SUPERCOMPUTER APPLICATIONS

SUPERCOMPUTER APPLICATIONS

Edited by

Robert W. Numrich

Control Data
Minneapolis, Minnesota

PLENUM PRESS • NEW YORK AND LONDON

Library of Congress Cataloging in Publication Data

Supercomputer Applications Symposium (1982–   ) (3rd: 1984: West Lafayette, Ind.)
  Supercomputer applications.

  ''Proceedings of the Supercomputer Applications Symposium cosponsored by the Purdue University Computing Center, the Purdue Center for Parallel and Vector Computing, and Control Data, held October 31–November 1, 1984, in West Lafayette, Indiana''—T.p. verso.
  Bibliography: p.
  Includes index.
  1. Supercomputers—Congresses. 2. CYBER 205 (Computer)—Congresses. I. Numrich, Robert W. II. Purdue University. Computing Center. III. Purdue University. Center for Parallel and Vector Computing. IV. Control Data Corporation. V. Title.
QA76.5.S8944   1984a   001.64   85-9394

ISBN-13: 978-1-4612-9514-3   e-ISBN-13: 978-1-4613-2503-1
DOI: 10.1007/978-1-4613-2503-1

Proceedings of the Supercomputer Applications Symposium cosponsored by the
Purdue University Computing Center, the Purdue Center for Parallel
and Vector Computing, and Control Data, held October 31–November 1, 1984,
in West Lafayette, Indiana

PREFACE

For the past three years, Control Data has cosponsored an applications symposium at one of its CYBER 205 customer sites. Approximately 125 participants from North America and Europe attended each of the three symposia. The Institute for Computational Studies at Colorado State University hosted the first symposium at Fort Collins, Colorado, August 12-13, 1982. The second annual symposium took place in Lanham, Maryland, and was hosted by the NASA Goddard Space Flight Center. This volume contains the proceedings of the Supercomputer Applications symposium held October 31-November 1, 1984, at Purdue University, West Lafayette, Indiana.

The purpose of this volume is to provide a forum for users of Control Data's CYBER 205 supercomputer to exchange common experiences and to discuss results of research projects performed on the computer. The unifying theme across the many disciplines is the development of methods and techniques to exploit the computational power of the CYBER 205. Somewhat surprisingly, these techniques are quite similar and apply to a wide range of problems in physics, chemistry, and engineering.

Selected papers are grouped into four main categories:

* Numerical Methods and Software Development

* Engineering and Petroleum Applications

* Fluid Dynamics and Weather

* Computational Physics and Chemistry

The categories roughly correspond to the four major areas where supercomputers have been most widely applied. The first paper in each group sets the tone for its section. Because of the interdisciplinary nature of the volume, the organizers asked the speakers to give brief overviews of their research projects before discussing specific results or techniques for using the CYBER 205.

Access to supercomputers is still limited for academic researchers. Nonetheless, the community of scientists routinely using supercomputers increases yearly, and research projects performed on these are ever more ambitious and sophisticated. Supercomputers have become friendly tools for solving problems once thought impractical. They have opened whole new areas of research in the traditional computational sciences, and have expanded the scope of applications of computational techniques.

Robert W. Numrich

CONTENTS

COMPUTATIONAL PHYSICS AND CHEMISTRY

IMPLEMENTATION OF LANCZOS ALGORITHMS

ON VECTOR COMPUTERS*

B.N. Parlett, B. Nour-Omid and J. Natvig

Center for Pure and Applied Mathematics
University of California, Berkeley, California 94720 USA

1. Introduction

We report on two recent studies [17],[18] in which the Lanczos algorithm was transported to a vector machine. The results suggest that our present implementations cannot exploit the full power of Class VI machines, nor even half of it. Yet the nature of the Lanczos algorithm lends itself to vectorization.

It seems as though sophisticated applications programs will need extensive revision in order to exploit these machines satisfactorily. An otherwise excellent algorithm can become unacceptably expensive quite suddenly, as soon as the problem size demands an out-of-core solution technique. Large page faults come to dominate the cost.

One purpose of the first study was to examine the performance of two different methods for the solution of some generalized eigenvalue problems when implemented on a CYBER 205 vector computer. We investigated the Subspace Iteration Method [13],[1] (called SUBIT hereafter), which is widely used on traditional serial computers for medium sized and large eigenproblems, and the Lanczos' Method [8],[10] (called LANC hereafter), which recently has been shown to be an order of magnitude faster than SUBIT [9]. It was of interest to see how the methods compare on a vector computer.

Standard in-core versions of the two algorithms were modified to take advantage of the special features of the CYBER 205 [4]. The algorithms were not redesigned, but wherever possible, a CYBER 205 vector function was substituted for the original FORTRAN code.

The test problems were derived from the dynamic analysis of idealized 3-dimensional structures, as modeled by the finite element program FEAP [16, Ch.24]. For various problem sizes the two methods were timed extensively, both before and after the explicit vectorization took place.

The second study was made at Boeing Computer Services (Seattle). Some relevant aspects are summarized in Section 8.

*Work made possible by a grant of computer time from CDC (Grant #82CSU22). Partial support by the U.S. Office of Naval Research under contract N00014-76-C-0013 is gratefully acknowledged.

2. The nature of the eigenvalue problem

We are interested in the solution of Eq. (1)

$$(A - \lambda M)x = 0 \qquad\qquad (1)$$

when A and M are large, sparse, n×n, symmetric matrices. In many applications the matrices have a banded form, i.e. $a_{ij} = 0$ for all $|i-j| > b_m$ where a_{ij} is the (i,j)-element of A and b_m is the half bandwidth. For typical problems in structural dynamics b_m is (5-10)% of n, cfr. Section 5. M must be positive semi-definite. In some cases M (or A) may be diagonal, which leads to a significant savings in the computational effort with either method.

"Large" today means $n > 10^3$, but ten years ago 10^2 was considered large. In these large problems all eigenpairs (λ_i, x_i) in a given interval may be sought; these are usually quite few, perhaps between 10 to 50. The interval may be at either end of the eigenspectrum and may contain the origin. For best convergence properties it is best to perform a shift of origin to this interval before the actual eigen computation takes place. For more details, see Section 4, as well as [10].

3. Special features on CYBER 205

The CYBER 205 is capable of attaining a rate of several hundred million floating point operations (add or multiply) per second, depending on the actual machine configuration and also on the precision of the arithmetic. We have used a 205 with two pipes in ordinary single precision (64 bits), and an asymptotic rate of 100 megaflops where a linked multiply-add is counted as one operation [4],[7]. One megaflop (mflop) is a rate of 1 million floating point per second.

3.1 Vectorization

We assume that readers have some familiarity with vector computers and make brief comments. Top performance can only be obtained by those parts of the program that operate on vectors, where a "vector" is a set of consecutive memory cells all treated in the same way. Vectorization of code may be delegated to the computer or it may be forced by using vector instructions. In either case a vector instruction is composed of a start-up phase and an execution phase with two operations (because of two pipes) per CPU-cycle. One CPU cycle equals 20 nanoseconds on the CYBER 205. The longer the vector the better for speed up. One needs $N \geqslant 100$ to achieve vector performance.

Linear combinations vectorize well. In particular we use repeatedly

$$Y(1;N) = Y(1;N) + A * X(1;N) \quad .$$

This is often called a SAXPY (single precision a times x plus y) although CDC calls it a linked triad. We found a rate of 50 mflops is achieved with N about 160.

The product (element by element) of two vectors goes even faster than a SAXPY.

The dot (or scalar) product of two vectors is not a vector function on the CYBER 205 but it has been implemented efficiently in a special function Q8SDOT. It can reach half the speed of a SAXPY for very long vectors, $N > 1000$. It follows that a matrix product VP, with V n×m, P m×m, and n>m is best coded as m^2 SAXPYs each of length n rather

2

than mm dot products of length m. More details are given in [18].

3.2 Memory management

The CYBER 205 has virtual memory (theoretical upper bound 2×10^{12} words per user). The real memory on the machine we used was 2 million words, and there was a 36.5 Mbit/s link, to a total of 450×10^6 words on disk. A program, with instructions and data, will be organized on pages, for which there are two choices, either small pages (equal to 512 words) or large pages (equal to 65,536 words). The paging system will seek to keep the most recently used pages in the primary storage. When the program references data (or instructions) that do not at that moment reside in the primary storage, a "page fault" occurs. CYBER 205 will halt the execution of the program until the page that contains the requested data has been transferred from the secondary storage, usually at the expense of another page being put out to the secondary storage. While this swapping takes place, the CYBER 205 may execute other programs that are allowed to occupy part of central memory. There is a small overhead in CPU-time when a vector crosses a page boundary, and also during a page fault. When a program references data in an "orderly" fashion, as when consecutive columns of a matrix are used consecutively and not at random, it is more efficient to use large pages. This is indeed the case with both our eigenvalue methods.

The accounting system assesses a cost penalty for a large page fault of .156 SBU (System Billing Units), equivalent to .156 CPU-seconds [12]. For large problems that cannot be fitted into primary storage, this penalty might actually be larger than the CPU time for the whole computation.

4. Description of the Lanczos method

We are interested in some of the eigenvalues closest to a specified value, σ. We perform initial calculations:

$$\text{shift the } A \text{ matrix by } \sigma: \bar{A} = A - \sigma M$$

$$\text{factor } \bar{A} \text{ into } L \Delta L^T$$

Here L is a lower triangular matrix with diagonal elements equal to one, Δ is a diagonal matrix, and L^T is the transpose of L. Incidentally the number of negative elements in Δ equals the number of eigenvalues that are smaller than σ.

We use a standard active column profile solver (called PROFIL). The upper triangular part of $\bar{A}$ is stored and gradually overwritten with L^T. Consider the computation of a typical column of L^T of "height" h (above the diagonal). Each element will require a dot product and the lengths of the vectors involved will vary from 1 to h-1. Altogether nb dot products are needed for L^T and the average length is b/2, where b is the average half bandwidth of $\bar{A}$.

Of course PROFIL could be reorganized to compute L by columns and so replace dot products by SAXPYs. This helps, but not much. The significant fact is that b/2 is small compared to n in most structural problems and the factorization of narrow banded matrices cannot exploit the full vector mflop rate of the CYBER 205 since it is manipulating vectors of (average) length b/2 rather than n.

As we shall see, the factorization process dominates the iteration to a greater extent on the CYBER 205 than in serial machines.

The remaining part of either method, SUBIT or LANC, is formulated in such a way as to solve the following transformed eigenvalue problem

$$((L \Delta L^T)^{-1} Mx) = x\alpha \quad . \tag{2}$$

The largest α's correspond to the eigenvalues λ closest to σ (these are the smallest λ's when $\sigma = 0$) according to

$$\alpha_i = \frac{1}{\lambda_i - \sigma} \quad . \tag{3}$$

The eigenvectors, x, in Eq. (2) are the same as in Eq. (1). See [5] for more on this transformation of the problem.

We used a simple Lanczos' algorithm. A single random starting vector, r_0, is iterated according to the scheme presented in [9]. Initializations include setting $q_0 = 0$, and computing $p_0 = Mr_0$ and $\beta_1 = \sqrt{r_0^T p_0}$.

In each step j (j = 1,2,...), the following tasks are done:

(a) Orthogonalization. r_{j-1} will be orthogonalized against the previous Lanczos vectors when needed [15]. (This is called selective orthogonalization.) A maximum of (j−2) SAXPY's and dot products with vector length n.

(b) Compute $q_j = \dfrac{r_{j-1}}{\beta_j}$ and $\bar{p}_{j-1} = \dfrac{p_{j-1}}{\beta_j}$. This is two vector operations, vector length n.

(c) Solve $\bar{A}r_j = \bar{p}_{j-1}$ for r_j.
     "Solve (A)", i.e. (2b+1)n operations, n dot products and n SAXPYs with average vector length equal to b, and one vector Schur multiply with vector length n.

(d) Compute $r_j = r_j - q_{j-1}\beta_j$. Single SAXPY, vector length n.

(e) Compute $\alpha_j = r_j^T \bar{p}_{j-1}$. Single dot product, vector length n.

(f) Compute $r_j = r_j - q_j \alpha_j$. Single SAXPY, vector length n.

(g) Compute $p_j = Mr_j$ and $\beta_{j+1} = \sqrt{r_j^T p_j}$.
     "mult M", i.e. a vector operation (Schur product) of length n (when M is diagonal); and a dot product, vector length n.

If β_{j+1} is small compared to $|\alpha_j|$ and β_j, (e), (f) and (g) will be repeated once. This is found to take place in fewer than 1/4 of the steps.

(h) Analysis of the symmetric tridiagonal matrix T_j which has the α's as diagonal elements and the β's as off-diagonal elements. $\approx 80j$ scalar operations [11].

(i) For converged eigenvalues compute eigenvectors of T_j and then compute eigenvectors x. j SAXPYs with vector length n per computed x.

Typically, the first eigenvalue will converge in 5-10 iterations, and 20 eigenvalues will converge in 40-50 iterations. For longer runs it is a good assumption that $\ell/2$ eigenvalues will have converged in ℓ iterations.

The operations count for a LANC run, when the M matrix is diagonal (cfr. Section 5), includes

(1) Factorization: $nb^2/2$ operations; dot products, average vector length $b/2$.

(2) Orthogonalization, total for ℓ steps: for selective orthogonalization we have found that $\approx n\ell^2/5$ operations are required; equally divided between SAXPYs and dot products, each of vector length n. For a full reorthogonalization, $n(\ell^2-\ell)$ operations are required; equally divided between SAXPYs and dot products, each of vector length n.

(3) ℓ Lanczos steps

   a) $(2b+1)n\ell$ operations; dominated by SAXPYs and dot products, average vector length b (solve $(\bar{A})$)

   b) $\frac{5}{4}\,n\ell$ operations, vector length n (mult (M))

   c) $6n\ell$ operations; equally divided between dot products and scalar vector products, vector length n

(4) Analysis of tridiagonal matrix, T, total for ℓ steps $\approx 40\,\ell^2$ operations; nonvectorizable.

(5) Computation of $\ell/2$ eigenvectors $\approx \frac{3}{8}\,n\ell^2$ operations; estimated for the case that one eigenvalue/eigenvector converges in each of the $\ell/2$ last steps. This is an overestimate, more typical would be $\frac{1}{8}\,n\ell^2$ to $\frac{1}{4}\,n\ell^2$ operations. These operations are SAXPYs with vector length n.

A comparison of operation counts shows that a LANC run with fewer than $b/4$ steps does not permit the initial cost of factorization, (1) above, to be amortized. For longer runs ($\ell > b/4$) the Lanczos' steps (3) will require more operations than the factorization, and then (selective) orthogonalization (2) and the analysis of T (4) become significant parts of a LANC step. Because of the poor vector length during factorization as compared with the other parts, we shall expect to need even more steps than $b/4$ to amortize the factorization on the CYBER 205.

The operation count break-even between parts (4) and (5) above is for $n \approx 100$, but because (5) is vectorized, the two modules will have equal CPU-time on the CYBER 205 for a much higher value of n. Note also that (5) will always be less than half the cost of (2).

5. Test examples

We used test examples that typically occur in structural dynamics. A version of the finite element program FEAP [16, Ch.24] was converted to run on CYBER 205; this program was then used to generate stiffness matrices, A, and mass matrices, M, for the test examples. The eigenvalues, λ_i, are the squares of the frequencies of free vibration of the structures, ω_i, i.e. $\lambda_i = \omega_i^2$.

We generated a total of four sets of matrices, with the order of the matrices ranging from 150 to 7296 (i.e. $n = 150$ to 7296). In all four examples we chose to have a diagonal M. There is no loss of generality with this assumption. Our intention was only to keep the total computational cost down, and yet be able to examine large problems.

Example (i): $n = 150$, $b = 17$. This is a simple truss structure.

Example (ii): $n = 468$; $b = 60$. This is a structure, first presented in [2], which we have used extensively during previous testing [8].

Examples (iii) and (iv) were generated with a 3-dimensional beam element. The model is an idealization of a multistorey structure, see Fig. 1. Each storey had the same geometry, with $(N_x-1)N_y+(N_y-1)N_x$ elements parallel to the x- and y-axes, and also N_xN_y elements parallel to the z-axis (connecting the storey to the next lower storey). Of the total number of nodes, $N_xN_yN_z$, all that belonged to the bottom storey were held fixed. Each node had six variables.

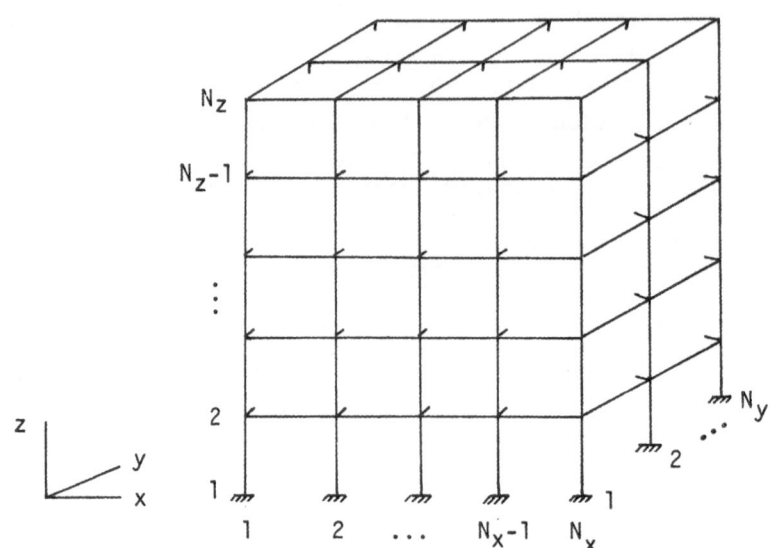

Figure 1. *Model used in examples (iii) and (iv).*

Example (iii): With $N_x = N_y = 4$, $N_z = 20$ we get $n = 1824$, $b = 95$.

Example (iv): With $N_x = N_y = 8$, $N_z = 20$ we get $n = 7296$, $b = 370$.

The model was so constructed that whenever $N_x = N_y$, due to symmetry there was a set of double eigenvalues corresponding to the transverse vibration of the structure. Both Example (iii) and Example (iv), therefore, contain double eigenvalues.

6. Test runs

While we ran our four test examples on LANC, the CPU-time consumed by all major parts of the programs was measured by the CDC FORTRAN function SECOND [3], and we kept track of the step at which eigenvalues were accepted. In all cases a number of the smallest eigenvalues/eigenvectors were computed, i.e. those closest to the shift $\sigma = 0$.

7. Results and comparisons

7.1 Effects of vectorization

7.1.1 <u>Performance improvements through vectorization</u>. As pointed out in section 3.1 vectorization may be achieved through an optimization option for the FORTRAN compiler or through the replacement of instructions by explicit vector instructions. Typically, the automatic vectorization will only affect quite obviously vectorizable code. More complicated sequences of instructions will not be vectorized through the compiler option, although they might well be vectorizable. We wanted to see how much various subroutines improved from the original scalar version to the explicit vectorized version, and made test runs under the following four regimes:

Condition (A): The code was as before vectorization, compiled with no optimization options.

Condition (B): The code was as before vectorization, and it was compiled with the scalar optimization option.

Condition (C): The code was as before vectorization, and it was compiled with scalar and vector optimization options.

Condition (D): The code was explicitly vectorized as outlined in section 3, and it was compiled with scalar and vector optimization options.

In Table 1 we show the CPU-times and mflop rates for the important factorization subroutine (PROFIL) as a function of the problem size.

TABLE 1. Performance of vectorized factorization subroutine PROFIL.

Example	Average vector length	Operations [millions]	CPU-time [seconds]	Mflop rate
(i)	9	.022	.0148	1.49
(ii)	30	.84	.1689	5.0
(iii)	48	8.23	1.091	7.6
(iv)	185	499	24.59	20

Note that these rates are well below vector rates. The reason is that PROFIL contains various conditional statements and also some scalar arithmetic (e.g. on indices) in addition to the vector operations, that will slow down the code accordingly. This is also the case for other subroutines.

TABLE 2. Performance of factorization subroutine PROFIL under various optimization options.

Example	Mflop rates			
	Cond A	Cond B	Cond C	Cond D
(i)	.53	.63	1.03	1.49
(ii)	1.04	1.64	3.5	5.0
(iii)	.94	1.70	4.4	7.6

The values in Table 1 are for condition (D), i.e. explicitly vectorized code. Table 2 shows how much was gained in this subroutine by the various compiler options.

In LANC the orthogonalization task may be coded as a full Gram-Schmidt orthogonalization, as in our subroutine GSORT, or as a selective orthogonalization, as in our subroutine PRORT. We shall comment on the choice of method in section 7.3, and at this point include performance data for GSORT with Table 3.

TABLE 3. *Orthogonalization performance GSORT.*

Example	Vector length	Mflop rate			
		Cond A	Cond B	Cond C	Cond D
(i)	150	1.9	2.1		13
(ii)	468	2.0	2.2	5.1	29
(iii)	1824	2.0		5.3	42
(iv)	7296				50

GSORT contains equal amounts of SAXPYs and dot products, all with vector length n. Table 3 illustrates that the automatic vectorization did not vectorize perfectly vectorizable code (low performance for condition (C)); this is because the original GSORT contained an external reference to a general purpose dot product function, which was coded in a way the automatic vectorizer did not recognize.

7.1.2 __The importance of vector length.__ So far we have seen that the performance of a given subroutine improves dramatically with longer vectors. However, this improvement is not as big as the direct measurements of pure vector indicate. For subroutines like GSORT, that consist almost completely of vector instructions, we have observed a markedly lower performance than expected. Typically, a subroutine — or a whole program for that matter — will include slow parts that will degrade the overall performance even further. Scalar operations, dot products, and/or operations with short vector lengths are examples of such slow parts. In order for the slow parts to have any significant influence on the overall performance, however, they must represent a significant fraction of the total work. Below we shall give a detailed discussion of an important subroutine that has certain operations on short vectors and other operations on long vectors. The overall performance will be in between what would have been expected for the short vector length alone and for the long vector length alone.

Within LANC, the tasks (b)-(g) were coded as one subroutine, LANSIM. During task (c) there are vectors of average length b, but in the other tasks the vector length is equal to n. In an average step there are about nine vector operations of length n (6.5 (fast) vector multiply's or SAXPYs, and 2.5 (slow) dot products) as opposed to 2n vector operations (n SAXPYs and n dot products) with average vector length b. Thus the shorter vector length occurs much more often than the longer vector length in typical examples, e.g. 100 times more often in Example (ii), and nearly

TABLE 4. *Lanczos' step performance LANSIM.*

Example	Vector length		No of steps ℓ	Operations [millions]	CPU-time [seconds]	Mflop rate
	Short b	Long n				
(i)	17	150	80	.507	.1183	4.3
(ii)	60	468	100	6.00	.497	12
(iii)	95	1824	100	36.3	2.08	17
(iv)	370	7296	100	546	13.66	40

400 times more often in Example (iii). As a result the overall mflop rate for LANSIM will be determined by the shorter vector length. Again the problem must be very large, with short vector length > 500 (if we ignore the effect of non-vectorized parts), to give an overall mflop rate in LANSIM that is higher than 50. Table 4 shows the performance that we measured for LANSIM for our more typical examples.

7.1.3 <u>Modules that were not vectorized</u>. Within LANC there is an unvectorized subroutine, ANALZT, that performs the analysis of the tridiagonal matrices. The operations count for ANALZT that we have given is quite approximate. Taken literally it indicates a performance rate of about 0.9 - 1.5 mflops. We recorded an improvement of about 10% for this subroutine through the scalar optimization option for the compiler.

7.2 <u>Overall LANC performance</u>

A LANC run consists mainly of the following subroutines: PROFIL and STPONE performed once; and GSORT, LANSIM, ANALZT, and VECT performed repeatedly (see section 4). GSORT, LANSIM, and ANALZT are performed in each step; VECT only in those steps when an eigenvalue has converged and its eigenvector is to be computed.

Our intention was to run LANC with our latest version of selective orthogonalization. However, our selective code, PRORT, which worked perfectly on our serial computer, began to misbehave on the CYBER 205 on the large example (Example (iv)). For this reason we switched to full reorthogonalization, GSORT, for all our tests. Separately we compared the two techniques on Examples (i), (ii), and (iii) and found that PRORT required about one-third the CPU-time of GSORT. As we shall see (e.g. Table 5) this is not a significant advantage for PRORT since GSORT vectorizes so well and is a minor part of the LANC loop. This is in strong contrast to the situation with serial computers in which full orthogonalization comes to dominate unless the LANC runs are kept short.

Because all of GSORT, ANALZT, and VECT gradually involve more operations as the iteration progresses, whereas LANSIM has the same number of operations in every step, the relative importance of GSORT, ANALZT, and VECT will increase with the number of iterations.

Table 5 gives cumulative CPU-times for the subroutines within the LANC loop after some typical number of steps, ℓ. The number of eigenvalues, c, that have converged to machine precision (1.42108×10^{-14}) is also given. The VECT column contains estimates on the CPU-time for the eigenvector computation. The estimates are taken to be one-half of the CPU-time measured for GSORT; this is most certainly an over-estimation. The column for ANALZT in the largest example also contains estimates, taken to be the time for $40 \ell^2$ operations at 1 mflops. The results are presented separately for Examples (i), (ii), (iii), and (iv).

In small problems the unvectorized ANALZT will take a significant portion of the CPU-time in the LANC loop. Even in Example (iii) ANALZT takes about as much time as the orthogonalization subroutine, GSORT, while there are about 45 times as many operations in GSORT. After 44 LANC steps (Example (iii)) ANALZT has consumed about 7% of the LANC loop time; its share rises to about 15% after 100 steps, and it will eventually take more time than LANSIM (after some 400 steps). If on a scalar computer we can achieve 1 mflops for all operations, we can expect ANALZT to catch up with LANSIM at a rate that is about 17 times slower than this, cfr. Table 4. GSORT would have been slowed down by a factor of 42, and would have taken about half of the time taken by LANSIM after 100 LANC steps with Example (iii) on this fictitious computer.

TABLE 5. *Cumulative CPU-times for subroutines in the LANC loop. (All times are in seconds.)*

Example	l	c	GSORT	LANSIM	ANALZT	VECT
(i)	10	3	.0020	.0135	.0070	(.0010)
n = 150	20	5	.0062	.0287	.0177	(.0031)
b=17	40	17	.0282	.0613	.0623	(.0141)
	80	40	.0771	.1183	.2575	(.0386)
(ii)	10	1	.0013	.0453	.0060	(.0007)
n = 468	20	5	.0058	.0963	.0182	(.0029)
b=60	40	15	.024	.196	.054	(.012)
	80	47	.099	.396	.188	(.050)
	100	54	.156	.497	.271	(.078)
	150	69	.352	.748	.545	(.176)
(iii)	10	2	.0034	.1908	.0074	(.0017)
n = 1824	20	5	.0153	.4053	.0197	(.0077)
b=95	44	20	.081	.987	.082	(.040)
	100	46	.418	2.080	.362	(.209)
(iv)	10	2	.0115	1.241	(.0040)	(.0058)
n = 7296	20	3	.053	2.621	(.016)	(.027)
b=370	40	17	.226	5.382	(.064)	(.113)
	60	26	.520	8.142	(.144)	(.260)
	80	59	.934	10.902	(.256)	(.467)
	100	72	1.469	13.664	(.400)	(.735)

The full LANC run also includes initializations, most of which is the factorization done by PROFIL. Table 6 shows that the initialization cost may be quite substantial compared to the cost of the LANC loop; e.g. in Example (iii) first after some 60 steps does the LANC loop take more CPU-time than the initialization, and in Example (iv) this does not happen in 100 steps.

For this method as well as the number of converged eigenvalues seems to be independent of problem size (n). For all examples about 30 steps were necessary for the first ten eigenvalues to converge. Table 7 shows the exact step when each of the first ten eigenvalues converged to machine precision. However, the eigenvalues will not necessarily be found in increasing order. This is particularly the case when there are double eigenvalues, as in Examples (iii) and (iv). E.g. in Example (iii), when 30 eigenvalues have converged, they are in fact found to be all of λ_1 to λ_{26}, then λ_{28} through λ_{30} and λ_{34}. The "missing" eigenvalue, λ_{27}, is equal to λ_{26}, and it was found in a few more steps (along with λ_{31}, λ_{33}, λ_{36}, and λ_{37}). In practice it is reassuring to do one extra factorization simply to check that there are no missing eigenvalues among the ones which have been computed. This represents an added cost of $nb^2/2$ operations; that has not been included in our tables or figures.

7.3   Comparison with SUBIT (Subspace Iteration)

The overall performance of SUBIT is described in [18]. Examples (i), (ii), and (iii) could be solved within the available CYBER 205 primary storage, and a comparison of the two methods may then justly consider only the CPU-time that was needed for computation. Figures 2, 3 and 4 show a graph of the total CPU-time vs. the number of converged eigenvalues for these examples.

TABLE 6. *Initialization vs. cumulative LANC steps,*
floating point operations, and CPU-time.

Example	No. of steps	Initialization		LANC steps	
		ops [millions]	CPU-time [seconds]	ops [millions]	CPU-time [seconds]
(i)	10	.027	.0211	.090	.0235
n = 150	20			.233	.0557
b=17	40			.678	.166
	80			2.20	.492
(ii)	10	.900	.193	.675	.0533
n = 468	20			1.50	.159
b=60	40			3.59	.286
	80			9.55	.733
	100			13.42	1.00
	150			25.7	1.82
(iii)	10	8.58	1.50	3.90	.204
n = 1824	20			8.34	.448
b=95	44			21.3	1.19
	100			63.9	2.44
(iv)	10	505	26.6	55.7	1.26
n = 7296	20			114	2.72
b=370	40			236	5.79
	60			367	9.07
	80			507	12.56
	100			655	16.3

TABLE 7. *Initial convergence in LANC.*

Eigenvalue no.	Iteration no. when eigenvalue converges			
	Example (i)	Example (ii)	Example (iii)	Example (iv)
1	6	9	6	5
2	8	16	9	10
3	9	17	11	12
4	15	18	12	21
5	18	20	18	21
6	21	22	21	22
7	21	23	26	22
8	24	24	27	24
9	25	27	27	25
10	29	29	29	26

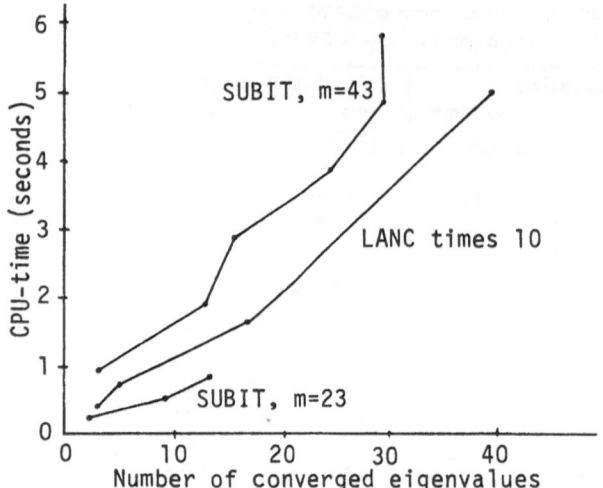

Figure 2. SUBIT vs. LANC, Example (i), n=150.

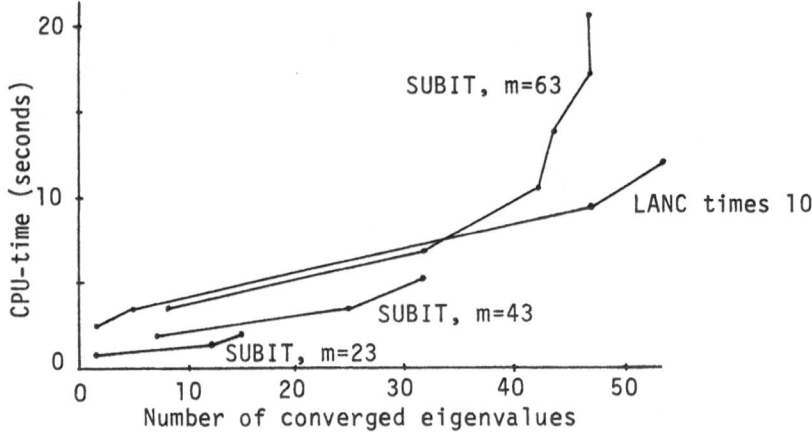

Figure 3. SUBIT vs. LANC, Example (ii), n=468.

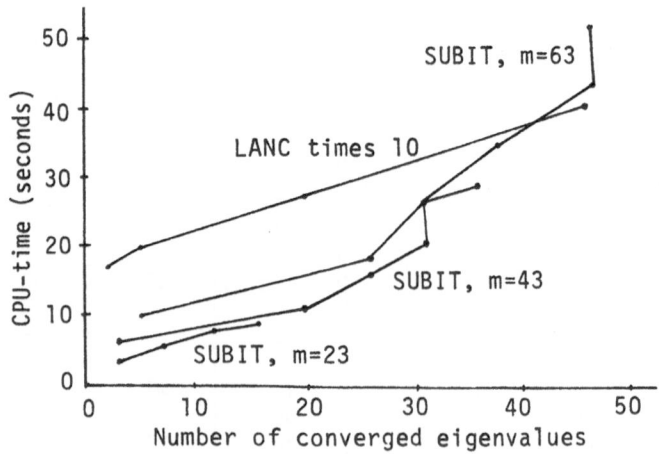

Figure 4. SUBIT vs. LANC, Example (iii), n=1824.

Note that the CPU-times for LANC have been multiplied by 10 in Figures 2, 3, and 4. We feel it is fair to say that LANC is an order of magnitude faster than SUBIT on these examples, but the trend is for SUBIT to gain relative to LANC as the problem size n increases. This is of course due to the fact that the cost of the initial factorization is more dominant in LANC. It is also to be seen that SUBIT is at its best when very few eigenpairs are sought; then a small number m of iteration vectors is sufficient. LANC is definitely superior also in these cases, although not by an order of magnitude.

Note also that we are comparing the computation of the same fixed number of eigenpairs, up to about 50, irrespective of problem size n. But it may be more common to look for more eigenvalues in a larger example than in a small one. From the tendencies that we have observed, e.g. in Figs. 2-4, we may argue that when we are looking for a number of eigenpairs that is the same fixed fraction of the problem size, SUBIT is no longer seen to gain relative to LANC with increasing problem size. However, we do not want to stretch this argument too far, because the best way to compute a large number of eigenpairs (with either method) will ordinarily include the use of repeated shifts and factorizations, which is not being discussed in the present report.

Our largest example, Example (iv), required almost 3 million words to store the A (or L) matrix. Because a typical step both in SUBIT and in LANC involves two passes through the L matrix and there are less than 2 million words of primary storage, it is evident that extensive swapping did take place during the course of the programs. Although it is possible to insert commands in the program that may advise the paging system about pages that can be removed from the primary storage and pages that are needed in the primary storage, we relied solely on the standard scheduling.

Figure 5 shows the comparison of SUBIT and LANC as far as CPU-time is concerned for Example (iv). To compute 30 eigenpairs in this case required about four times as much CPU-time in SUBIT as in LANC.

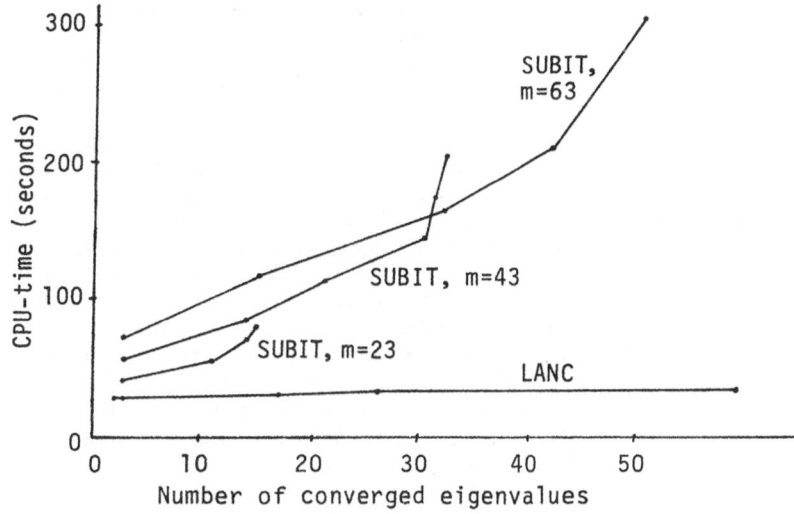

Figure 5. SUBIT vs. LANC, Example (iv), n=7296.

Table 8 summarizes some typical runs of Example (iv) and compares the cost of input/output to the cost of computation. ℓ is the number of iteration steps, m is the subspace size (in SUBIT), LPF is the number of

TABLE 8. *Page fault penalties, Example (iv)*.

Method	ℓ	m	LPF	LPF/ℓ	Penalty [SBU]	Comp.time [CPU-sec]
SUBIT	19	23	1036	55	161.6	83.3
	30	43	2010	67	313.6	206.6
	30	63	2251	75	351.2	308.1
LANC	20		885	45	138.1	29.5
	40		1710	43	266.8	32.4
	60		2624	44	409.3	35.5
	80		3629	45	566.1	38.7
	100		4714	47	735.4	41.9

large page faults, "Penalty" is the SBU corresponding to the large page faults, and "Comp.time" is the total CPU-time for the computation. The load module for both programs was about 61 large pages.

When we are searching for quite a few eigenvalues, SUBIT may be run with a small subspace (m) and get away with fewer page faults per iteration step. In LANC it is seen that the number of page faults increases slightly in the later steps of an iteration. This is due to the fact that more vectors will take part in the orthogonalization in a later step. Because the first steps of LANC only involves a few vectors, whereas in SUBIT all vectors participate in each iteration, LANC requires fewer page faults than SUBIT.

Essentially it is the two passes through L that lead to the about 70 large page faults in each step of SUBIT and the about 45 large page faults in each step of LANC. The penalty for these page faults is larger than the CPU-time for the computation. (And when M is not diagonal, we shall expect a substantial increase in the number of page faults, because we shall also need to make a pass through M.) Because SUBIT works on several vectors at the same time and converges to a required number of eigenvalues and eigenvectors in fewer steps than does our simple LANC, SUBIT may perform better than (simple) LANC in cases like Example (iv), where the problem is too big for the primary storage and extensive paging takes place.

It is possible to work with more than one vector in each step of the Lanczos method, using Block Lanczos [14] (called BLANC hereafter). When the block size is 8 (i.e. 8 vectors), each BLANC step will involve about 8 times as many operations and 8 times as much CPU-time as a step of LANC. At the same time the number of steps required for a certain number of eigenpairs to converge in BLANC will be only a fraction (roughly 1/8) of that in LANC, at least for fairly long runs. The net effect is that the computation's total CPU-time will remain about the same. However, the overall number of page faults with BLANC will be greatly reduced (by a factor of 8) compared to LANC. This is because still only two passes through L are needed in each step of the algorithm. Table 8 shows that BLANC with $\ell=10$, m=8 would incur only 443=885/2 page faults to deliver as many eigenvalues as does LANC with $\ell=80$.

When BLANC is used for finding very few eigenpairs and/or with a large block size, the number of steps is reduced by a factor that is less than 8, and the total CPU-time will increase compared with simple LANC. Further, the saving in the number of page faults will be smaller because of the increase in the number of steps.

We conclude that for problems that are too big for the primary storage, BLANC would be more efficient than LANC because of the smaller penalties for page faults. Both LANC and BLANC have smaller CPU-time than SUBIT. The number of page faults occurring in LANC is greater than that for SUBIT. With a proper choice of block size BLANC will remain at the same low level for the CPU-time as simple LANC, and at the same time have fewer page faults than SUBIT.

At present we are experimenting with different ways of adapting ANALZT to block tridiagonal matrices.

8. The role of I/O

We summarize here one aspect of the transportation of a Lanczos code from a CDC 6600 to a CRAY-1S carried out by John Lewis and Horst Simon at Boeing Computer Services. See [17] for many more details. The FORTRAN package STLM was developed by Ruhe and Ericsson in Umeå, Sweden, and is perhaps the best available implementation of the simple Lanczos process.

STLM was developed on a computer system where I/O was free and core memory was expensive. This situation strongly influenced the structure of the package. On the CRAY-1S it is just the opposite: core memory is cheap and I/O is expensive. The consequences can be dramatic.

Lewis and Simon compared two extreme implementations on some medium and large sized problems. One was a completely in-core solution, the other wrote all Lanczos vectors out to the disc. The out-of-core solution cost is between 3 and 10 times more than the in-core version.

The 2M words of core memory on the CRAY-1S permit an in-core solution of some interesting problems but the trend is to larger and larger problems. When $N \geq 10,000$ out-of-core versions become necessary and I/O costs will dominate CPU costs.

To economize such costs it becomes necessary to make fundamental changes in the structure of the algorithm. Another way to put this is that a different version of the Lanczos algorithm might be needed for each type of billing algorithm on the system. This is not a happy situation.

Having stressed that execution time is not an accurate measure of cost it is still of interest to identify the computationally intensive parts of the STLM package. In reading the table below it should be mentioned that the algorithm in STLM for analyzing T is quite different from the corresponding part of LANC.

TABLE 9. Percentage of CPU-time in parts of STLM.

N	$L\Delta L^t$	Solve	$v \leftarrow Mv$	Analyze T	Reorthog.	No. of λ's
48	9	26	1	16	21	24
420	17	22	20	20	9	50
1074	73	18	0	5	3	20
362	11	27	0	22	9	2×15
1922	39	34	0	12	9	200

Two conclusions can be drawn from these numbers:

- For very large problems the solution of $(A - \sigma M)r = Mq$ dominates the other parts of a Lanczos step.

- There is some advantage to be gained from halving the execution time for reorthogonalization and the analysis of T.

Although Lewis and Simon do not study the degree to which STLM was successfully vectorized, the circumstantial evidence is that it was low.

9.   Acknowledgments

We want to thank Professor R. L. Taylor for valuable discussions and for making a copy of the FEAP program available to us. We also thank John Lewis and Horst Simon for making their studies available to us. The first author takes this opportunity to thank R.Numrich for organizing a gift from Control Data Corporation of 10 hours of CPU time on the CYBER 205. It made possible the first of these studies.

REFERENCES

[1] K.J.Bathe and E.L.Wilson, "Numerical Methods in Finite Element Analysis", (Prentice Hall, Englewood Cliffs, New Jersey, 1977).
[2] K.J.Bathe and E.L.Wilson, Large eigenvalue problems in dynamic analysis, A.S.C.E., Journal of Engineering Mechanics Division **99** (1973), 467-479.
[3] CDC CYBER 200 FORTRAN Version 2 Reference Manual, Publication No. 60485000 (CDC, 1981).
[4] CDC CYBER 200 Model 205 - Technical Description (CDC, 1980).
[5] T.Ericsson and A.Ruhe, The spectral transformation Lanczos method for the numerical solution of large sparse generalized symmetric eigenvalue problems, Math. Comp. **34** (1980) 1251-1268.
[6] B.S.Garbow et al., "Lecture Notes in Computer Science", vol. 51, (Springer-Verlag, 1977).
[7] M.J.Kascic, Jr., "Vector Processing on the CYBER 205", (CDC CYBER 205 Workshop, Fort Collins, Colorado, February 1983).
[8] C.Lanczos, An iteration method for the solution of the eigenvalue problem of linear differential and integral operators, J. Res. Nat. Bur. Standard **45** (1950) 255-282.
[9] B.Nour-Omid, B.N.Parlett, and R.L.Taylor, Lanczos versus subspace iteration for solution of eigenvalue problems, International Journal for Numerical Methods in Engineering **19** (1983) 859-871.
[10] B.N.Parlett, "The Symmetric Eigenvalue Problem", (Prentice-Hall, Englewood Cliffs, N.J., 1980).
[11] B.N.Parlett and B.Nour-Omid, The use of refined error bounds when updating eigenvalues of tridiagonals, Technical Report PAM-175, (Center for Pure & Applied Math., Univ. of California, Berkeley, 1983).
[12] Price Schedule for CYBER 205 Services, (CSU, Computer Center, Fort Collins, Colorado, 1982).
[13] H.Rutishauser, Simultaneous iteration method for symmetric matrices, in: J.H.Wilkinson and C.H.Reinsch, eds., "Handbook for Automatic Computation (Linear Algebra)", (Springer-Verlag, New York, 1971) 284-302.
[14] D.S.Scott, Block Lanczos software for symmetric eigenvalue problems, Report ORNL/CSD-48, UC-32 (Union Carbide Corporation, 1979).
[15] H.D.Simon, The Lanczos algorithm for solving symmetric linear systems, Technical Report PAM-74 (Center for Pure & Applied Math., Univ. of California, Berkeley, 1982).
[16] O.C.Zienkiewics, "The Finite Element Method", 3rd edition, (McGraw-Hill, London, 1977).

[17] John G. Lewis and Horst D. Simon, Numerical experience with the
      spectral transformation Lanczos method, Technical Report MM-TR-16,
      Boeing Computer Services, Seattle, WA 98124 (April 1984).
[18] Jon Natvig, B.Nour-Omid, and B.N.Parlett, Effect of the CYBER 205
      on methods for computing natural frequencies of structures, Technical
      Report PAM-218 (Center for Pure & Applied Math., Univ. of California,
      Berkeley, March 1984).

(Jon Natvig was on leave from Rogaland Regional College, N-4000 Stavanger,
Norway. Dr. B. Nour-Omid is presently consultant to the Applied Mechanics
Group, Lockheed Palo Alto Research Laboratory, Palo Alto, California.)

INCOMPLETE CHOLESKI CONJUGATE GRADIENT ON THE CYBER 203/205

Eugene L. Poole and James M. Ortega

University of Virginia

1. Introduction

We consider in this paper the Incomplete Cholesky Conjugate Gradient (ICCG) method on the CDC Cyber 203/205 vector computers for the solution of an NxN system of linear equations $Ax=b$. We assume that the matrix A is large, sparse, and symmetric positive definite with non−zero elements lying along a few diagonals of the matrix, such as arises in the solution of elliptic partial differential equations by finite difference or finite element discretizations. Results are given for two model problems, run on a Cyber 203 at NASA−Langley Research Center. In the sequel, we shall refer to the Cyber 203 and 205 as the Cyber 200 unless there is a reason to differentiate between them.

Since Meijerink and van der Vorst [1977], several authors have considered ICCG. Most of this work has been directed towards serial computers although Kershaw [1982] gives an implementation for the CRAY−1 using a cyclic reduction technique applied to block tridiagonal matrices, and Lichnewsky [1984] discusses parallel and vector implementations for ICCG but, as in Kershaw, mainly gives algorithms with vector lengths better suited for the CRAY computers.

Efficient use of the Cyber 200 requires algorithms that consist mainly of operations on long vectors. In particular, our goal is vectors of length $O(N/p)$, where N is the number of unknowns and p is a small constant independent of N. This is achieved in the basic conjugate gradient algorithm (with $p = 1$) and the problem is to achieve it also in the incomplete Cholesky preconditioning. To this end, we adopt and extend a suggestion of Schreiber and Tang [1982] to use a multicolor ordering of the grid points. This is combined with diagonal storage of the matrix and it is shown that by judicious choice of the multicolor ordering, vector lengths of $O(N/p)$ are achieved, where p is the number of colors (p=4 and 6 in the model problems). There are many incomplete Choleski fill−in strategies and we have chosen the simplest, ICCG(0), in which the non−zero elements in the incomplete Choleski factors are in the same diagonals as the original matrix; that is, no fill is allowed outside these diagonals. We will briefly discuss, however, how other fill strategies can be implemented. We also consider an m−step ICCG(0) method in the sense of Adams and Ortega [1982]; m=1 is then ICCG(0).

Our first model problem is the differential equation: $U_{xx} + \frac{1}{2}U_{xy} + U_{yy} = 4$ on the unit square with boundary conditions chosen so that the solution is $x^2 + y^2$. The usual finite difference approximations are used and a typical equation is

$$4u_{i,j} - u_{i,j-1} - u_{i,j+1} - u_{i-1,j} - u_{i+1,j} - \frac{1}{8}u_{i-1,j-1} + \frac{1}{8}u_{i-1,j+1} + \frac{1}{8}u_{i+1,j-1} - \frac{1}{8}u_{i+1,j+1} = -4h^2$$

The second model problem is a plane stress problem as described in Adams [1983a,b]. The domain is a rectangular plate discretized using triangular finite elements on which linear basis functions are defined (see Figure 4). At each node of the plate there are two unknowns, the u and v displacements. The plate is constrained on one side and loaded on the other. Results for these model problems are given in **7**.

2. The ICCG(0) Method

The preconditioned conjugate gradient algorithm is described in a number of places (see, e.g. Adams [1983b]). The basic loop is

a) $\alpha_k = \dfrac{(\hat{r}^k , r^k)}{(p^k , A\, p^k)}$, $\quad x^{k+1} = x^k + \alpha_k p^k$, $r^{k+1} = r^k - \alpha_k A p^k$

b) solve: $M\hat{r}^{k+1} = r^{k+1}$

c) $\beta_k = \dfrac{(\hat{r}^{k+1} , r^{k+1})}{(\hat{r}^k , r^k)}$, $\quad p^{k+1} = \hat{r}^{k+1} + \beta_k p^k$

with suitable initialization and convergence test. The steps a) and c) vectorize well, in principle, and our main concern is with step b), which defines the preconditioning. For ICCG(0), $M = LDL^T$ is the incomplete Choleski decomposition, where D is diagonal and L is unit lower triangular with the same diagonal sparsity pattern as A.

We assume that the grid points have been ordered by a suitable multicolored ordering (Adams and Ortega [1982]) so that A has the form

$$
A = \begin{vmatrix}
A_{11} & A_{12} & . & . & . & A_{1p} \\
A_{12}^T & A_{22} & & & & . \\
. & & . & & & . \\
. & & & . & & . \\
. & & & & . & . \\
A_{1p}^T & . & & . & . & A_{pp}
\end{vmatrix}
\tag{2.1}
$$

where p is the number of colors and the diagonal blocks A_{ii} are diagonal. A matrix with this structure is called a p-colored matrix. (Note that the classical red/black ordering is the case p=2). This is illustrated in Figure 1 for the first model problem in which p=4.

It is convenient to view the incomplete Choleski decomposition as derived from block Choleski with the partitioning of (2.1):

$$
L_{i,j} = \left| A_{i,j} - \sum_{k=1}^{j-1} L_{i,k} D_k L_{jk}^T \right| D_j^{-1} \quad i,j=1,2,...,p \qquad 0 < i-j \leqslant p
\tag{2.2}
$$

$$
D_i = A_{ii} - \sum_{j=1}^{i-1} L_{i,j} D_j L_{i,j}^T
\tag{2.3}
$$

These equations define a complete block Choleski decomposition and we obtain the incomplete scalar decomposition by constraining each D_i to be diagonal and each L_{ij} to have the same non-zero diagonal structure as the corresponding A_{ij}. This is accomplished by modifying the matrix multiplication by diagonals algorithm and is discussed in **6**. We note that the vector lengths are N/p or slightly less. Although, in general, block Choleski is not equivalent to scalar Choleski, for the ICCG(0) no fill strategy on the p-colored matrix (2.1), our version of a block incomplete Choleski decomposition is identical to the scalar incomplete Choleski decomposition we desire.

To carry out the block incomplete decomposition, we adapt a row-wise scalar algorithm of Lambiotte [1974, p. 35]. In the calculation of L_{ij}, the quantity in parenthesis in (2.2) is stored in a temporary location to be used in (2.3) before being multiplied by D_j^{-1}; this avoids the multiplication by D_j in (2.3). Of course, all calculations which would give non-zero elements in other than the diagonal of D_i and specified diagonals of the L_{ij} are suppressed. We

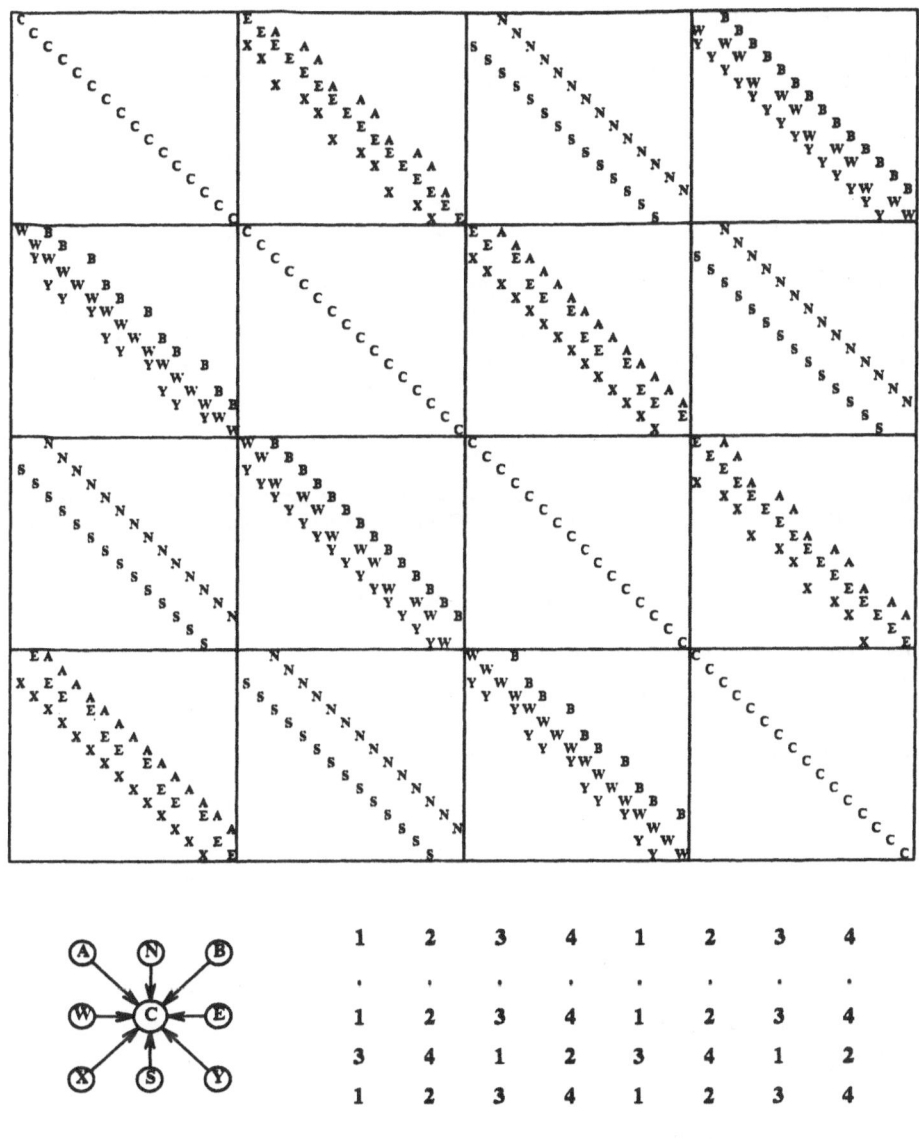

Grid Stencil

Grid Multi–Color Ordering

1	2	3	4	1	2	3	4
.	.	.	.	.	.	.	.
1	2	3	4	1	2	3	4
3	4	1	2	3	4	1	2
1	2	3	4	1	2	3	4

Multicolor Ordering for First Model Problem

Figure 1

note that other partial fill strategies could be used in our algorithm as long as no fill is allowed in the diagonal matrices D_i and is allowed only along certain diagonals in the L_{ij}.

The matrix A is, of course, required in the CG algorithm and separate storage is needed for L and D. The storage requirement for L is considerably less than that for A for small p since the first column of blocks of L is the same as in A; for example, for a 4 color matrix, the only blocks altered during the decomposition are $L_{3,2}$, $L_{4,2}$, and $L_{4,3}$. For p colors, storage requirements are $\frac{p^2-3p+2}{2}$ blocks for the L_{ij} and one N vector for D. The amount of storage for each L_{ij} depends on the number of diagonals it contains.

3. Forward and Back Substitution

The incomplete decomposition is done only once at the beginning of the ICCG(0) process while the solution of $LDL^T \hat{\underline{f}}^{k+1} = \underline{r}^{k+1}$ is done at each iteration. This is the usual 3-step process:

1) Solve $L\underline{z} = \underline{r}^{k+1}$;   2) $\hat{\underline{z}} = D^{-1}\underline{z}$;   3) Solve: $L^T \hat{\underline{f}}^{k+1} = \hat{\underline{z}}$

Our incomplete decomposition of A ensures that each diagonal block of L is the identity matrix, and this is crucial in order to achieve vector operations that are O(N/p) in steps 1) and 3) above. The first step, the forward solve, is illustrated in Figure 2 for the 4-color matrix of the first model problem. Note that the entire process consists of subtracting the products of two vectors (a diagonal of $L_{i,j}$ times the corresponding subvector of $\underline{z}$) from the appropriate subvector of $\underline{r}$. Step 3) is done in a similar way. Further details on the multiplication process are discussed in **6.** Note that the vector lengths in both 1) and 3) are the desired O(N/p), and step 2) is just a vector multiply of length N.

4. M-step ICCG(0)

Incomplete factorization of A can also be viewed as a splitting of A, thus defining an iterative method, and we consider m-step ICCG methods defined as follows (see Adams and Ortega [1982]). If

$$A = M - R, \quad M = LDL^T \tag{4.1}$$

then the inner iterative method at the (k+1)st stage is:

$$LDL^T \hat{r}_{i+1}^{k+1} = R\hat{r}_i^{k+1} + r^{k+1} \ , \ i = 0,1,...,m-1 \tag{4.2}$$

which, using R = M − A, becomes

$$LDL^T \hat{r}_{i+1}^{k+1} = (M-A)\hat{r}_i^{k+1} + r^{k+1} = M\hat{r}_i^{k+1} + (r^{k+1}-A\hat{r}_i^{k+1}) \ , \ i=0,1,...,m-1 \tag{4.3}$$

The vector $r^{k+1}-A\hat{r}_i^{k+1}$ is formed at the end of the ith step and added to the previous right hand side, $M\hat{r}_i^{k+1}$, to form the right hand side for the $i + 1^{st}$ step of the inner iteration. Note that if we take $\hat{r}_0^{k+1}=0$ and m=1, we obtain the ICCG(0) method already discussed. The

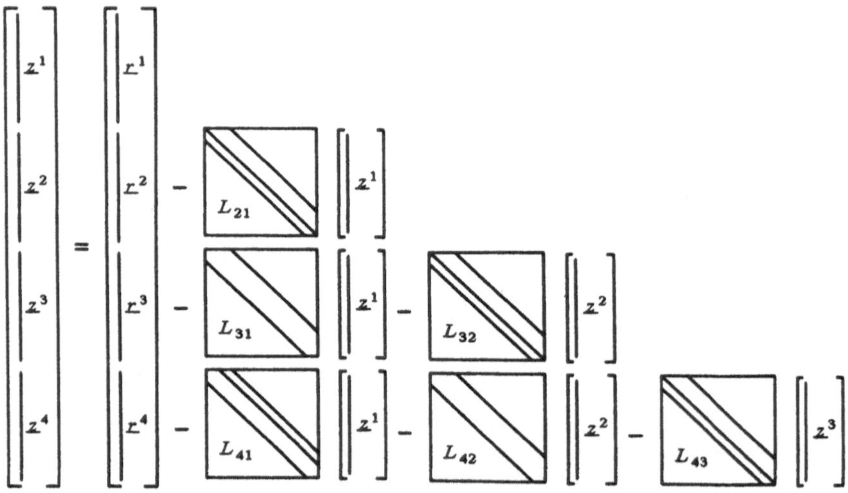

Forward Substitution, $L\underline{z} = \underline{r}$, for 4 x 4 Block Matrix

Figure 2

disadvantage of the splitting (4.1) is that R is not known explicitly and consequently $A\hat{r}_i^{k+1}$ must be calculated at each step, requiring an expensive matrix vector multiplication at each step when m is greater than 1. Results for this m—step process are discussed in **7**.

5. The Coloring Problem

Our algorithm is predicated on the coefficient matrix having the p—colored form (2.1). Determining the smallest number of colors p of a general grid for an arbitrary grid stencil is a graph coloring problem which, in general, is NP—complete (see, e.g., Horowitz and Sahni [1978]). For many problems of interest, however, it is apparent how to achieve such a coloring. Indeed, there are usually many different p—colorings and we wish to choose those which minimize storage requirements and execution time while maximizing vector lengths whenever possible.

In Figure 3 , we give two different colorings of a 6x6 grid for the first model problem. In both cases, the grid points are numbered left to right, bottom to top within each color class. Each of these colorings gives a 4—colored matrix but the detailed properties of the matrices are quite different. In a), there are 25 non-zero diagonals while in b) there are only 17 diagonals, each in general almost full with a few gaps due to boundary values. Additionally, many of the vectors in a) require storage of groups of $O(N^{1/2}/p)$. zeros which, of course, grow as the problem size increases.

Note that the same coloring pattern is used in Figure 1 as in Figure 3 (b) but because there is an 8x8 grid some of the E, W, A, B and X,Y points do not line up nicely and the matrix of Figure 1 has 23 non-zero diagonals rather than 17. This illustrates the fact that in order to obtain the minimum number of non-zero diagonals we not only need the correct coloring pattern but also the correct number of grid points. For this model problem, if the grid consists of an even number of rows and 4i + 2 columns then the coloring pattern of

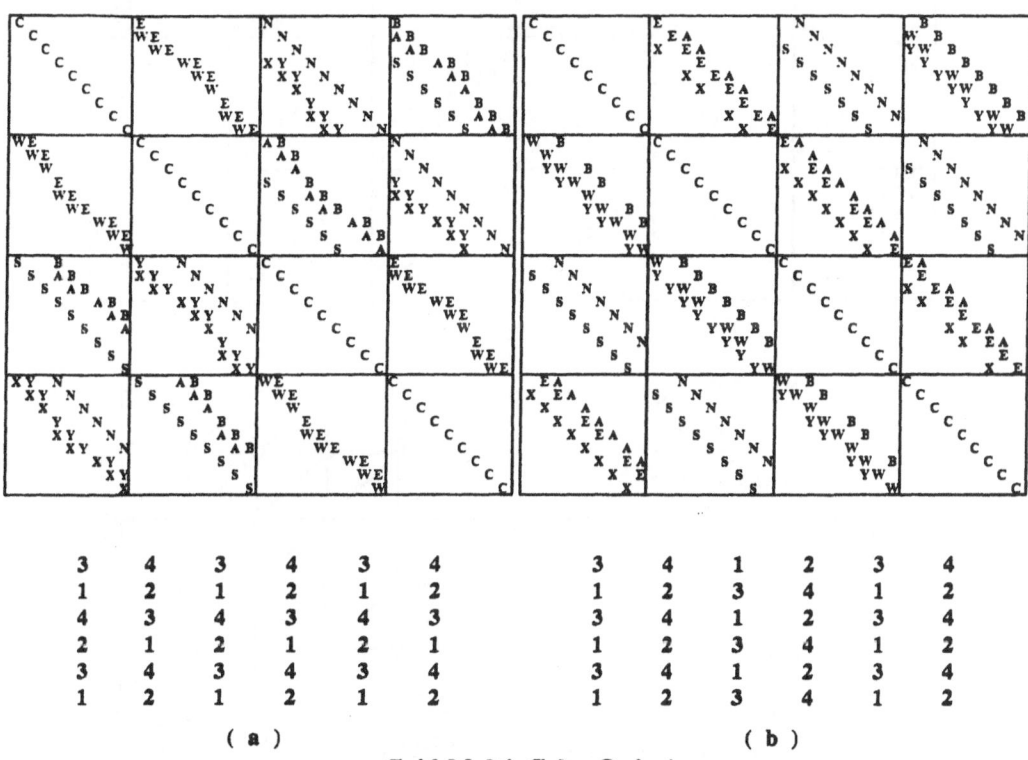

3	4	3	4	3	4
1	2	1	2	1	2
4	3	4	3	4	3
2	1	2	1	2	1
3	4	3	4	3	4
1	2	1	2	1	2

(a)

3	4	1	2	3	4
1	2	3	4	1	2
3	4	1	2	3	4
1	2	3	4	1	2
3	4	1	2	3	4
1	2	3	4	1	2

(b)

Grid Multi—Color Orderings

Multi—Color Matrices for First Model Problem

Figure 3

Figure 3(b) gives the minimum number of diagonals. In general, the coloring pattern must be replicated in a continuous fashion as one goes through the grid.

In addition to saving storage, an advantage in using an ordering in which the number of non-zero diagonals is minimized is that longer vector lengths can be used in the matrix vector multiplication done at each iteration of the outer CG loop. However, one must then add zeros where appropriate; for example, in Figure 3(b), $N^{1/2}/4+1$ zeros must be added to the "B"

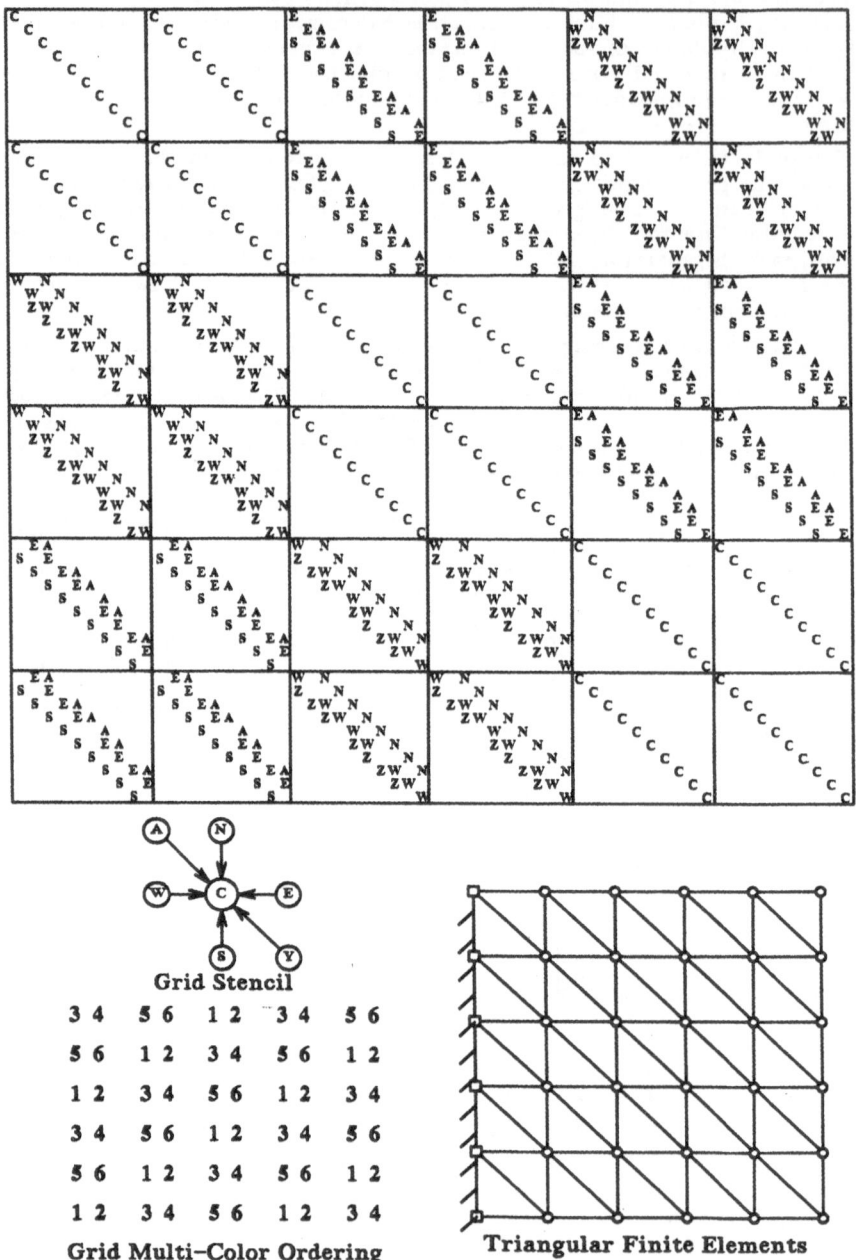

Grid Stencil

3 4	5 6	1 2	3 4	5 6
5 6	1 2	3 4	5 6	1 2
1 2	3 4	5 6	1 2	3 4
3 4	5 6	1 2	3 4	5 6
5 6	1 2	3 4	5 6	1 2
1 2	3 4	5 6	1 2	3 4

Grid Multi-Color Ordering

Triangular Finite Elements

Plane Stress Problem

Figure 4

24

vector between blocks $A_{2,1}$ and $A_{3,2}$. Execution time will be decreased as long as the number of zeros added to eliminate a startup is less than the number of calculations that could be performed during a startup. For the first model problem the conjugate gradient method without preconditioning was run on a Cyber 203 using the diagonals that line up as one vector and without doing so. The time saved was approximately 7 to 9 percent on problems that were as large as 22000 unknowns. For the Cyber 205 the startup overhead is significantly reduced and so the savings realized should not be as great. The savings is less important also as the size of the problem increases since the length of the vectors within each block becomes sufficiently large that the startup times become less significant. This lining up of vectors may however prove to be more important in three dimensional problems since even relatively small numbers of points per row generate very large matrices. Of course we still must access the vectors in each block independently for the preconditioning steps.

For our second model problem, the plane stress problem, there are 2 unknown displacements, u and v, at each grid point and although 3 colors decouple the points in the stencil the resulting matrix does not have the p–colored structure we desire. Instead 6 colors must be used to decouple the u and v unknowns at each grid point. Again we color the grid in a continuous fashion. In Figure 4 are shown the grid stencil and domain of the problem, the multi–color ordering of the gridpoints and the resulting 6–colored matrix. In general for this problem, we wish a grid of 3i + 2 columns and 3j rows. This ensures that each block of the p–colored matrix is square and of dimension N/p. Note that some of the vectors still do not line up in this ordering although most adjacent block vectors do. For example, the E vector in block $A_{5,2}$ does not line up with the W vectors of blocks $A_{4,1}$, and $A_{6,3}$. This suggests that it is not always possible to achieve the goal of lining up all adjacent vectors even when the continuous pattern rule is followed. The coloring pattern used and the constraint on the number of grid points per row is obviously dependent on the grid stencil and the number of unknowns associated with each grid point. Further research is necessary to determine what strategies are best for other problems with different stencils, especially three dimensional problems.

6. Matrix Multiplication by Diagonals

We adapt an algorithm for matrix multiplication by diagonals given by Madsen et al. [1976] for banded matrices. Recall that the main computation performed in the block incomplete factorization is incomplete matrix–matrix multiplication while matrix–vector multiplication is the main computation in the forward and back solves in each iteration as well as the formation of Ap in the CG algorithm. In all cases the matrices are stored by diagonal. The matrix–vector multiplication is the easier of the two operations and Figure 5 (a) illustrates the process. Here, a subdiagonal multiplies the corresponding first positions of w while a superdiagonal multiplies the corresponding last positions of w. These contributions are added to the correct positions of the result vector z.

The more difficult problem is the incomplete multiplication AB = C where A and B are any q x q matrices. Each diagonal of the product is of the form

$$c_k = \sum a_i b_j$$

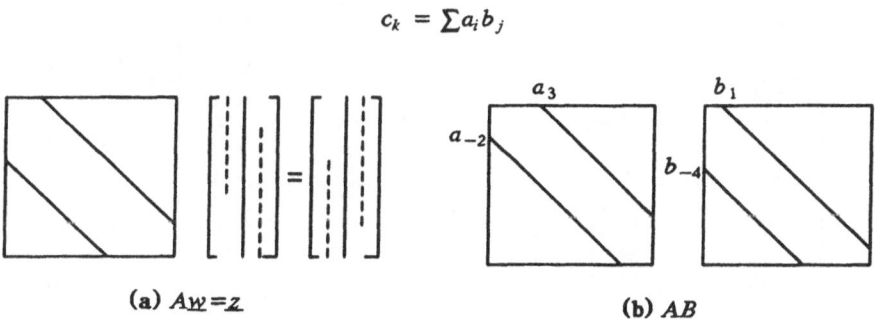

(a) $A\underline{w}=\underline{z}$ (b) AB

Matrix Multiplication by Diagonals

Figure 5

For each **k** do

    if ($i+j = k$) then do

$$L_{min} = \max(1, 1-i, 1-i-j); \quad L_{max} = \min(q, q-i, q-i, j); \quad L = L_{max} - L_{min} + 1$$

$$L_a = \max(1, 1-i); \quad L_b = \max(1-i, 1-i-j); \quad L_c = \max(1, 1-i-j)$$

$$\mathbf{a}_{start} = L_{min} - L_a + 1; \quad \mathbf{b}_{start} = L_{min} - L_b + 1; \quad \mathbf{c}_{start} = L_{min} - L_c + 1$$

Figure 6

where the a_i and b_j are diagonals of A and B and the summation is over all products which contribute to the product diagonal c_k. For the incomplete multiplication, we do only those calculations for diagonals c_k which are allowed to be non–zero.

The programming problem is to determine the diagonals of A and B which contribute to a diagonal of C, the starting positions and lengths of the diagonal operands and the start-ing position of the result in the product diagonal. For example, in Figure 5 (b), suppose that c_{-1}, the first subdiagonal of C, is an allowed non–zero diagonal. Then $a_{-2}b_1$ and a_3b_{-4} con-tribute to c_{-1} in the following way. The q–2 long vector a_{-2} multiplies the first q–2 posi-tions of the q–1 long vector b_1 and is stored in c_{-1} beginning in the second position. Then the last q–4 elements of the q–3 long vector a_3 multiply the q–4 elements of b_{-4} and the pro-duct vector is added to the first q–4 position of c_{-1}.

Our program uses three sets of positive and negative integers { i }, { j }, { k }, called offsets, to specify the non–zero diagonals of A and B and the allowable non–zero diagonals of C. A positive i denotes a non–zero diagonal i units above the main diagonal, a negative i denotes a diagonal below the main diagonal, and i=0 is the main diagonal. Similarly for j and k. A pseudo–code is given in Figure 6 which indicates how these offsets are used to determine the starting locations in diagonals of A and B which contribute to allowable diag-onals of C, the length, L, of each vector operation and the starting location in the diagonal of C for the product.

7. Results for the Model Problems

Figures 7 and 8 give results for the two model problems obtained from runs on the Cyber 203 at NASA–Langley Research Center. In each table, m=0 means that no

Model Problem No. 1: Differential Equation m–steps of Incomplete Cholesky Preconditioner					
n pts/row	N Unknowns	m Steps	# of iterations	time (sec)	time per iteration
130	16900	0	315	9.03	.029
130	16900	1	124	4.69	.037
130	16900	2	73	4.37	.060
130	16900	3	63	5.13	.081
114	12996	0	278	6.06	.022
114	12996	1	109	3.16	.029
114	12996	2	64	2.94	.046
114	12996	3	56	3.50	.063
98	9604	0	239	3.82	.016
98	9604	1	95	2.03	.021
98	9604	2	56	1.89	.034
98	9604	3	49	2.24	.046

Results for First Model Problem

Figure 7

Model Problem No. 2: Plane Stress m—steps of Incomplete Cholesky Preconditioner					
n pts/row	N Unknowns	m Steps	# of iterations	time (sec)	time per iteration
80	12960	0	744	20.19	.027
80	12960	1	298	11.54	.039
80	12960	2	172	11.35	.066
80	12960	3	155	14.36	.092
62	7812	0	580	9.77	.017
62	7812	1	235	5.57	.024
62	7812	2	134	5.34	.040
62	7812	3	121	6.74	.056
41	3444	0	393	3.21	.008
41	3444	1	158	1.83	.012
41	3444	2	90	1.73	.019
41	3444	3	82	2.18	.027
20	840	0	194	.59	.003
20	840	1	78	.36	.005
20	840	2	45	.35	.008
20	840	3	41	.46	.011
11	264	0	105	.202	.002
11	264	1	43	.135	.003
11	264	2	25	.137	.005
11	264	3	23	.179	.008

Results for Plane Stress Model Problem

Figure 8

preconditioning step was used so that the method is just the conjugate gradient iteration. The timing results are for the solution of $Ax=b$ and do not include the time for assembling the matrices or outputting results, etc. For the first problem with N = 16,400, maximum vector lengths of approximately 4200 are used in the preconditioning calculations while for the plane stress problem, with N = 12,960, maximum vector lengths in the preconditioning are about 2160. Although the number of CG iterations decreases as a function of m, for both problems the best timing results were obtained using two steps of the inner iteration. The improvement in execution time between $m=1$ and $m=2$ is slight, however, when compared to the improvement between m=0 and m=1. The matrix—vector multiplication necessary for m greater than 1 adds a large time penalty to the inner iteration.

The plane stress problem was identical to the one presented in Adams [1983b] except for the addition of one more row of nodes so that there were equal numbers of nodes of each color. For comparison, for the plane stress with 12,800 unknowns, Adams' parametrized SSOR PCG method required 395 iterations and 17.2 seconds with a one—step method and 55 iterations and 11.6 seconds with a 10—step method, the best time achieved for her method. (All runs were also on the Cyber 203 at NASA–Langley). Thus the running times for her method and ours, both for the optimal number of interior iterations, were about identical for this problem. The number of outer iterations decreases much more rapidly with m—step ICCG(0) as m increases which indicates that the formation of the incomplete factors of A gives a much better approximation to A than SSOR and if the iterations of the inner loop in ICCG for m greater than 1 could be performed without the costly additional matrix vector multiplications, m—step ICCG(0) would achieve a much greater speedup.

8. Summary

We have described an m—step incomplete Choleski preconditioned conjugate gradient method for solving large sparse symmetric positive definite systems of equations when the sparsity pattern of the matrix allows storage by diagonals after reordering the unknowns to obtain a p—colored matrix. We have shown how to obtain multi—color orderings which give good vector lengths for the two model problems. noting that the optimal orderings depend on the coloring pattern, the grid stencil, the number of grid points in each row and column and the number of unknowns.

Acknowledgement

We are indebted to Dr. J. Lambiotte of NASA–Langley Research Center for many enlightening discussions and assistance in using the CYBER 203. This research has been partially supported by NASA–Langley Research Center under grants NAG1–242 and NAG1–46.

References

Adams, L. [1983a], "Iterative Algorithms for Large Sparse Linear Systems on Parallel Computers," Ph.D. Dissertation, Applied Mathematics, University of Virginia.

Adams, L. [1983b], "An M–Step Preconditioned Conjugate Gradient Method for Parallel Computation," Proc. 1983 Intern. Conf. Par. Proc., pp. 36–43.

Adams, L. and **Ortega, J.** [1982], "A Multi–Color SOR Method for Parallel Computation," Proc. 1982 Intern. Conf. Par. Proc., pp. 53–56.

Horowitz, E. and **Sahni, S.** [1978], *Fundamentals of Computer Algorithms*, Computer Science Press, Rockville, Maryland.

Kershaw, D. [1982], "Solution of Single Tridiagonal Linear Systems and Vectorization of the ICCG Algorithm on the CRAY–1," in *Parallel Computations*, G. Rodrigue, Editor, Academic Press, New York, N.Y. pp. 85–99.

Madsen, N. , Rodrigue, G. and **Karush, J.** [1976], "Matrix Multiplication by Diagonals on Vector/Parallel Processors," Infor. Proc. Letters 5, pp. 41–45.

Lambiotte, J. [1974], "The Solution of Linear Equations On A Vector Computer," Ph.D. Dissertation, Applied Mathematics, University of Virginia.

Lichnewsky, A. [1984], "Some Vector and Parallel Implementations for Preconditioned Conjugate Gradient Algorithms." Proceedings of the NATO Workshop on High–Speed Computations. NATO ASI Series, v. F7, pp. 343–359.

Meijerink, J. and **Van der Vorst, H.** [1977], "An Iterative Solution for Linear Systems of Which the Coefficient Matrix is a Symmetric M–Matrix," Math. Comp. 31, pp. 148–162.

Meijerink, J. and **Van der Vorst, H.** [1981], "Guidelines for the Usage of Incomplete Decompositions in Solving Sets of Linear Equations," *J. Comp. Phys. 44, pp. 134–155.*

Schreiber, R. and **Tang, W.** [1982], "Vectorizing the Conjugate Gradient Method, " Proceedings Symposium Cyber 205 Applications. Ft. Collins, CO.

COMPARISON OF METHODS AND ALGORITHMS FOR TRIDIAGONAL

SYSTEMS AND FOR VECTORIZATION OF DIFFUSION COMPUTATIONS*

I. K. Abu-Shumays

Bettis Atomic Power Laboratory
P.O. Box 79, W. Mifflin, PA 15122-0079

ABSTRACT

   This work focuses on development and testing of alternative numerical
methods and computational algorithms specifically designed for the solution of tridiagonal systems and for the vectorization of diffusion computations on a Control Data Corporation Cyber 205 vector computer.

   Solution of tridiagonal systems of linear equations is a central part
for several efficient numerical methods for multidimensional diffusion
computations and is also essential for fluid flow and other physics and
engineering problems. The first part of this paper deals with the numerical solution of linear symmetric positive definite tridiagonal systems.
Among the methods tested, a combined odd-even cyclic reduction and modified Cholesky factorization algorithm is found to be the most effective
for these systems on a Cyber 205. For large tridiagonal systems, computation with this algorithm is an order of magnitude faster on a Cyber 205
than computation with the best algorithm for tridiagonal systems on a
CDC-7600.

   The above mentioned algorithm for solving tridiagonal systems is also
utilized as a basis for a new hyper-line method for implementing the red-
black cyclic Chebyshev iterative method to the solution of two-dimensional
diffusion problems. The hyper-line method is found to be competitive with
other alternative options developed in this work. This hyper-line method
has an attractive feature of being compatible with so called "concurrent"
iteration procedures whereby iterations $n+1,...,n+k$, can be started before
the completion of iteration n. This feature is very effective in balanc-
ing computations and data transfer requirements for very large diffusion
problems. Consequently implementation of the hyper-line method is suitable for certain iterative procedures used to solve large three-
dimensional diffusion problems.

   Some experience gained with Cyber 205 vector syntax statements related to diffusion computations is discussed in an Appendix to this paper.

*Work supported by the Department of Energy under Contract
DE-AC11-76PN00011.

I. INTRODUCTION

One of our main objectives is to test numerical methods and to develop algorithms to utilize features of new computers. This work is part of a continuing effort to improve the efficiency and expand the applicabilities of the Bettis PDQO8 multi-dimensional diffusion program[1,10].

Solution of tridiagonal systems of linear equations[3,4,6-9,11,12,14,15] arises in one-dimensional (1-D) diffusion computations and is a central part for several efficient numerical methods for multidimensional diffusion computations. The first part of this paper deals with the solution of linear symmetric positive definite tridiagonal systems. Several algorithms have been tested. Of these algorithms, a variant of odd-even cyclic reduction is found to be the most suitable for a Cyber 205. For large tridiagonal systems, computations with this algorithm is an order of magnitude faster on a Cyber 205 than computation with the best algorithm for tridiagonal systems on a CDC-7600. One advantage of the algorithm discussed here, relative to those given elsewhere, is the full utilization of matrix symmetry.

The second part of this paper summarizes numerical results obtained from a Cyber 205 computer program DXY constructed to compare alternative vectorization formulations of point Chebyshev and red-black line cyclic Chebyshev iterative methods[3] for the solution of a 7-point discretization of the 2-D diffusion equation. Numerical results demonstrate that the point Chebyshev iterative method, contrary to initial expectations and in spite of its being highly vectorizable, is inferior to the line cyclic Chebyshev method. The line cyclic Chebyshev iterative method involves repeated solution of tridiagonal symmetric positive definite linear systems. Moreover, the cyclic Chebyshev method permits the simultaneous solution of the tridiagonal systems and this presents an opportunity for additional vectorization.

Relative to the red-black line cyclic Chebyshev method, two alternative algorithms for vectorizing the simultaneous solution of multi-tridiagonal systems were tested: (a) a parallel line method where the factorization and subsequent solutions of all tridiagonal systems corresponding to the odd lines are carried out simultaneously, and the same is done for the even lines; and (b) a hyper-line (multi-odd, multi-even line) method where the tridiagonal systems for all the odd lines are grouped together to form a single long tridiagonal system, and the same is done for the tridiagonal systems corresponding to the even lines. For the hyper-line method, the resulting tridiagonal systems are solved by the odd-even cyclic reduction direct method described above. Of these methods, the hyper-line method is to be preferred for systems with $N_x M$ spatial mesh whenever $M < 250$ while the parallel line method is better for $M > 250$. The parallel line method is not compatible with the concurrent iteration strategy for three-dimensional computations in the PDQO8 program[3,10]. The hyper-line method is compatible with concurrent iterations and can be generalized to dynamically combine as many of the odd-numbered (even-numbered) mesh lines into hyper-lines, as would be suitable for improving the overall efficiency of three-dimensional diffusion computations.

An outline of the rest of this paper is as follows. Section II deals with solutions of tridiagonal linear systems. Section III considers the simultaneous solution of several tridiagonal systems. The contents of Section III are relevant to the vectorization of the solution of the systems of equations corresponding to a red-black ordering of mesh line as part of an overall cyclic Chebyshev iterative solution procedure. Section IV com- to the solution of 3-D diffusion problems. Figures and Tables summarizing numerical results are distributed throughout the chapter. Appendix A summarizes experience with Cyber 205 vector syntax statements.

II. SOLUTION OF TRIDIAGONAL SYSTEMS

Considerable information is available in the open literature regarding alternative solution strategies for tridiagonal systems and comparison of performance on various computers[4,6-8,11,12,14,15]. The tradition by some has been (a) to consider either symmetric positive definite or diagonally dominant tridiagonal systems but (b) to ignore symmetry in the solution strategy and note that[8] "nothing is gained computationally or storagewise in Gauss Elimination for the case when the coefficient matrix A is a symmetric matrix." The above argument, leading authors to ignore symmetry, is not necessarily valid, in the opinion of this author, when vector computers are concerned. A subroutine utilizing symmetry in cyclic reduction for solving positive definite tridiagonal systems on a Cyber 205 is developed in this work and is found to require a factor of 1.5 less storage and to be practically as efficient for very large systems as the best machine language subroutine TRID which was developed by control data for symmetric and non-symmetric tridiagonal systems (TRID is currently available on Cybernet).

The main solution methods reported in the literature for tridiagonal systems are (a) standard Gauss-Elimination[15], (b) recursive doubling[7,11,12], (c) a parallel partition method[14] and (d) odd-even cyclic reductions[6-8,11,12]. Among these methods, the odd-even cyclic reduction algorithm proved to be the superior approach for solving tridiagonal systems on the STAR-100 family of computers[7] and its characteristics suggest that it would also prove to be highly competitive on a Cyber 205 and on other vector computers. As a consequence of a literature search, the author decided to restrict the present work to odd-even cyclic reduction and to the modified LDL^T Cholesky decomposition variant of Gauss-Elimination.

The system of equations of interest here is the following

$$\underline{A}\underline{x} = \begin{bmatrix} a_1 & b_1 & & & & \\ b_1 & a_2 & b_2 & & & \\ & & & \ddots & & \\ & & & a_{n-1} & b_{n-1} \\ & & & b_{n-1} & a_n \end{bmatrix} \begin{bmatrix} x_1 \\ x_2 \\ \vdots \\ \\ x_n \end{bmatrix} = \begin{bmatrix} k_1 \\ k_2 \\ \vdots \\ \\ k_n \end{bmatrix} = \underline{k} \, , \quad (1)$$

where the tridiagonal coefficient matrix A is symmetric positive definite.

A. $\underline{LDL^T \text{ Factorization}}$

Matrix A of Eq. (1) can be factored into a lower triangular L, a diagonal D and an upper triangular matrix L^T (L transpose);

$$A = LDL^T, \tag{2}$$

where

$$L \equiv \begin{bmatrix} 1 & & & & \\ \ell_1 & 1 & & & \\ & \ell_2 & 1 & & \\ & & \ddots & \ddots & \\ & & & \ell_{n-1} & 1 \end{bmatrix} , \ D = \begin{bmatrix} d_1 & & & \\ & d_2 & & \\ & & \ddots & \\ & & & d_n \end{bmatrix} . \tag{3}$$

The main advantage of the factorization in Eq. (2) over the standard Cholesky decomposition is the fact that no square roots are involved. Square root extraction is relatively expensive on most computers.

Substituting the expressions for L, L^T and D given in Eq. (3) into Eq. (2), performing the matrix multiplications involved and equating the result to the expression for A in Eq. (1) yields

$$a_1 = d_1, \quad a_{i+1} = d_{i+1} + \ell_i d_i \ell_i, \qquad \qquad \text{(4a, b)}$$

$$b_i = \ell_i d_i, \qquad\qquad\qquad\qquad i = 1, \ldots, n-1. \qquad \text{(4c)}$$

Note from Eq. (4) that a_i may also be expressed in the form

$$a_i = d_i + b_{i-1}^2 / d_{i-1}, \qquad i > 1. \qquad\qquad\qquad \text{(5)}$$

It is clear from Eqs. (4) or (5) that the calculations required to obtain d_i are recursive in nature and that a special effort[9] is needed in order to optimize the computation for vector machines. For simplicity, only the following two algorithms for completing the factorization of Eq. (2) are examined:

(i) The algorithm based on Eqs. (4) with scalar computations for ℓ_i

$$d_1^{-1} = 1/a_1 , \qquad\qquad\qquad\qquad\qquad\qquad \text{(6a)}$$

$$t = b_{i-1}, \quad \ell_{i-1} = t \, d_{i-1}^{-1},$$

$$d_i^{-1} = 1/(a_i - t\ell_{i-1}), \qquad i=2,\ldots,n. \qquad \text{(6b)}$$

The same computer storage locations for a_i and b_i can be used for d_i^{-1} and ℓ_i. The total storage required here is $2n + 1$.

(ii) The algorithm based on Eqs. (4a, c) and (5) with vector computations for ℓ_i

$$t_i = b_i \cdot b_i, \qquad i=1, \ldots, n-1, \text{ (vectorizable)}, \qquad \text{(7a)}$$

$$d_1^{-1} = 1/a_1, \qquad d_i^{-1} = 1/(a_i - t_{i-1} d_{i-1}^{-1}), \qquad i=2, \ldots, n, \text{(7b)}$$

$$\ell_i = b_i d_i^{-1} , \qquad i=1,\ldots,n-1, \qquad \text{(vectorizable)}. \qquad \text{(7c)}$$

The same storage locations for a_i and b_i can be used for d_i^{-1} and ℓ_i. Extra $(n - 1)$ working storage locations are needed here for the dummy variable t_i.

Once the factorization is completed, the solution to the system Eq. (1), given by

$$\underline{x} = A^{-1} \underline{k} = L^{-T} D^{-1} L^{-1} \underline{k} = L^{-T}(D^{-1}(L^{-1} \underline{k})) \, , \tag{8}$$

can be accomplished by the following successive computations:

$$y_1 = k_1, \qquad y_{i+1} = k_{i+1} - \ell_i y_i, \qquad i=1,\dots,n-1. \tag{9a}$$

$$z_i = y_i \, d_i^{-1}, \qquad i=1,\dots,n. \tag{9b}$$

$$x_n = z_n, \qquad x_i = z_i - \ell_i x_{i+1}, \qquad i = n-1, n-2, \dots, 2, 1. \tag{9c}$$

Note that the computations in Eqs. (9a) and (9c) are recursive; these computations can be optimized by using special Cyber 205 STACKLIB routines[2].

The treatment in this section is aimed primarily at symmetric positive definite tridiagonal systems which are solved repeatedly, possibly with several right hand sides. When solutions of the same tridiagonal systems are required a large number of times, as in the solution of the eigenvalue problem

$$A\underline{x} = \lambda B\underline{x} \, , \tag{10}$$

considerable additional gain can be realized by a normalization to eliminate the step Eq. (9b) from the computations. For this case the desired normalization can be obtained by multiplying Eq. (10) by $D^{-1/2}$, making the following changes in variables

$$\underline{y} \equiv D^{1/2} \underline{x} \, , \qquad \hat{B} \equiv D^{-1/2} B \, D^{-1/2} \, , \tag{11}$$

and modifying the factorization in Eq. (2) to transform Eq. (10) into the form

$$\hat{L} \, \hat{L}^T \underline{y} = \lambda \hat{B} \underline{y} \, , \tag{12}$$

where $\hat{L}$ can be deduced from L by replacing ℓ_i by

$$\hat{\ell}_i \equiv \ell_i \, d_i^{1/2}/d_{i+1}^{1/2} \, . \tag{13}$$

Each step of the solution of the system Eq. (12) with a fixed right-hand side (such as in a power method) would involve steps analogous to Eqs. (9a, c) without the need for repetitive multiplications as in Eq. (9b). The subject normalization is an integral part of the Bettis PDQ08 diffusion program[10]. In summary, the advantage of the normalization as outlined above is expected to be significant for most eigenvalue computations. This advantage was not numerically tested in this study, which was limited by choice to source problems where the additional cost of the normalization is not justified.

B. Odd-Even Cyclic Reduction

Cyclic reduction has proved to be a very effective method to vectorize the solution of tridiagonal systems by basically carrying out repetitive block (rather than point) factorization steps, followed by block solution steps[6-8,11,12]. Each block step is preceded by a permutation step to locally separate odd and even numbered unknowns. For illustration, a permutation or regrouping of the unknowns in Eq. (1) into odd-numbered unknowns

$x_1, x_3, x_5 \ldots$, followed by even-numbered unknowns $x_2, x_4, x_6, \ldots$, would transform Eq. (1) into the form

$$A_1 \, \underline{x}_1 = \begin{bmatrix} a_1 & & & & & \vline & b_1 & & & & \\ & a_3 & & & & \vline & b_2 & b_3 & & & \\ & & \cdot & & & \vline & & b_4 & \cdot & & \\ & & & \cdot & & \vline & & & \cdot & \cdot & \\ & & & & \cdot & \vline & & & & \cdot & \\ \hline & & & & a_{n'} & \vline & a_2 & & & & \cdot \\ b_1 & b_2 & & & & \vline & & a_4 & & & \\ & b_3 & b_4 & & & \vline & & & \cdot & & \\ & & \cdot & \cdot & & \vline & & & & \cdot & \\ & & & \cdot & \cdot & \vline & & & & & \cdot \\ & & & & & \vline & & & & & a_{n''} \end{bmatrix} \begin{bmatrix} x_1 \\ x_3 \\ \cdot \\ \cdot \\ \cdot \\ x_{n'} \\ x_2 \\ x_4 \\ \cdot \\ \cdot \\ x_{n''} \end{bmatrix} = \begin{bmatrix} k_1 \\ k_3 \\ \cdot \\ \cdot \\ \cdot \\ k_{n'} \\ k_2 \\ k_4 \\ \cdot \\ \cdot \\ k_{n''} \end{bmatrix} = \underline{k}_1 \quad (14)$$

where,

$$n' \equiv n, \qquad n'' \equiv n-1 \quad \text{for n odd,} \tag{15a}$$

$$n' \equiv n-1, \qquad n'' \equiv n \quad \text{for n even.} \tag{15b}$$

The coefficient matrix A_1 in Eq. (14) can be block factored into the form

$$A_1 = \begin{bmatrix} A_{11} & A_{12} \\ A_{12}^T & A_{22} \end{bmatrix} = \begin{bmatrix} I & 0 \\ A_{12}^T A_{11}^{-1} & I \end{bmatrix} \begin{bmatrix} A_{11} & 0 \\ 0 & (A_{22} - A_{12}^T A_{11}^{-1} A_{12}) \end{bmatrix} \begin{bmatrix} I & A_{11}^{-1} A_{12} \\ 0 & I \end{bmatrix}, \tag{16}$$

which can be abbreviated as

$$A_1 = L_1 D_1 L_1^T \, . \tag{17}$$

Note the following from Eqs. (14)-(17):

(a) A_{11} is a diagonal matrix whose elements are the odd diagonal elements $a_1, a_3, \ldots$ of the original matrix A of Eq. (1). Thus the inverse of A_{11}, denoted by

$$A_{11}^{-1} = \text{Diag} \, [a_1^{-1}, \, a_3^{-1}, \, \ldots, \, a_{n'}^{-1}] \, ,$$

$$\equiv \text{Diag.} \, [d_1^{-1}, \, d_2^{-1}, \, \ldots, \, d_{n_1}^{-1}], \quad n_1 = (n' + 1)/2 \, , \tag{18}$$

can be easily computed on a vector computer by a vector division operation.

(b) The matrix product $A_{12}^T A_{11}^{-1}$ is an upper bi-diagonal matrix whose elements are given by

$$\text{DIAGONAL } (A_{12}^T A_{11}^{-1}) = \{b_1 a_1^{-1},\ b_3\ a_3^{-1},\ \ldots,\ b_{n''-1} a_{n''-1}^{-1}\},$$

$$\equiv \{\ell_1,\ \ell_2,\ \ldots,\ \ell_{n_2}\},\qquad n_2 = n''/2 \qquad (19a)$$

$$\text{UPPER BI-DIAGONAL } (A_{12}^T A_{11}^{-1}) = \{b_2 a_3^{-1},\ b_4 a_5^{-1},\ \ldots,\ b_{n'-1} a_{n'}^{-1}\}$$

$$\equiv \{\ell_{n_2+1},\ \ell_{n_2+2},\ \ldots,\ \ell_{n-1}\}. \qquad (19b)$$

Note that the elements of $A_{12}^T A_{11}^{-1}$ in Eqs. (19a, b) can be computed by vector multiplications.

(c) The matrix

$$\tilde{A}_{22} \equiv A_{22} - A_{12}^T A_{11}^{-1} A_{12}, \qquad (20a)$$

is a symmetric positive definite tridiagonal matrix whose elements are given by

$$\text{DIAGONAL } (\tilde{A}_{22}) = \{(a_2 - b_1 \cdot b_1 a_1^{-1} - b_2 \cdot b_2 a_3^{-1}),\ (a_4 - b_3 \cdot b_3 a_3^{-1} - b_4 \cdot b_4 a_5^{-1}),$$

$$\ldots,\ (a_{n''} - b_{n''-1} \cdot b_{n''-1} a_{n''-1}^{-1} - b_{n''} \cdot b_{n''} a_{n''+1}^{-1})\}$$

$$\equiv \{\tilde{a}_1,\ \tilde{a}_2,\ \ldots,\ \tilde{a}_{n''/2}\}, \qquad (20b)$$

$$\text{UPPER BI-DIAGONAL } (\tilde{A}_{22}) = \text{LOWER BI-DIAGONAL } (\tilde{A}_{22})$$

$$= \{b_2 \cdot b_3 a_3^{-1},\ b_4 \cdot b_5 a_5^{-1},\ \ldots,\ b_{n''-2} \cdot b_{n''-1} a_{n''-1}^{-1}\}$$

$$\equiv \{\tilde{b}_1,\ \tilde{b}_2,\ \ldots,\ \tilde{b}_{n''/2-1}\}. \qquad (20c)$$

It is assumed here that

$$a_{n''+1}^{-1} \equiv 0 \qquad \text{for} \qquad n'' > n'. \qquad (20d)$$

It follows from the form of Eqs. (20a-d) that the computation of the elements of the tridiagonal matrix $\tilde{A}_{22}$ involves vector multiplications and vector subtractions.

The permutation applied to Eq. (1) to deduce Eq. (14), and the subsequent block factorization in Eqs. (16)-(20) can be denoted by

$$A = P_1 A_1 P_1^T = P_1 L_1 D_1 L_1^T P_1^T \qquad (21)$$

where P_1 is a permutation matrix which separates the odd and even unknowns, L_1 is the lower triangular matrix defined in Eqs. (16)-(19) and D_1 is the matrix with block diagonal elements A_{11} and $\bar{A}_{22}$. The above permutation and factorization procedure can now be applied to the symmetric tridiagonal positive definite principal sub-matrix $\bar{A}_{22}$ and can also be repeatedly applied to subsequent smaller and smaller tridiagonal systems to result in a global factorization represented symbolically as

$$A = P_1 L_1 \ P_2 L_2 \ \cdots \ P_r L_r D_r L_r^T P_r^T \ \cdots \ L_2^T P_2^T L_1^T P_1^T \ , \tag{22}$$

where D_r is a diagonal matrix. Note that the permutation matrices are orthogonal so that $P_i^T = P_i^{-1}$.

The above is a summary of the cyclic reduction algorithm. Computer storage allocation for this algorithm is as follows: (i) N storage locations for array D containing inverses of successive groups of odd-numbered diagonal elements in D_i, (ii) 2N storage locations for array L containing all arrays L_i corresponding to the appropriately transformed lower (upper) bidiagonal elements, (iii) N/2 temporary working storage locations used to enhance the vectorization, and (iv) 18 storage locations for an integer array M whose elements $M(i)=N_i$ are the successive orders of tridiagonal systems resulting at each stage of the cyclic reduction algorithm. M(16) is reserved for storage of the number of distinct steps of the odd-even cyclic reduction algorithm.

Once the cyclic reduction of a tridiagonal symmetric positive definite matrix is completed, the repeated solution of systems of the form of Eq. (1) can be accomplished by a series of successive permutations and forward elimination steps followed by successive permutations and backward substitution steps. This solution process would be made clearer by combining Eqs. (1) and (22) to represent the system under consideration in the form

$$P_1 L_1 P_2 L_2 \ \cdots \ P_r L_r DL_r^T P_r^T \ \cdots \ L_2^T P_2^T L_1^T P_1^T \ \underline{x} = \underline{k} \ . \tag{23}$$

Symbolically, the solution $\underline{x}$ can be achieved by multiplying Eq. (23) from the left successively by P_1^T, L_1^{-1}, ..., P_r^T, L_r^{-1}, D^{-1} as successive steps of forward elimination, and subsequently multiplying the resulting system from the left, successively by L_r^{-T}, P_r, ..., L_1^{-T}, P_1 as successive steps of backward substitution. Details of the solution of Eq. (23) are omitted for brevity. It suffices to say that the forward elimination and backward solution steps involved are all vectorizable with vector length varying from N, the order of the tridiagonal system, [N/2], [N/4], ..., to 1.

REMARK. The primary objective above is to solve tridiagonal positive definite linear systems as efficiently as possible on vector computers. For eigenvalue problems, as in Eq. (10), a normalization similar to that summarized in Eqs. (11)-(13) for the LDL^T factorization can be accomplished to reduce the diagonal matrix D_r in Eq. (22) to the identity matrix. Details of this normalization are omitted for brevity. The extra cost of the normalization is fully justified by the expected saving in overall eigenvalue computations.

C. Modified Odd-Even Cyclic Reduction

In the cyclic reduction method given above, one starts with a symmetric positive definite tridiagonal system, and reduces it by a permutation and block factorization into a new sub-system which is also a symmetric positive definite tridiagonal system of approximately half the size of the original system. The permutation and subsequent block factorization steps are vectorizable with an effective vector length half the order of the tridiagonal system involved. This process of permutation followed by block factorization is then repeatedly applied to the resulting smaller and smaller tridiagonal symmetric positive definite sub-systems until the order of the last sub-system is reduced to one. Observe that the effective vector length of the permutation and factorization process is initially half the order of the original tridiagonal system and this effective vector length is cut in half in each of the above successive steps. Once the vector length drops below a certain value, say 4, 8, 16, etc., this value to be determined based on the properties of the computer used (based on the relation between scalar and vector computations), it is advisable to terminate the cyclic reduction cycle and solve the remaining tridiagonal sub-system by a variant of Gaussian Elimination. Here the modified odd-even cyclic reduction of symmetric positive definite tridiagonal systems refers to the procedure of (a) applying odd-even cyclic reduction for systems and sub-systems of size larger than $N_c = 2^m$, m being a user-specified number and (b) applying the LDL^T factorization and subsequent solution as described above, to sub-systems of size 2^m or less.

D. Numerical Results

A number of subroutines which utilize matrix symmetry to solve tridiagonal positive definite symmetric linear systems were constructed and applied to problems with between 2 and 64000 unknowns. These subroutines include (i) LDL^T of Eqs. (6a, b), (ii) $LDL^T V$ of Eqs. (7a-c), (iii) $SLDL^T$ of Eqs. (9a-c), (iv) CYRED Cyclic Reduction of Eqs. (14)-(23), (v) CYSOL Cyclic Solution of Eq. (23), (vi) CYREDV Modific Cyclic Reduction combining CYRED and LDL^T, and (vii) CYSOLV Modified Cyclic Solution combining CYSOL and $SLDL^T$ as indicated in Section IIC above. The latest versions of the last four subroutines mentioned above employ Cyber 205 Q8 vector compress and merge instructions Q8VCMPRS and Q8VMERG[2].

All the subroutines developed in this work were verified for model problems and were shown to yield accurate answers. A sample timing comparison for the subroutines for symmetric positive definite tridiagonal systems is presented in Tables Ia, b. The timings presented here are for runs on the Cybernet Cyber 205 with the FORTRAN 2.0 compiler dated October 26, 1983. Tests for large tridiagonal systems showed that the computation on the Cybernet Cyber 205 is up to 16% more efficient than on the Bettis Cyber 205 with the FORTRAN 2.1 compiler dated October 26, 1983.

The results quoted in the tables for the modified cyclic reduction correspond to the cutoff value between vector and scalar computations of $N_c = 2^5$ which is regarded here as the reference case for a Cyber 205. The entries in the row and column labeled RATIOS in the tables indicate the time ratios for the various methods as compared to this reference case for the modified cyclic reduction/solution method.

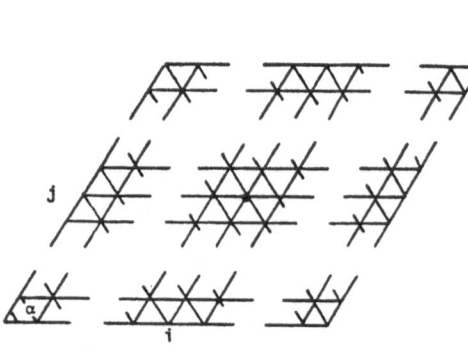

Fig. 1. A Typical Triangulated Parallelogram Mesh

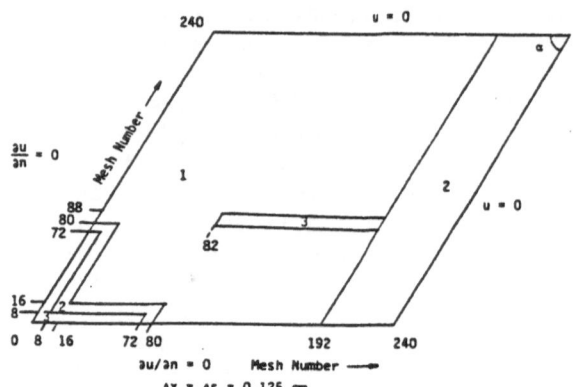

Fig. 2. Modified Wachpress Problem See Ref. 15, Fig. 8.7 and Tables 8.7 and 8.10

Table I.a

Comparison of Factorization Time for Symmetric Tridiagonal Systems

Order N	CDC-7600[a]			CYBERNET CYBER 205					
	LDL^T (μ-sec)	LDL^Tv (μ-sec)	Cyclic Reduction (μ-sec)	LDL^T (μ-sec)	LDL^Tv (μ-sec)	Cyclic Reduction (μ-sec)	Modified Cyclic $N_C=2^5=32$ (μ-sec)	TRID (μ-sec)	(ratios)[b]
16	26	41	123	37	40	75	48	43	(.90)
32	51	81	220	70	74	94	71	57	(.80)
64	99	161	408	136	141	116	83	74	(.89)
128	196	319	772	268	274	145	108	100	(.93)
256	389	637	1490	531	540	185	145	141	(.97)
512	776	1270	2917	1059	1072	248	206	211	(1.02)
1024	1553	2538	5740	2114	2138	357	312	341	(1.09)
2048	3102	5074	11405	4224	4270	560	512	589	(1.15)
4096	6000	10000	95000	8446	8532	951	902	1075	(1.19)
8192	12500	22500	190000	16884	17054	1710	1659	2035	(1.23)
16384	25000	45500	380000	33764	34105	3215	3164	3946	(1.25)
20000	31000	55000	239000			3865	3820	4799	(1.26)
32000	49000	75000	687000			6056	6030	7573	(1.26)
40000	61000	110000	610000			7530	7486	9435	(1.26)
64000				131874	132173	11992	11962	14993	(1.25)
Ratios[b]	.54– 8.1	.85– 14.7	2.56– 81.5	.77– 10.7	.83– 10.8	1.002– 1.56	1.0	.80– 1.26	

[a]Each entry for the CDC-7600 is the average of several runs; Large Core Memory (LCM) was used for systems of order N > 8192.

[b]Time ratios for the various methods as compared to the modified cyclic reduction/solution method.

38

Table I.b

Comparison of Solution Times for Already Factored Tridiagonal Systems

Order N	CDC-7600[a] SLDL^T (μ-sec)	Cyclic Solution (μ-sec)	CYBERNET CYBER 205 SLDL^T (μ-sec)	Cyclic Solution (μ-sec)	Modified Cyclic $N_c=2^5=32$ (μ-sec)	TRID (μ-sec)	(ratios)
16	38	117	50	66	30	42	(1.40)
32	73	205	58	82	46	56	(1.22)
64	143	368	73	100	63	72	(1.14)
128	284	686	113	124	86	93	(1.08)
256	566	1312	191	158	119	124	(1.04)
512	1130	2556	340	212	171	177	(1.04)
1024	2257	5015	637	305	264	268	(1.02)
2048	4515	9942	1240	477	436	440	(1.01)
4096	9000	55000	2445	808	767	771	(1.01)
8192	19000	111000	4848	1456	1414	1420	(1.00)
16384	39000	223000	9654	2738	2696	2704	(1.00)
20000	47000	246000		3313	3276	3280	(1.00)
32000	88000	408000		5232	5210	5198	(1.00)
40000	94000	427000		6428	6381	6373	(.999)
64000			37610	10411	10384	10164	(.979)
Ratios	1.27–14.7	3.90–66.9	1.16–3.62	1.003–1.78	1.0	.98–1.40	

[a]See footnote to Table Ia.

Table II.a

Comparison of Algorithms for Factorization of Multi N_xM Symmetric Tridiagonal Systems

N_xM	LDL^T-2D + TPMOV (μ-Sec)	LDL^{T}1-2D + TPMOV (μ-Sec)	LDL^{T}1-2D + GATHR SCATR (μ-Sec)	TRID (μ-Sec)	Modified Cyclic $N_c=2^5=32$ (μ-Sec)
200x100=20000	4240	4784	4782	4720 (1.11)[a]	3816 (0.90)[a]
256x125=32000	6434	6604	9926	7488 (1.16)	6023 (0.94)
200x200=40000	7378	7526	7576	9353 (1.27)	7502 (1.02)
400x160=64000	11902	11218	11528	14916 (1.25)	11973 (1.01)
320x200=64000	11800	11243	11187	14906 (1.26)	11964 (1.01)
256x250=64000	11448	10978	17720	14869 (1.30)	11951 (1.04)
200x320=64000	11056	10714	10950	14891 (1.35)	11968 (1.08)
20000x1=20000				4799	3820
32000x1=32000				7573	6030
40000x1=40000				9436	7486
64000x1=64000				14993	11962

[a]Ratio TIME (•)/TIME (LDL^T-2D + TPMOV)

Table II.b

Comparison of Algorithms for Solution of Factored Multi N_xM Symmetric Tridiagonal Systems

N_xM	SLDL^T-2D + GATHR SCATR (μ-Sec)	SLDL^{T}2D + TPMOV (μ-Sec)	TRID (μ-Sec)	Modified Cyclic $N_c=2^5=32$ (μ-Sec)
200x100=20000	3305	3364	3218 (0.96)[a]	3274 (0.97)[a]
256x125=32000	8436	5109	5129 (1.00)	5207 (1.02)
200x200=40000	5842	5908	6285 (1.06)	6388 (1.08)
400x160=64000	9828	9520	10097 (1.06)	10378 (1.09)
320x200=64000	9284	9365	10101 (1.08)	10386 (1.11)
256x250=64000	15892	9210	10064 (1.09)	10379 (1.13)
200x320=64000	8890	8914	10064 (1.13)	10378 (1.16)
20000x1=20000			3280	3276
32000x1=32000			5198	5210
40000x1=40000			6373	6381
64000x1=64000			10164	10384

[a]Ratio TIME (•)/TIME(SLDL^T + TPMOV)

The columns in these tables under the heading TRID refer to timing results obtained when applying the Control Data cyclic reduction/solution subroutine TRID available on Cybernet.

The tables presented and other results omitted indicate the following: Cyclic reduction/solution is considerably worse (factors of up to 14) than the standard LDL^T factorization and subsequent solution on a CDC-7600. The reverse is true for large tridiagonal systems on a Cyber 205. The modified cyclic reduction/solution on a Cyber 205 is better than the LDL^T factorization/solution for systems of order 64 or more and the advantage of the modified cyclic reduction solution becomes more pronounced for very large systems of order 1000-64000 where the vectorization is most effective.

Of particular interest are the following: (a) for the factorization part of the modified odd-even cyclic reduction procedure, subroutine CYREDV is better than the machine language subroutine TRID for systems of order $N > 512$ and is asymptotically 25% better than TRID for systems with $N = 64,000$; (b) for the solution part, the modified cyclic solution subroutine CYSOLV is better than TRID for systems with $N < 20,000$ and is asymptotically worse than TRID by less than 3% for very large systems. The main advantage of the reference modified cyclic reduction/solution method implemented in the subroutines CYREDV and CYSOLV is that it utilizes matrix symmetry and requires a factor of 1.5 less storage than subroutine TRID.

III.   SOLUTION OF MULTI-TRIDIAGONAL SYSTEMS

Often in practice one is confronted with the requirement to solve many tridiagonal linear systems of the same order N, where N is of a moderate size. This situation arises for example when implementing red-black line methods for solving multi-dimensional diffusion equations.

Subroutine TRID has the option of solving multi-tridiagonal systems of the same order N. The efficient subroutines for the modified odd-even cyclic reduction/solution method developed in this work and discussed in the previous section can also be applied to solve several symmetric positive definite tridiagonal systems. Here, the coefficient matrices of the given tridiagonal systems are grouped together and regarded as separate diagonal blocks of a single large tridiagonal matrix whose order is the sum of the orders of the separate systems.

Another efficient algorithm to solve M symmetric positive definite tridiagonal systems, each of order N, is to apply the LDL^T factorization/solution procedure to all of these systems simultaneously, computing the factorization, forward elimination and backward substitution for the corresponding unknowns of the M systems in parallel. Thus while the LDL^T factorization/solution procedure for a single system is predominantly scalar, applying the method to M systems simultaneously results in converting every step of the method to a vector computation with vector length M. The following subroutines to apply the LDL^T factorization/solution procedure for M systems were constructed: LDL^T-2D, LDL^{T}1-2D and SLDL^T-2D generalize Eqs. (6a, b), (7a-c) and (9a-c), respectively to treat M symmetric positive definite tridiagonal systems.

Tables IIa, b present a timing comparison for alternative methods to solve M tridiagonal positive definite systems, each system having N unknowns. The reordering of the M systems required to vectorize LDL^T can be accomplished on a Cyber 205 by applying the Q8 gather-scatter vector operations[2] or applying the Control Data matrix transpose subroutine TPMOV. Because of the particular choice of computer storage for the M tridiagonal systems adopted in this work, the reordering required here can be shown to be equivalent to matrix transpose operations (details are omitted for brevity).

A peculiar behavior of the gather-scatter operations as compared to TPMOV for certain combinations of N and M is exhibited in Tables IIa, b for example, for N = 256, M = 125 or N = 256, M = 250.

Tables IIa, b and other results omitted for brevity indicate that the combination modified odd-even cyclic reduction and solution is better than the combination LDL^T-2D, SLDL^T-2D and TPMOV for M < 125 and that the reverse is true for M > 125.

Finally, note from Tables IIa, b that there is no advantage to using TRID for a large number M of tridiagonal systems, except possibly when the individual systems lack symmetry.

IV. TWO-DIMENSIONAL DIFFUSION COMPUTATIONS

Of primary interest in this work is the vectorization of multidimensional diffusion computations. The current phase of this work is restricted to the solution of the fixed source one energy group two-dimensional diffusion equation

$$- \nabla \cdot D \nabla u(\underline{x}) + \sigma u(\underline{x}) = S , \qquad (24)$$

over a parallelogram domain in cartesian x-y geometry. Zero flux u or zero current $\partial u/\partial n = 0$ is allowed on each boundary. Parallelogram mesh subdivisions as shown in Figure 1 are allowed where each parallelogram mesh is further subdivided along its shortest diagonal. The diffusion coefficient D, the cross-section σ and the source term S in Eq. (24) are assumed to be constant over each parallelogram mesh subdivision.

The standard low order finite difference, finite element and pseudo finite element discretizations[1,10] of the diffusion equation (24) over a mesh subdivision such as shown in Figure 1, yield systems of linear algebraic equations with so called 7-point couplings. A typical equation corresponding to the mesh point i,j is given by

$$a_{ij} u_{ij} + a_{i-1,j} u_{i-1,j} + a_{i+1,j} u_{i+1,j} + a_{i,j-1} u_{i,j-1}$$

$$+ a_{i+1,j-1} u_{i+1,j-1} + a_{i-1,j+1} u_{i-1,j+1} + a_{i,j+1} u_{i,j+1} = b_{ij} . \quad (25)$$

The system of linear equations under consideration can also be written in the compact form

$$A \underline{u} = \underline{b} , \qquad\qquad (26)$$

where the coefficient matrix A is a symmetric positive definite block tridiagonal matrix

$$A = \begin{bmatrix} A_{11} & A_{12} & & & & \\ A_{12}^T & A_{22} & A_{23} & & & \\ & A_{23}^T & A_{33} & A_{34} & & \\ & & \cdot & \cdot & \cdot & \\ & & & \cdot & \cdot & \cdot \\ & & & & \cdot & \cdot & \cdot \\ & & & & & A_{J-1,J}^T & A_{JJ} \end{bmatrix} , \qquad (27)$$

The diagonal blocks A_{jj} are symmetric positive definite tridiagonal matrices and the off diagonal blocks $A_{j,j+1}$ are lower bidiagonal matrices. A symbolic representation of the matrix A is as follows

$$A = \qquad\qquad . \qquad\qquad (28)$$

Note from Eqs. (25)-(28) and from the above description that the matrix A has a regular sparse structure. Furthermore, because of symmetry, the matrix A can be represented by the following four arrays.

(i) A_0 is an array of coefficients a_{ij} of Eq. (25) corresponding to diagonal elements of A.

(ii) A_1 is an array of coefficients $a_{i+1,j}$ of Eq. (25) which constitute the first upper (or lower) diagonal elements of A.

(iii) A_2 is an array of coefficients $a_{i-1, j+1}$ of Eq. (25) containing the second non-zero upper (or lower) diagonal elements of A. A_2 contains the lower bidiagonal elements of the submatrices $A_{\ell,\ell+1}$ of Eq. (27).

(iv) A_3 is an array of coefficients $a_{i,j+1}$ of Eq. (25) containing the third and last non-zero upper (or lower) diagonal elements of A. A_3 contains the diagonal elements of the submatrices $A_{\ell,\ell+1}$ of Eq. (27).[3]

The seven point discretization of the diffusion equation described above was adopted in this work for the sake of illustration. The relative behavior of the various solution methods described below for a seven point coupling discretization is expected to remain approximately the same for a nine point coupling discretization resulting from more general triangulations.

A Cyber 205 experimental computer program DXY to solve the Diffusion equation in XY geometry was constructed. The program DXY allows for a mesh, such as in Figure 1, accepts any number of distinct mesh regions with constant diffusion coefficients D, cross-section σ and source term S, and generates the corresponding discretized equation $A\underline{u} = \underline{b}$.

The program DXY has as options the solution of the resulting system of equations by one of the following iterative methods: (a) Point Chebyshev, (b) Red-Black Line Cyclic Chebyshev with the equations for all the red (black) lines solved in parallel, (c) Red-Black Line Cyclic Chebyshev with the separate equations for all the red (black) lines combined to form a single equation with a tridigonal coefficient matrix for a hyper-line corresponding to the red (black) lines combined, (d) Red-Black Line Cyclic Chebyshev with the equations for the distinct red (black) lines solved separately using scalar recursive relations and (e) the same method as in (d) above but with the recursive relations replaced by Q8 Cyber 205 vector syntax STACKLIB routines[2]. Methods (d) and (e) above simulate the method used in PDQ08. The above alternative methods are described briefly in the following sub-sections.

A. Point Chebyshev Iterative Method

The aim here is to outline the steps of the point Chebyshev iterative method as they are applied in the model DXY program to solve the discretized two-dimensional diffusion equation embodied in Eqs. (25)-(28) above. Eq. (25) can be solved for u_{ij} in terms of the other unknowns by transferring all terms other than the first term in Eq. (25) to the right-hand side and dividing by a_{ij} to get

$$u_{ij} = \frac{b_{ij}}{a_{ij}} - \frac{a_{i-1,j}}{a_{ij}} u_{i-1,j} - \frac{a_{i+1,j}}{a_{ij}} u_{i+1,j} - \frac{a_{i,j-1}}{a_{ij}} u_{i,j-1}$$

$$- \frac{a_{i+1,j-1}}{a_{ij}} u_{i+1,j-1} - \frac{a_{i-1,j+1}}{a_{ij}} u_{i-1,j+1} - \frac{a_{i,j+1}}{a_{ij}} u_{i,j+1}$$

$$\equiv k_{ij} - \hat{a}_{i-1,j} u_{i-1,j} - \hat{a}_{i+1,j} u_{i+1,j} - \hat{a}_{i,j-1} u_{i,j-1}$$

$$- \hat{a}_{i+1,j-1} u_{i+1,j-1} - \hat{a}_{i-1,j+1} u_{i-1,j+1} - \hat{a}_{i,j+1} u_{i,j+1} . \quad (29)$$

The notation $\hat{a}$ in Eq. (29) denotes normalized coefficients. The normalization in Eq. (29) is vectorizable as follows: (a) the elements a_{ij} of A_0 are replaced by their inverses by a vector operation to yield the array A_0^{-1} and (b) each element of A_i, $i=1,2,3$, is multiplied by the corresponding element of A_0^{-} to yield the corresponding arrays of normalized coefficients $\hat{A}_i$. The normalization as described here is very effective as it saves a considerable number of arithmetic operations. The system of Eqs. (29) corresponding to all the mesh points can be written in matrix notation as

$$\underline{u} \equiv G\underline{u} + \underline{k} . \quad (30)$$

The optimal point Chebyshev iterative method as applied to the system Eq. (30) is given by[3]

$$\underline{u}^{(n+1)} = \overline{\rho}_{n+1} \left\{ \overline{\gamma} (G\underline{u}^{(n)} + \underline{k}) + (1 - \overline{\gamma})\underline{u}^{(n)} \right\} + (1 - \overline{\rho}_{n+1})\underline{u}^{(n-1)} , \quad (31)$$

43

where

$$\overline{\gamma} = 2/\{2 - M(G) - m(G)\} \ , \tag{32}$$

$$\overline{\sigma} = \{M(G) - m(G)\}/\{2 - M(G) - m(G)\} \ , \tag{33}$$

$$\overline{\rho}_1 = 1, \qquad \overline{\rho}_2 = 1/(1 - \overline{\sigma}^2/2) \ ,$$

$$\overline{\rho}_{n+1} = 1/(1 - \overline{\sigma}^2 \ \overline{\rho}_n/4) \ , \qquad n > 2 \ , \tag{34}$$

and where $m(G)$ and $M(G)$ denote, respectively, the algebraically smallest and largest eigenvalues of G.

The optimal acceleration parameters in Eqs. (31)-(34) are not known in advance. The adaptive Chebyshev procedure of Hageman[3], whereby the various iteration parameters are estimated and whereby the estimates are updated and improved as the solution steps proceed, is indeed very effective in the present situation. His subroutine CHEBY[3] is applied in this work to estimate the acceleration parameters and to monitor convergence of the point Chebyshev iterations.

The subroutine CHEBY[3] accepts as input estimates m_E and M_E of the eigenvalues $m(G)$ and $M(G)$ of the iteration matrix G. Theoretically $m_E > - 3$ for the 7-point discretization of the diffusion equation under consideration, and thus $m_E = - 3$ can be used. However, a better approach here is to start with $m_E = -1$, and to let subroutine CHEBY update this estimate if necessary.

B. Red-Black Line Cyclic Chebyshev Iterative Method

The solution algorithm in the Bettis PDQ08 diffusion program for 2-D problems is based on the line cyclic Chebyshev method which, theoretically and in practice on scalar computers, is highly superior to the point Chebyshev method. The only reason for considering the point Chebyshev method in this study is the fact that this method is highly vectorizable with very long effective vector lengths. One objective here is to study alternative ways to vectorize and optimize the line cyclic Chebyshev method and to compare the relative performance on a Cyber 205 of the line Chebyshev method and the highly vectorizable point Chebyshev method.

A brief description of the red-black line cyclic Chebyshev iterative method now follows. Recall from the discretized system of diffusion equations (25)-(27) that (a) each diagonal block A_{jj} in Eq. (27) represents the coupling of mesh line j to itself, (b) each off-diagonal block $A_{j,j+1}$ represents the coupling of line j to the next adjacent line j+1, and (c) each mesh line is coupled only to itself and to its two immediate neighboring lines. Assume now that the lines are regrouped in such a way that (i) the odd-numbered lines, to be referred to as the red-lines, appear first in their natural ordering and that (ii) the even-numbered lines, to be referred to as the black-lines, appear last in their natural ordering. This regrouping or permutation transforms Eqs. (26) and (27) into the so called 2-cyclic form

$$
\left[\begin{array}{cccc:cccc}
A_{11} & & & & A_{12} & & & \\
& A_{33} & & & A_{23}^T & A_{34} & & \\
& & \cdot & & & A_{45}^T & \cdot & \\
& & & \cdot & & & \cdot & \cdot \\
& & & \cdot & & & & \cdot \\
& & & A_{J'J'} & & & & \\
\hdashline
A_{12}^T & A_{23} & & & A_{22} & & & \\
& A_{34}^T & A_{45} & & & A_{44} & & \\
& & \cdot & \cdot & & & \cdot & \\
& & & \cdot & \cdot & & & \cdot \\
& & & & \cdot & & & A_{J''J''}
\end{array}\right]
\left[\begin{array}{c}
\underline{u}_1 \\ \underline{u}_3 \\ \underline{u}_5 \\ \cdot \\ \cdot \\ \cdot \\ \hline \underline{u}_2 \\ \underline{u}_4 \\ \cdot \\ \cdot \\ \cdot
\end{array}\right]
=
\left[\begin{array}{c}
\underline{b}_1 \\ \underline{b}_3 \\ \underline{b}_5 \\ \cdot \\ \cdot \\ \cdot \\ \hline \underline{b}_2 \\ \underline{b}_4 \\ \cdot \\ \cdot \\ \cdot
\end{array}\right], \quad (35)
$$

or the condensed form

$$
\begin{bmatrix} D_R & H \\ H^T & D_B \end{bmatrix} \begin{bmatrix} \underline{u}_R \\ \underline{u}_B \end{bmatrix} = \begin{bmatrix} \underline{b}_R \\ \underline{b}_B \end{bmatrix} . \quad (36)
$$

The system Eq. (36) can also be written in the form

$$
D_R\, \underline{u}_R = \underline{b}_R - H\, \underline{u}_B, \qquad D_B \underline{u}_B = \underline{b}_B - H^T \underline{u}_R . \quad (37)
$$

The red-black line cyclic Chebyshev method as applied to Eqs. (37) is then given by[3]

$$
\underline{u}_R^n = \rho_R^n\, [D_R^{-1}\, (\underline{b}_R - H\underline{u}_B^{n-1}) - \underline{u}_R^{n-1}] + \underline{u}_R^{n-1} , \quad (38a)
$$

$$
\underline{u}_B^n = \rho_B^n\, [D_B^{-1}\, (\underline{b}_B - H^T\, \underline{u}_R^n) - \underline{u}_B^{n-1}] + \underline{u}_B^{n-1} , \quad (38b)
$$

where

$$
\rho_R^1 = 1, \qquad \rho_B^1 = 1/(1 - M^2/2) \quad (39a)
$$

$$
\rho_R^n = 1/(1 - M^2\, \rho_B^{n-1}/4) , \qquad \rho_B^n = 1/(1 - M^2\, \rho_R^n/4) , \qquad n \geq 2 , \quad (39b)
$$

and where $M = M(G)$, is the spectral radius of the Jacobi matrix associated with Eqs. (35)-(37). The adaptive subroutine CCSI of Hageman[3] is applied in this work to provide estimates of the spectral radius $M(G)$ and the acceleration parameters in Eqs. (39) and also to measure the iteration error for the red-black line cyclic Chebyshev iterative method.

C. Implementation of Red-Black Line Cyclic Chebyshev

The red-black line cyclic Chebyshev method Eq. (38a, b) involves at each step of the iteration solving systems of the form

$$D \, \underline{u} = \underline{k} \qquad\qquad\qquad\qquad\qquad (40)$$

where D is either D_R or D_B, as given in Eqs. (36)-(38). As an illustration, the expression in Eq. (40) corresponding to the red unknowns as deduced from Eqs. (35) and (36) is given by

$$\begin{bmatrix} A_{11} & & & & \\ & A_{33} & & & \\ & & \cdot & & \\ & & & \cdot & \\ & & & & A_{J'J'} \end{bmatrix} \begin{bmatrix} \underline{u}_1 \\ \underline{u}_3 \\ \cdot \\ \cdot \\ \underline{u}_{J'} \end{bmatrix} = \begin{bmatrix} \underline{k}_1 \\ \underline{k}_3 \\ \cdot \\ \cdot \\ \underline{k}_{J'} \end{bmatrix}, \qquad (41)$$

where A_{ii}, u_i and k_i refer here to coefficient matrices, unknowns and right-hand sides corresponding to the odd-numbered or red mesh lines.

It suffices here to specify how a system of the form of Eq. (41) can be solved. The program DXY developed in this work applies one of the following three methods to the solution of Eq. (41).

1. <u>PDQ08 Simulation.</u> The PDQ08 program considers separately each block system corresponding to a single odd (or even) line

$$A_{ii} \, \underline{u}_i = \underline{k}_i \, , \qquad\qquad\qquad\qquad (42)$$

and essentially applies the LDL^T factorization to the tridiagonal coefficient matrix A_{ii} and repeatedly solves (42) for varying right-hand sides, as was outlined in Section II.A above. The LDL^T solution procedure is basically recursive in nature and thus the computations are predominantly scalar. The recursive relations can be optimized by applying recursive doubling[11,12] or by using special Cyber 205 Q8 STACKLIB routines[2].

PDQ08 performs normalization to eliminate multiplication by diagonal matrices. This normalization is very desirable as was discussed in Section II.A above.

For the DXY program, normalization for line methods is ignored in order to simplify the logic. What is important is to evaluate the relative performance of alternative methods and this can be accomplished whether or not normalization is implemented.

Two options exist in DXY to simulate features of PDQ08. In both options the LDL^T factorization is done for each individual mesh line. In the first option the associated recursive relations are carried out in scalar mode. In the second option the recursive computations associated with every mesh line are optimized by applying Q8 STACKLIB routines[2].

2. <u>Parallel Line Method.</u> The parallel line method is based on the observation that the various steps in the factorization $A_{ii} = L_i D_i L_i^T$ for all odd (even) i in Eqs. (41) and (42) can be carried out in parallel. Also, corresponding subsequent forward elimination backward substitution steps for Eq. (42) for all odd (even) i can be carried out in parallel. Such parallel

computations are vectorizable with a vector length equal to the number of odd (even) mesh lines.

The efficient implementation of the parallel computations on a Cyber 205 requires renumbering the equations and unknowns in Eq. (41) [and in the equivalent equation corresponding to all A_{ii} for i even] in such a way that the first unknowns in $\underline{u}_1$, $\underline{u}_3$, etc., appear consecutively followed by the second unknowns in $\underline{u}_1$, $\underline{u}_3$, etc., and so on until the last set of unknowns. The reordering is needed only at the beginning of the solution process and also at the end in order to transform the answers for the unknowns into their natural ordering. This reordering is essential because of the relatively poor performance of a Cyber 205 computer for operations on nonconsecutive vector elements (in contrast, a Cray is very efficient for nonconsecutive vector operations when a fixed stride is involved). The reordering discussed here can be achieved either by applying a Cyber 205 Q8 gather-scatter vector statement or by applying efficient matrix transpose algorithms such as the Control Data Corporation Cyber 205 TPMOV subroutine.

Once the reordering is accomplished, the vectorization amounts to converting every scalar statement in the LDL^T factorization/solution procedure into a vector statement with a vector length equal to the number of odd (even) mesh lines.

3. Hyper-Line Method. In the present application in the DXY program all the odd-numbered (red) mesh lines are grouped together to simulate a single long line referred to here as a red hyper-line and all the even-numbered (black) mesh lines are grouped together to simulate a single long line referred to here as a black hyper-line. Specifically, for example, instead of regarding the system of Eq. (41) as having a block diagonal coefficient matrix with each block A_{ii} being a tridiagonal symmetric positive definite matrix, the entire coefficient matrix in Eq. (41) is regarded as a single long tridiagonal symmetric positive definite matrix. Program DXY applies the modified odd-even cyclic reduction method introduced in Section II.C to the solution of the hyper-line system in Eq. (41) and to the solution of the analogous system corresponding to the even-numbered (black) mesh lines.

Again, for simplicity, the program DXY does not apply normalization to the hyper-line method. Also in practice for large two and three-dimensional problems, it suffices for effective utilization of the hyper-line method to regroup the odd-numbered (even-numbered) mesh lines into sets of odd-numbered lines to be referred to as red (black) hyper-lines. The odd-even cyclic reduction direct solution method described above would then be repeatedly applied to the regrouped systems corresponding to the distinct hyper-lines.

D. Numerical Results

Several problems were utilized to compare the various methods applied in the DXY program for the two-dimensional diffusion equation. Tables III-VI exhibit sample results. Table V corresponds to the problem illustrated in Fig. 2. The last set of columns in the tables labelled ADVANTAGE FACTORS summarize (a) the numerically observed (T/T $\equiv$ Total time for a method/Total time for reference method; T/T = 1.00* for reference method) relative behavior of the various methods when the computations are carried out to maximum pointwise relative error of EPS = .005, and (b) the theoretically estimated asymptotic ratio[3] (log AER/log AER) of the number of iterations required by each method relative to the reference method for which this ratio is labelled 1.00*. Note that the reference method is different for the different columns under advantage factors.

47

Table III

L-Shaped Domain Problem with 60° Angle and Variable Mesh
D = 0.01   σ = 1.0   S = 1.0

	Setup Time (sec)	Solution Time T(sec)	Ave. Time Per Iter. (sec)	Iter.	AER	Advantage Factors T/T		Log AER/Log AER	
256x250[a]									
PT. CHEBY-3	.0427	34.2917	.0208	1652	.9937	3.40	1.23	3.16	1.33
PT. CHEBY-1	.0427	27.8322	.0208	1339	.9916	2.76	1.00*	2.37	1.00*
LINE CCSI 1	.0383	10.0886	.0168	599	.9802	1.00*		1.00*	
LINE CCSI 2	.0391	13.3631	.0223	599	.9802	1.32			
LINE PDQ(R)	.0383	55.7084	.0930	599	.9802	5.52	1.38		
LINE PDQ(Q8)	.0383	40.4454	.0675	599	.9802	4.01	1.00*		
160x400									
PT. CHEBY-3	.0447	31.6669	.0207	1529	.9934	3.06	1.27	3.02	1.37
PT. CHEBY-1	.0448	24.8997	.0207	1205	.9910	2.40	1.00*	2.21	1.00*
LINE CCSI 1	.0405	10.3612	.0160	648	.9802	1.00*		1.00*	
LINE CCSI 2	.0443	14.4276	.0223	648	.9802	1.39			
LINE PDQ(R)	.0404	63.3914	.0978	648	.9802	6.12	1.31		
LINE PDQ(Q8)	.0404	48.2369	.0744	648	.9802	4.66	1.00*		
400x160									
PT. CHEBY-3	.0418	31.5963	.0207	1529	.9934	7.91	1.27	6.81	1.37
PT. CHEBY-1	.0419	24.8740	.0206	1205	.9910	6.23	1.00*	4.99	1.00*
LINE CCSI 1	.0375	3.9925	.0178	224	.9559	1.00*		1.00*	
LINE CCSI 2	.0383	5.0110	.0223	225	.9559	1.26			
LINE PDQ(R)	.0376	20.1547	.0896	225	.9559	5.05	1.48		
LINE PDQ(Q8)	.0376	13.5745	.0603	225	.9559	3.40	1.00*		

[a]256 x 250 = mesh chosen, PT. CHEBY = Point Chebyshev method with an initial estimate $M_E = -3$ or $m_E = -1$.
LINE CCSI 1 = Parallel line method, LINE CCSI 2 = Hyper-line method, LINE PDQ ≡ PDQ08 simulation with (R) recursive scalar computation or (Q8) optimized Cyber 205 STACKLIB routines.
1.00 is reference case for respective columns. See text for an explanation of advantage factors.

Table IV

Four Region Problem for Group 1 with 90° Angle
D_1 = .0025, D_2 = .025, D_3 = .25, D_4 = 2.5
σ = S = 1.0 throughout

2	3	
1	4	

	Setup Time (sec)	Solution Time T(sec)	Ave. Time Per Iter. (sec)	Iter.	AER	Advantage Factors T/T		Log AER/Log AER	
256 x 250[a]									
PT. CHEBY-3	.0447	26.7322	.0210	1274	.9910	4.11	1.43	4.07	1.41
PT. CHEBY-1	.0429	18.6570	.0210	889	.9873	2.87	1.00*	2.88	1.00*
LINE CCSI 1	.0383	6.5118	.0175	372	.9639	1.00*		1.00*	
LINE CCSI 2	.0391	8.4423	.0227	372	.9639	1.30			
LINE PDQ(R)	.0340	35.5908	.0957	372	.9639	5.46	1.36		
LINE PDQ(Q8)	.0340	26.1083	.0702	372	.9639	4.01	1.00*		
160 x 400									
PT. CHEBY-3	.0448	31.5094	.0210	1502	.9925	3.23	1.29	3.20	1.40
PT. CHEBY-1	.0448	24.4782	.0210	1167	.9895	2.51	1.00*	2.28	1.00*
LINE CCSI 1	.0404	9.7591	.0166	588	.9762	1.00*		1.00*	
LINE CCSI 2	.0413	13.3234	.0227	588	.9762	1.37			
LINE PDQ(R)	.0405	59.0367	.1004	588	.9762	6.05	1.30		
LINE PDQ(Q8)	.0405	45.2730	.0770	588	.9762	4.64	1.00*		
400 x 160									
PT. CHEBY-3	.0432	30.7830	.0205	1503	.9925	7.07	1.29	7.60	1.40
PT. CHEBY-1	.0414	23.9088	.0205	1167	.9895	5.49	1.00*	5.42	1.00*
LINE CCSI 1	.0370	4.3512	.0179	243	.9444	1.00*		1.00*	
LINE CCSI 2	.0378	5.3848	.0222	243	.9444	1.24			
LINE PDQ(R)	.0374	22.5349	.0927	243	.9444	5.18	1.46		
LINE PDQ(Q8)	.0374	15.4250	.0635	243	.9444	3.54	1.00*		

[a]See footnotes to Table III.

48

Table V

Modified Wachpress Problem with 60° Angle
See Fig. 2 and Ref. 15 § 8.3 Fig. 8.7 and Tables 8.7, 8.10

	Setup Time (sec)	Solution Time T(sec)	Ave. Time Per Iter. (sec)	Iter.	AER	Advantage Factors T/T		Log AER/Log AER	
GROUP 1[a]									
PT. CHEBY-3	.133	19.8347	.0189	1050	.9884	4.04	1.39	3.22	1.40
PT. CHEBY-1	.133	14.2856	.0189	756	.9838	2.91	1.00*	2.30	1.00*
LINE CCSI 1	.129	4.9143	.0156	315	.9631	1.00*		1.00*	
LINE CCSI 2	.130	6.4717	.0205	315	.9631	1.32			
LINE PDQ(R)	.038	25.6801	.0815	315	.9631	5.23	1.40		
LINE PDQ(Q8)	.038	18.2933	.0581	315	.9631	3.72	1.00*		
GROUP 2									
PT. CHEBY-3	.133	11.0793	.0189	586	.9793	4.36	1.35	3.56	1.41
PT. CHEBY-1	.133	8.2103	.0189	434	.9709	3.23	1.00*	2.52	1.00*
LINE CCSI 1	.129	2.5417	.0157	162	.9282	1.00*		1.00*	
LINE CCSI 2	.130	3.3382	.0206	162	.9282	1.31			
LINE PDQ(R)	.038	13.3288	.0823	162	.9281	5.24	1.40		
LINE PDQ(Q8)	.038	9.5295	.0588	162	.9281	3.75	1.00*		
GROUP 3									
PT. CHEBY-3	.133	6.4550	.0189	341	.9531	4.11	1.35	3.20	1.41
PT. CHEBY-1	.133	4.7948	.0190	253	.9347	3.05	1.00*	2.27	1.00*
LINE CCSI 1	.129	1.5710	.0159	99	.8577	1.00*		1.00*	
LINE CCSI 2	.130	2.0490	.0207	99	.8577	1.30			
LINE PDQ(R)	.038	8.3240	.0849	98	.8582	5.30	1.38		
LINE PDQ(Q8)	.038	6.0248	.0615	98	.8582	3.84	1.00*		

[a]See footnotes to Table III.

Table VI

Model Hyper-Lines with 60° Angle
D = .01, σ = 1.0, S = 1.0, Δx = Δs = .002

	Setup Time (sec)	Solution Time T(sec)	Ave. Time Per Iter. (sec)	Iter.	AER	Advantage Factors T/T		Log AER/Log AER	
500 x 5[a]									
PT. CHEBY-3	.0072	.0273	.00074	37	.6942	2.44	1.19	3.70	1.43
PT. CHEBY-1	.0055	.0229	.00074	31	.5943	2.05	1.00*	2.60	1.00*
LINE CCSI 1	.0056	.0506	.00562	9	.2591	4.52		1.00*	
LINE CCSI 2	.0061	.0112	.00124	9	.2591	1.00*			
LINE PDQ(R)	.0054	.0304	.00338	9	.2591	2.72	1.36		
LINE PDQ(Q8)	.0054	.0224	.00249	9	.2591	2.00	1.00*		
500 x 9									
PT. CHEBY-3	.0065	.0785	.00133	59	.8059	3.21	1.27	3.32	1.39
PT. CHEBY-1	.0065	.0617	.00134	46	.7403	2.52	1.00*	2.38	1.00*
LINE CCSI 1	.0062	.0734	.00564	13	.4884	3.00		1.00*	
LINE CCSI 2	.0070	.0245	.00188	13	.4884	1.00*			
LINE PDQ(R)	.0063	.0800	.00615	13	.4884	3.26	1.40		
LINE PDQ(Q8)	.0063	.0573	.00220	13	.4884	2.34	1.00*		
500 x 17									
PT. CHEBY-3	.0085	.2333	.00256	91	.8976	2.84	1.24	3.49	1.41
PT. CHEBY-1	.0085	.1879	.00257	73	.8587	2.29	1.00*	2.47	1.00*
LINE CCSI 1	.0079	.1573	.00605	26	.6858	1.91		1.00*	
LINE CCSI 2	.0087	.0822	.00316	26	.6859	1.00*			
LINE PDQ(R)	.0080	.2974	.01144	26	.6859	3.62	1.45		
LINE PDQ(Q8)	.0080	.2058	.00792	26	.6859	2.50	1.00*		

[a]See footnotes to Table III.

Tables III-VI and the results omitted indicate the following:

(a) The vectorized red-black line cyclic Chebyshev methods are at least a factor of two better than the highly vectorized point Chebyshev method for 2-D diffusion computations on a Cyber 205.

(b) The parallel line method is superior to the hyper-line method when the number of red or black mesh lines is relatively large (> 125) and the reverse is true when the number of mesh lines is relatively small.

(c) The point Chebyshev method is insensitive to a row and column interchange of the mesh grid. In contrast, the red-black line cyclic Chebyshev method is very sensitive to interchange of rows and columns as is evident from a glance at the results in Tables III and IV for the choice of the 160 x 400 mesh versus the choice of the 400 x 160 mesh. In summary, while the efficiency of vectorization for the parallel-line method improves as the number of rows increases, this should be a secondary consideration relative to the choice of preferred orientation with relatively stronger couplings among row mesh cells and relatively weaker couplings along column mesh cells.

REFERENCES

1. I. K. Abu-Shumays and L. A. Hageman, Development of Comparison of Practical Discretization Methods for the Neutron Diffusion Equation Over General Convex Quadrilateral Partitions, Proc. ANS Conf. "Computational Methods in Nuclear Engineering," CONF-750413, Vol. I, pp. I-117 through I-165, Charleston, SC (April 15-17, 1975).
2. CDC "CYBER 200 FORTRAN Version 2 Reference Manual," Publication No. 60485000 (June 15, 1983).
3. L. A. Hageman and D. M. Young, "Applied Iterative Methods," Academic Press, New York (1981).
4. D. E. Heller, D. K. Stevenson and J. F. Traub, Accelerated Iterative Methods for the Solution of Tridiagonal Systems on Parallel Computers, Carnegie-Mellon University, Department of Computer Science Report, AD A006868 (December 1974).
5. M. Ishiguro and Y. Koshi, Vectorization for Solving the Neutron Diffusion Equation--Some Numerical Experiments, Nucl Sci Eng., 80, 322-328, (1982).
6. D. S. Kershaw, The Solution of Single Linear Tridiagonal Systems and Vectorization of the ICCG Algorithm of the CRAY 1, Lawrence Livermore Laboratory Report UCID-19085, (June 25, 1981).
7. J. J. Lambiotte, Jr. and R. G. Voigt, The Solution of Tridiagonal Systems on the CDC STAR-100 Computer, ICASE Report, July 19, 1974.
8. N. K. Madsen and G. H. Rodrigue, A Comparison of Direct Methods for Tridiagonal Systems on the CDC-STAR-100, Lawrence Livermore Laboratory Report UCRL-76993, Rev. 1 (May 28, 1976).
9. J. S. Nolen, D. W. Kuba, and M. J. Kascic, Jr., Application of Vector Processors to the Solution of Finite Difference Equations, SPE 7675, AIME Fifth Symposium on Rervervoir Simulation, Denver, Colorado, February 1-2, 1979, Petroleum Engineers Journal (August 1981).
10. C. J. Pfeifer and C. J. Spitz, PDQ-8 Reference Manual, Bettis Atomic Power Laboratory report WAPD-TM-1266 (May 1978).
11. H. S. Stone, An Efficient Parallel Algorithm for the Solution of a Tridiagonal Linear System of Equations, J. ACM, 20, No. 1, 27-38 (January 1973).
12. H. S. Stone, Parallel Tridiagonal Equation Solvers, ACM Transactions on Math. Software, 1, No. 4, 289-307 (December 1975).

13. W. A. Thomas and E. E. Lewis, Two Vectorized Algorithms for the Solution of Three-Dimensional Neutron Diffusion Equation, <u>Nucl. Sci. Eng.</u>, <u>84</u>, 67-71 (1983).

14. H. H. Wang, A Parallel Method for Tridiagonal Equation, <u>ACM Transactions on Math Software</u>, 7, No. 2, 170-183 (June 1981).

15. E. L. Wachspress, "Iterative Solution of Elliptic Systems," Prentice-Hall, Englewood Cliffs, New Jersey (1966).

APPENDIX A. GATHER, SCATTER AND RELATED ALGORITHMS FOR A CYBER 205

The objective here is to summarize some experience with vector Q8 syntax and related algorithms associated with diffusion computations on the Control Data Corporation CYBER 205 computers.

One of the successful methods described above for two dimensional diffusion computations is the parallel-line cyclic Chebyshev iterative method which involves at each step of the iterations, solving several tridiagonal systems (systems corresponding to the even-numbered lines or the odd-numbered lines) simultaneously in parallel. Simultaneous solution of several tridiagonal systems is also of interest in other science and engineering applications. An important step in this connection is to transform the various systems into a form whereby elements to be operated on at the same time correspond to consecutive elements of a vector array. This is very important for the Cyber 205 since the use of a stride other than 1 is inefficient. The transformations involved here amount essentially to the construction of a matrix transpose of a given rectangular matrix.

No efficient Cyber 205 subroutine existed at Bettis for matrix transpose when this work was initiated. A special program was constructed for this purpose based on the Q8 gather-scatter instructions. The general gather-scatter instructions require the construction of an index for assigning how elements are to be reordered. An efficient index subroutine was constructed based on a doubling algorithm illustrated by the following example.

A.I. INDEX MATRIX

Suppose that we are required to find the transpose of the $N_x M = 8_x 5$ index matrix

$$A = \begin{bmatrix} 1 & 9 & 17 & 25 & 33 \\ 2 & 10 & 18 & 26 & 34 \\ 3 & 11 & 19 & 27 & 35 \\ 4 & 12 & 20 & 28 & 36 \\ 5 & 13 & 21 & 29 & 37 \\ 6 & 14 & 22 & 30 & 38 \\ 7 & 15 & 23 & 31 & 39 \\ 8 & 16 & 24 & 32 & 40 \end{bmatrix} . \tag{A.1}$$

Note that the elements of A are stored by columns and in the order specified in Eq. (A.1). The first step in achieving the transpose of A is to construct an array whose elements are the top row of A, namely

$$A_1 = \begin{bmatrix} 1 \\ 9 \\ 17 \\ 25 \\ 33 \end{bmatrix} . \tag{A.2}$$

The elements of A_1 in Eq. (A.2) have a constant stride N=8, the row dimension of the matrix A, and thus the construction of A_1 can be vectorized by the Q8VINTL(1,N; M) Cyber 205 vector instruction. Now an M x 2 = 5 x 2 matrix is constructed based on A_1 as its first column and based on constructing a second column by adding 1 to each element of A_1 to get

$$A_2 = \begin{bmatrix} 1 & \vdots & 2 \\ 9 & \vdots & 10 \\ 17 & \vdots & 18 \\ 25 & \vdots & 26 \\ 33 & \vdots & 34 \end{bmatrix} . \tag{A.3}$$

The process of extending A_1 to construct A_2 is a vector addition with vector length M=5 in this example. Now A_2 is extended by adjoining two additional columns to it based on adding the number 2 to each element in A_2 to get

$$A_3 = \begin{bmatrix} 1 & 2 & | & 3 & 4 \\ 9 & 10 & | & 11 & 12 \\ 17 & 18 & | & 19 & 20 \\ 25 & 26 & | & 27 & 28 \\ 33 & 34 & | & 35 & 36 \end{bmatrix} . \qquad (A.4)$$

The process of extending A_2 to construct A_3 is a vector addition with vector length 2M=10 in this example. Now A_3 is extended by adjoining four additional columns to it based on adding the number 4 to each element in A_3 to get

$$A_4 = \begin{bmatrix} 1 & 2 & 3 & 4 & | & 5 & 6 & 7 & 8 \\ 9 & 10 & 11 & 12 & | & 13 & 14 & 15 & 16 \\ 17 & 18 & 19 & 20 & | & 21 & 22 & 23 & 24 \\ 25 & 26 & 27 & 28 & | & 29 & 30 & 31 & 32 \\ 33 & 34 & 35 & 36 & | & 37 & 38 & 39 & 40 \end{bmatrix} . \qquad (A.5)$$

The process of extending A_3 to construct A_4 is a vector addition with vector length 4M=20 in this example. Note that A_4 is the desired answer.

To summarize, the construction of the index of a matrix transpose is vectorizable with vector length initially equal to the number of columns (or rows). The vector length is doubled at each step of construction until the index matrix is completed. The third column in Table A.I shows the run times in microseconds (μ-sec = 10^{-6} sec) for constructing index matrices corresponding to a varying number of rows N and columns M for a fixed total of N·M = 64,000 matrix elements. This index construction is indeed efficient and the example given above is intended to demonstrate the use of a doubling method for vectorization. This doubling method is similar in nature to an algorithm that is commonly applied to optimize computations of recursive relations which are basically scalar on a Cyber 205.

A.II. GATHER-SCATTER

Assume that an N by M matrix A with NM total number of elements is given together with the corresponding transpose of the index matrix associated with A. The process of replacing the matrix A by its transpose can be achieved on a Cyber 205 by one of the following gather instructions:

(i) $\quad$ A(1; NM) = Q8VGATHR(A(1; NM), INDEX(1; NM); NM) $\qquad$ (A.6)

(ii) $\quad$ A(1; NM) = Q8VGATHR(A(1; NM), INDEX(1; NM); B(1; NM)) $\qquad$ (A.7)

or

$\quad$ B(1; NM) = Q8VGATHR(A(1; NM), INDEX(1; NM); B(1; NM)) $\qquad$ (A.8a)

$\quad$ A(1; NM) = B(1; NM) $\qquad$ (A.8b)

The easiest of the above steps to understand is the one illustrated in Eqs. (A.8a, b). The instruction in Eq. (A.8a) reorders the elements of A according to the INDEX array in order to construct A^T and store A^T in the array B. The second instruction Eq. (A.8b) copies the array B into the old array A. The instruction in Eq. (A.7) combines the two steps in Eqs. (A.8a, b) into one instruction. Actual timing on the Bettis Cyber 205 and on the

Cybernet Cyber 205 demonstrated that the single instruction (A.7) and the double instructions (A.8a, b) are essentially indistinguishable as is expected. In principle, the instruction in (A.6) should also be indistinguishable from those in Eqs. (A.7) and (A.8a, b) but this is not the case. In Eq. (A.6), the compiler assigns a dummy internal array, in place of the user specified array B in Eqs. (A.7) and (A.8a, b), and then copies this undeclared array into the array A.* Note that the statement

$$A(1; NM) = Q8VGATHR(A(1; NM), INDEX(1;NM); A(1;NM)) \qquad (A.9)$$

is unacceptable and would lead to wrong answers.

The converse process of Eqs. (A.6)-(A.8a, b) for transforming the transpose of A into A is given by

(i) $\qquad A(1; NM) = Q8VSCATR(A(1; NM), INDEX (1; NM); NM) \qquad (A.10)$

(ii) $\qquad A(1; NM) = Q8VSCATR(A(1; NM), INDEX(1;NM); B(1; NM)) \qquad (A.11)$

or

$$B(1; NM) = Q8VSCATR(A(1; NM), INDEX(1; NM); B(1; NM)) \qquad (A.12a)$$

$$A(1; NM) = B(1; NM) \qquad (A.12b)$$

Here again, Eq. (A.11) and Eqs. (A.12a, b) are found to be computationally indistinguishable. The timing performance of the statement in Eq. (A.10) is disappointing as will become evident below.*

Table A.I provides the run times for Eqs. (A.6)-(A.8a, b), (A.10)-(A.12a, b) on the Cybernet Cyber 205, for systems of order N by M with NM = 64,000 elements. The results clearly indicate that the vector instructions in Eqs. (A.7) and (A.11), or their equivalent instructions in Eqs. (A.8a, b) and (A.12a, b), are to be preferred to the instructions in Eqs. (A.6) and (A.10). In fact the instruction in (A.10) increases the computing time by factors ranging from 2 to 8.*

The lesson learned from Table A.I is that seemingly equivalent vector operations may sometimes behave differently in practice, and thus, in the absence of guidelines as was the case above, the proper use of vector syntax should be tested and ascertained numerically.*

A.III. COMPARISON OF GATHER-SCATTER AND OTHER INSTRUCTIONS

Reference to the gather and scatter instruction below will refer to the formulations in Eqs. (A.7), (A.8a, b), (A.11), (A.11a, b). Control Data Corporation has installed on their Cybernet Cyber 205, an efficient machine language subroutine TPMOV to transpose a rectangular matrix and simultaneously move it from one storage location to another.

Table A.II presents a comparison of the gather-scatter instructions and TPMOV carried out on a Cybernet Cyber 205. In this table, the timings are for the matrix transpose followed by the copy instruction needed in order to replace a given matrix in computer memory by its transpose and thus econo-

---

*As pointed out by Ron Selva of CDC, Eqs. (A.6) and (A.10) become indistinguishable respectively from Eqs. (A.7), (A.8a, b) and (A.11), (A.12a, b) if the option RLP=1 is selected in the program card to assign a large page to the dynamic stack.

mize on overall storage. Note that the entries in Table A.II corresponding to matrices with N=256 rows and M=125 columns (total NM=32,000 elements) and to matrices with N=256 rows and M=250 columns (total NM=64,000 elements) exhibit a peculiar unexpected behavior of the gather-scatter instructions as compared to similar size problems, and result in these cases in a factor of more than two degradation in run time as compared to TPMOV. This unexpected behavior is a result of memory bank conflicts which arise whenever N is an exact multiple of 64.

In summary, the random gather-scatter Q8VGATHR, Q8VSCATR instructions do not always behave as advertised.

The above discussion relates to algorithms involved in the parallel-line method for solving two and three-dimensional diffusion problems. The competitive hyper-line method requires separating and later combining odd and even array elements. Table A.III compares 4 different approaches to separate odd and even array elements (i) random gather, (ii) periodic gather, (iii) matrix transpose subroutine TPMOV and (iv) double compress operations. Of these choices, the period gather is the best overall and is preferred over the double compress operations because the latter operations require the generation of a BIT vector corresponding to odd or even array elements. However, this author expects the BIT vector involved here (101010...) to be useful for other operations in a general diffusion program. Thus, if the cost of the BIT vector is excluded, Table A.III shows that the compress operation Q8VCMPRS would become the most economical approach to separate odd and even array elements.

Table A.III also shows that if the cost of the construction of the BIT vector is ignored, then the merge operation Q8VMERG is decisively the preferred approach for later recombining separated odd and even array elements.

The cyclic reduction algorithms discussed in the text were implemented using the different options for separating odd and even array elements given in Table A.III. The preferred version of the algorithms applies the compress and merge instructions.

TABLE A.I

Time Comparison (µ-Sec.) For Alternative Expressions[a] For
Q8VGATHR and Q8VSCATR For Vector Length N*M = 64,000
(CYBERNET 205 FORTRAN 2.0 10/26/83)

N	M	Index	GATHR1 + Copy	GATHR2 + Copy	SCATR1 + Copy	SCATR2 + Copy
80	800	683	2,657	2,594	6,643	2,322
160	400	673	2,650	2,534	10,791	2,444
200	320	668	2,620	2,565	13,050	2,387
320	200	668	2,655	2,494	19,358	2,384

[a]GATHR1 = Eq. (A.6),   GATHR2 = Eq. (A.7)
  SCATR1 = Eq. (A.10),   SCATR2 = Eq. (A.11)

Table A.II

Time Comparison (μ-Sec.) for Q8VGATHR, Q8VSCATR and TPMOV
With Vector Copy Instructions Added to These Expressions
(CYBERNET 205 FORTRAN 2.0 10/26/83)

N	M	Index	GATHR + Copy	TPMOV + Copy	SCATR + Copy	TPMOV + Copy
500	2	84	44	44	41	45
500	4	46	84	96	78	81
500	8	66	163	189	154	174
500	10	76	205	210	190	198
500	20	126	339	367	373	384
500	40	226	791	729	744	770
500	64	346	1278	1157	1192	1202
500	128	666	2554	2642	2367	2753
200	100	225	802	804	745	803
200	200	426	1599	1579	1488	1577
256[a]	125	346	2913	1252	2910	1244
256	250	666	5823	2455	5822	2512
320	200	668	2491	2522	2384	2434
400	160	668	2537		2579	2428

[a]The underlined numbers highlight an unexpected adverse behavior of the
gather and scatter instructions for N an exact multiple of 64.

Table A.III

Time Comparison (μ-Sec.) of Alternative Algorithms for
Separating and Later Combining Odd and Even Array Elements
(CYBERNET 205 FORTRAN 2.0 10/26/83)

N	Bit[a]	Index	GATHR	GATHP	TPMOV	CMPRS	SCATR	MERG	TPMOV
8	4	21	5	5	7	5	4	4	8
16	4	18	5	5	7	5	5	4	8
32	4	17	5	5	8	5	4	5	9
64	5	18	6	6	8	6	5	4	9
128	6	19	8	7	10	7	6	5	11
256	9	21	11	11	14	10	10	7	13
512	14	25	19	18	21	15	17	11	20
1024	24	33	33	31	34	25	31	17	32
2046	44	48	62	58	61	46	58	29	58
4096	86	79	122	112	114	87	113	55	107
8192	168	140	238	219	222	169	221	105	207
16384	331	263	474	434	437	335	438	207	407
20000	404	321	575	529	532	405	534	247	495
32000	643	498	917	844	847	645	853	463	788
40000	804	617	1150	1055	1058	814	1066	492	983
64000	1283	982	1830	1685	1687	1285	1704	794	1567

[a]Data for this column for constructing bit vectors was obtained on Bettis
Cyber 205 after the Cybernet system became unavailable to the author.

A MODIFIED CONJUGATE GRADIENT SOLVER

FOR VERY LARGE SYSTEMS

D. Barkai

Control Data Corporation at the
Institute for Computational Studies at CSU
P.O. Box 1852
Fort Collins, Colorado 80522, U.S.A.

K.J.M. Moriarty

Institute for Computational Studies
Department of Mathematics, Statistics
and Computing Science
Dalhousie University
Halifax, Nova Scotia   B3H 4H8
Canada

C. Rebbi

Theory Division
CERN
CH-1211, Geneva 23
Switzerland
    and
Department of Physics
Brookhaven National Laboratory
Upton, New York   11973, U.S.A.

ABSTRACT

A modified Conjugate Gradient method is derived which requires only
one pass through the coefficients and the temporary vectors. The method
is applicable to problems which may be complex and non-symmetric. The
method is implemented on a vector processor (the CDC CYBER 205) and
applied to a high-energy physics lattice gauge theory problem, though the
implementation methodology is quite general.

1. INTRODUCTION

In the course of establishing a suite of programs for measuring
physical observables predicted by the lattice formulation of Quantum
Chromodynamics (QCD), it became necessary to develop an efficient solver
for very large systems of linear equations. The outcome of this search
turned out to be quite general and the detailed structure of the coefficients matrix to be inverted is incidental to the method. The only

assumptions made here regarding this matrix are: (i) the size of the problem is large enough to warrant an iterative method, and (ii) there exist simple rules for applying the "operator" (i.e. the coefficients matrix) to a solution vector, without decompositions etc., so as to make a method like the Conjugate Gradient (CG) attractive.

The problem in the application of the conjugate gradient method to systems of very large extent is that all the relevant variables cannot be kept in fast memory; in the course of the iterations the data must then be repeatedly brought in and out of mass storage device. In the standard implementation of the algorithm two passes through the data are required at each iteration and this makes the whole computation strongly I/O bound. The main purpose of this paper is to illustrate a modification of the CG method which dramatically reduces the I/O requirements (by 50%), thus making the procedure substantially more efficient.

In the next section we derive a modified CG procedure based on a fairly simple-minded conventional CG algorithm, which also allows for a complex non-symmetric coefficients matrix, and such that only one pass through the data is needed for each iteration. This is the essence of this paper - a procedure that halves the "cost" of solving very large system of linear equations utilizing a CG method. Section 3 contains general considerations and details for implementing the modified algorithm on a powerful vector processor. The discussion there is also relevant to other parallel computer architectures. Finally, in section 4 we present a specific high-energy physics problem, to which the new method has been applied, its structure and performance figures on the CDC CYBER 205.

2. DERIVATION OF THE MODIFIED ALGORITHM

The CG method proposed by Hestenes and Stiefel[1] may be regarded as a family of methods, each corresponding to an acceleration processes for a particular basic iterative procedure. In the literature one can find slightly different presentations of the method, almost always restricted to the case of a symmetric-positive-definite (SPD) coefficients matrix [2,3,4]. It is, therefore, useful to establish our notation and formulation. We wish to find a vector f satisfying

$$Mf = b \qquad\qquad (2.1)$$

where M is $N{\times}N$ matrix and f and b are vectors of length N ; M and b are given. Lowercase variables, with and without a superscript, will denote vectors; lowercase variables with a subscript are scaler quantities. Uppercase letters are reserved for matrices. Let M be SPD and initialize the "residual" vector $u^{(0)} = b$, the "direction" vector $p^{(0)}$ to an arbitrary initial guess of the solution and the "solution" vector $f^{(0)} = 0$. Then iteration n of the CG method is given by

$$\alpha_n = \frac{(u^{(n-1)},u^{(n-1)})}{(p^{(n-1)},Mp^{(n-1)})} \quad \text{for} \quad n > 1 \ ; \ \alpha_1 = 1$$

$$f^{(n)} = f^{(n-1)} + \alpha_n \cdot p^{(n-1)}$$

$$u^{(n)} = u^{(n-1)} - \alpha_n \cdot Mp^{(n-1)} \qquad\qquad (2.2)$$

$$\beta_n = \frac{(u^{(n)},u^{(n)})}{(u^{(n-1)},u^{(n-1)})} \quad \text{for} \quad n > 1 \ ; \ \beta_1 = 0$$

$$p^{(n)} = u^{(n)} + \beta_n \cdot p^{(n-1)} .$$

With a few renamings and simple manipulations one can establish the equivalence between this formulation and the one given e.g. in Ref. 2. We wish to consider very large systems, such that the vectors used and the non-zero elements of M cannot all fit into the real memory of the computer system. The problem has, then, to be "sliced", i.e. the variables (vector and matrix elements) must be subdivided in sets (slices), according to some reasonable partitioning of the matrix, and the computations must be carried out for one slice at a time. It is apparent, however, that it is not possible to execute a whole iteration with a single loop over slices in the algorithm above. For example, $u^{(n)}$ has to be known for all slices before β_n can be computed, and β_n itself is required in this iteration for computing each slice of $p^{(n)}$. The reader will notice that α_n may be computed at the end of the $(n-1)$ iteration, and, therefore, only two (rather than three) passes through all the slices are needed for the completion of each iteration.

We wish to derive a formulation where each iteration can be done with just one pass through the various vectors. Before getting into that let us note that the coefficients in M (representing, typically, most of data to be accessed) are required only once - for computing the vector $Mp^{(n-1)}$. This is true, however, only when M is SPD. The formulation for the case when M is not SPD is achieved by the following modifications to Eqns. (2.2): Introduce an additional vector $r = Mp$; change the denominator of α_n to $(r^{(n-1)}, r^{(n-1)})$; change u in the expressions for α_n , β_n and $p^{(n)}$ to M^+u , and denote $|d|^2 = (d,d)$. We also allow M to be a complex matrix. With these changes the algorithm may be written as:

$$\text{initialization:} \quad f^{(0)} = 0 \ , \ u^{(0)} = b \ , \ p^{(0)} = x$$

$$r^{(n)} = Mp^{(n-1)}$$

$$\alpha_n = \frac{|s^{(n-1)}|^2}{|r^{(n)}|^2} \quad \text{for} \ n > 1 \ , \ \alpha_1 = 1$$

$$f^{(n)} = f^{(n-1)} + \alpha_n \cdot p^{(n-1)}$$

$$u^{(n)} = u^{(n-1)} - \alpha_n \cdot r^{(n)} \tag{2.3}$$

$$s^{(n)} = M^+ u^{(n)}$$

$$\beta_n = \frac{|s^{(n)}|^2}{|s^{(n-1)}|^2}$$

$$p^{(n)} = s^{(n)} + \beta_n \cdot p^{(n-1)} .$$

One can easily verify that when M is real and symmetric Eqns. (2.3) can be converted back to Eqns. (2.2). The changes introduced here are equivalent to considering the problem

$$(M^+ M)f = M^+ b \tag{2.4}$$

without evaluating $M^{\dagger}M$, which may, in general, reduce the sparsity of the coefficients matrix. In any event, the amount of data to be accessed is nearly double for this more general case than it is for when M is SPD, since M is required in each of two passes - once for computing $r^{(n)}$, and again for $s^{(n)}$.

We will now proceed to derive the "one-pass" algorithm from Eqns. (2.3). The main idea is to foresee the inner-products needed for the next iteration so that all the computations can be done slice-by-slice without having to compute an inner-product in the course of the current iteration. As we shall see, an incomplete initial iteration (or, more extensive initializ- ation) is added before getting into the modified one-pass iterations. The modified algorithm will be constructed in several stages. First, multiply the expression for $u^{(n)}$ from the left by $M^{\dagger}$, and use $s^{(n)} = M^{\dagger}u^{(n)}$. Also, define $q^{(n)} = M^{\dagger}r^{(n)}$. Eqns. (2.3) then become

$$f^{(0)} = 0 \ , \ p^{(0)} = x \ , \ s^{(0)} = M^{\dagger}b \ \ \text{and}$$
$$r^{(n)} = Mp^{(n-1)}$$

$$\alpha_n = \frac{|s^{(n-1)}|^2}{|r^{(n)}|^2} \quad \text{for} \ n > 1 \ ; \ \alpha_1 = 1 \tag{2.5}$$

$$f^{(n)} = f^{(n-1)} + \alpha_n \cdot p^{(n-1)}$$
$$q^{(n)} = M^{\dagger}r^{(n)}$$
$$s^{(n)} = s^{(n-1)} - \alpha_n \cdot q^{(n)}$$

$$\beta_n = \frac{|s^{(n)}|^2}{|s^{(n-1)}|^2} \quad \text{for} \ n > 1 \ ; \ \beta_1 = 0$$

$$p^{(n)} = s^{(n)} + \beta_n \cdot p^{(n-1)} \ .$$

Next, compute $s_n = |s^{(n)}|^2$ prior to computing the current slice of $s^{(n)}$ by noting that

$$s_n = |s^{(n-1)} - \alpha_n \cdot q^{(n)}|^2$$
$$= |s^{(n-1)}|^2 - 2\alpha_n \ \text{Re} \ \bar{s}^{(n-1)} \cdot q^{(n)} + \alpha_n^2 |q^{(n)}|^2 \tag{2.6}$$

so that β_n can be computed before $s^{(n)}$ is known.

The $f^{(n)}$ computation can be moved down to the lower half of the procedure. What we have now is an equivalent representation of Eqns. (2.5), namely:

initialization: $f^{(0)} = 0 \ , \ p^{(0)} = x \ , \ s^{(0)} = M^{\dagger}b$
$$r^{(n)} = Mp^{(n-1)}$$
$$r_n = |r^{(n)}|^2$$
$$\alpha_n = s_{n-1}/r_n \quad \text{for} \ n > 1 \ , \ \alpha_1 = 1$$

$$q^{(n)} = M^\dagger r^{(n)}$$

$$s_n = |s^{(n-1)}|^2 - 2\alpha_n \, \text{Re} \, \bar{s}^{(n-1)} \cdot q^{(n)} + \alpha_n^2 |q^{(n)}|^2$$

$$\beta_n = s_n/s_{n-1} \quad \text{for} \quad n > 1 \; , \; \beta_1 = 0 \tag{2.7a}$$

$$f^{(n)} = f^{(n-1)} + \alpha_n \cdot p^{(n-1)}$$

$$s^{(n)} = s^{(n-1)} - \alpha_n \cdot q^{(n)}$$

$$p^{(n)} = s^{(n)} + \beta_n \cdot p^{(n-1)} \; . \tag{2.7b}$$

The last step involves swapping Eqns. (2.7a) and (2.7b), so that Eqns. (2.7a) will serve to compute α_n and β_n for the next iteration. Of course, the initialization will need to be more extensive (or, one may refer to it as an "incomplete" first iteration). The final procedure is:

$$f^{(0)} = 0$$

$$p^{(0)} = x$$

$$s^{(0)} = M^\dagger b$$

$$r^{(0)} = Mp^{(0)}$$

$$q^{(0)} = M^\dagger r^{(0)}$$

$$\alpha_0 = 1$$

$$\beta_0 = 0 \tag{2.8a}$$

begin loop over "regular" iterations
begin loop over slices of data

$$f^{(n)} = f^{(n-1)} + \alpha_{n-1} \cdot p^{(n-1)}$$

$$s^{(n)} = s^{(n-1)} - \alpha_{n-1} \cdot q^{(n-1)}$$

$$p^{(n)} = s^{(n)} + \beta_{n-1} \cdot p^{(n-1)}$$

$$s_n = |s^{(n)}|^2$$

$$r^{(n)} = M \cdot p^{(n)}$$

$$r_n = |r^{(n)}|^2$$

$$q^{(n)} = M^\dagger r^{(n)}$$

$$q_n = |q^{(n)}|^2$$

$$sq_n = \text{Re} \, \bar{s}^{(n)} q^{(n)} \tag{2.8b}$$

end loop over slices,

$$\alpha_n = s_n/r_n$$

$$\beta_n = (s_n - 2\alpha_n \cdot sq_n + \alpha_n^2 \cdot q_n)/s_n \qquad\qquad (2.8c)$$

end loop over iterations.

The procedure expressed in Eqns. (2.8a, b and c) is perfectly equivalent to Eqns. (2.3), as indeed was verified by implementing the two procedures as computer programs. However, since in Eqns. (2.8) the full moduli s_n, r_n, q_n and the scalar product sq_n are needed only at the end of the loop over the iterations, the manipulations on the vectors s, q, f, p and r (Eqns. (2.8b)) can be done in a "sliced" form with a single pass through M and the previous values of these vectors. Eqns. (2.8c) set α_n and β_n for the next iteration.

3. IMPLEMENTATION

In the previous section we have presented an alternative formulation to a conjugate gradient method. Similar strategies may be applied to different, traditional variants of the CG family of algorithms (e.g., various acceleration schemes for the basic CG method). All one has to do is manipulate the given expressions so that results of inner-products are needed only at the end of the iteration for quantities to be used at the next iteration. A "look-ahead" approach is adopted whereby one employs scaling factors computed at the previous iteration and accumulates values to be used at the next one. Our discussion so far has been quite general, in the sense that we have not assumed any special structure of the coefficients matrix M ; in fact, we have allowed for M not to be SPD. The result is a formulation which is not very interesting for small-size problems. The amount of computations remains the same (it is assumed, as is almost always the case, that most of the computational effort goes in the multiplications of the vectors by M or M^T), and the convergence rate has not changed. The advantage of our proposed formulation pertains to problems so large that the coefficients and the vectors to be manipulated cannot all be simultaneously kept in memory. When this is the case the problem has to be implemented on a computer system in a sliced fashion; for example, for a three-dimensional problem one might choose one definite coordinate, e.g. the one along the z-axis, as a slicing parameter, and each slice would contain all the x-y plane values for a fixed z value. During the computation of an iteration the slices are being rolled in and out of memory, as required. If we knew (as will be shown below) that the arithmetic can be done very fast compared to the time it takes to swap slices in and out of memory, then an algorithm requiring only half the amount of data motion becomes very attractive. It may, as was the case in our implementation, cut the real cost of executing the program in half, since one is normally charged for the clock time of keeping real memory (even if only to wait for input/output operations to complete) and for the amount of data read in and written out.

Before discussing a specific application of this method, let us mention some computer characteristics which are relevant to tackling a large CG problem. When treating a large multi-dimensional problem, supercomputers come to mind; both due to their big memories and their ability to perform very fast arithmetic. The word "supercomputers" refers today to vector processors or other powerful computers with parallel architectures, equipped with large, fast memories. From Eqns. (2.8) one can easily deduce that the CG method is likely to be very suitable for vector or parallel processing. The element of uncertainty is due to the fact that the structure, sparsity and storage mode of the matrix M is specific to each application, but is crucial to the parallelism of the procedure. Multiplying a matrix by a vector is easily vectorizable, of course, but it is assumed that M is sparse and that a set of rules is given for the product M•(vector). The storage mode and slicing mechanism for the non-zero elements of M will determine the vectorazibility of this operation. We shall return to this issue when describing a specific implementation. In the meantime let us make the following observations:

(i) Using vector arithmetic will be efficient since one can expect long vectors, if one assumes a three-, or even four-, dimensional problem. The vectors in Eqns. (2.8) will then be of length m^2 or m^3, where m is the grid size, having already accounted for slicing; i.e., there are m "sections" or slices of the length given above for each of the vectors f, p, s, r and q . Long vectors are useful for efficient processing, because in their manipulations the effect of the fixed-time vector operation start-up time becomes quite negligible.

(ii) The updating of the vectors f, s and p in Eqns. (2.8) is accomplished by using only values from the previous iteration or from other vectors. There is no dependence of elements in any vector on elements of the same vector which are computed in the current iteration; in other words, there are no recursive relationships for any individual vector in the lifetime of an iteration. In fact, the combination of multiplying by a scalar and adding to, or subtracting from, another vector can be executed at the rate of a single operation on current vector processors. The two operations may be "linked" in such a way that a result element from the first operation can be used as input operand as soon as it becomes available.

(iii) There are more floating-point operations needed for performing inner-products than for the vector updating (we do not consider the matrix multiplications yet). The four scalars s_n , r_n , q_n and sq_n in Eqns. (2.8) are the result of accumulating, slice by slice, the sum of products of two vectors. This operation, when considered as progressing in lexicographic order, is recursive by nature - a previous partial sum is needed before the next product can be added to it. It is possible, however, to accumulate partial sums and products in parallel from different segments of the vectors, and, after several stages, get the final result. Such a logic has been implemented on the CDC CYBER 205, for example, in a way which enables the inner product to be performed at full streaming rate. It may not seem significant, but the reader should realize that when only, say, 5% of the operations perform 10 times slower than the other 95%, they will amount to about one-third of the final computation time.

The overall efficiency of the arithmetic really depends on how fast the two "matrix by vector" multiplies in Eqns. (2.8) can be performed. This will determine the CPU time for executing an iteration. Our experience has shown, however, that the true crucial factor is how fast one can restructure and move data to and from memory, and not how fast the computations are done. The arithmetic is fairly simple, and when the computations of one slice of data can be done concurrently with fetching the next slice, the data motion is more time-consuming. Input/output bandwidth seems to be the most relevant factor for the CG method on present supercomputers. The implementation effort should, therefore, be directed at minimizing the time spent in moving data in and out of fast memory. Some or all of the following may be done when implementing the algorithm:

(a) Hardware and software features supporting concurrency of input/output with computations should be used. In addition to the slices needed for updating the current slice, there should be some space allocated for swaping the slice which is not needed anymore with the one which will be needed for updating the next slice. Then, at least part of the data fetching operation will be overlapped with computations.

(b) There are two major data items which may be treated separately. One is the set of coefficients in the matrix M ; the second is the set of vectors being updated each iteration. It is important, of course, that the slicing mechanism used for the two be consistent; e.g., that the coefficients in slice i of M will be those needed for slice i of the vectors f, p, s, r and q . If the vectors are sliced by planes or cubes, for example, so should the non-zero elements in M be arranged.

(c) It is possible to reduce the data swapping time by utilizing more than one input/output stream (or I/O channel). The obvious thing to do is to create two files residing on storage media associated with different channels or memory ports, one for elements of M and one for the elements of the vectors above. In this way two I/O requests may be initialized at the same time, and the total elapsed time corresponds then to the larger of the two slices of data files. This will be, typically, the one containing the elements of M , therefore, there is no point in further subdividing the file containing the updated vectors. Depending on the structure of M it may make sense to split it into sections within the same slice.

(d) There are five vectors being computed each iteration. However, only four of them (f,p,s and q) must be preserved for the next iteration. These should, then, be part of the file of slices. The vector r , on the other hand, needs to be computed only for the current iteration, and does not have to be stored at all. This is better even if a couple of r slices may have to be computed twice in an iteration.

(e) The time spent performing I/O operations is a function of the number of bits transferred. On some computer systems one can choose different size words. Halving the size of words in the data being sliced cuts in half the time spent performing I/O . One can truncate the word size before writing out and extending it after reading in, or just work with the smaller size word. We have found, on the CDC CYBER 205, that the CG method is stable enough to permit the latter procedure. Using 64-bit or 32-bit arithmetic on this system resulted in values identical to machine accuracy and no deterioration of convergence rates. The cost of executing the code, and its elapsed time, were halved, of course.

4. AN APPLICATION

In this section we shall give some details regarding a specific problem to which the modified CG method described before has been applied. This is where the structure of the matrix M , or the rules for applying the "operator" to a given vector, have to be spelled out. The application considered is a stage in a suite of programs for evaluating physical observables in Quantum Chromodynamics (QCD) (the theory describing strong interactions among subnuclear particles), discretized on a four-dimensional lattice. Monte Carlo simulation techniques are used to create space-time configurations of the gauge dynamical variables on the lattice. These configurations amount to collections of 3×3 complex unitary-unimodular matrices (corresponding to the SU(3) gauge symmetry group of QCD) each associated with an oriented link (i.e. site and direction) on the four-dimensional lattice, as described in Ref. 5. A configuration contains (in our case) $16^3 \times 32$ SU(3) matrices, in each of four directions, 16 being the spatial size, and 32 the size along the time axis; i.e. more than 9.4 million real variables. In this lattice we place a "source" representing a quark (the fundamental building block of hadronic particles in the QCD theory), and then compute at each lattice site the value of the Green's function (or propagator) describing the propagation of the quark. This requires the solution of the lattice equivalent of the Dirac equation with a point-like source, namely

$$\sum_\mu S_x^\mu (\Delta^\mu \psi)_x + m\psi_x = \delta_{x,x_0} s_0 \; . \tag{4.1}$$

In Eqn. (4.1) x represents a generic lattice point, x_0 the position of the source, μ one of the 4-directions, s_0 a vector describing the orientation of the source in unitary space ($\delta_{x,x_0} s_0$ is the b vector of the CG algorithm). S_x^μ are suitable spin factors, reproducing on the lattice the algebra of Dirac's γ-matrices, and ψ_x is the quark field propagating from x_0 (ψ_x is also a 3-dimensional complex vector, in unitary space, and all the components of ψ constitute the vector f in the notation previously used). Δ^μ is a gauge-covariant central difference operator defined by

$$(\Delta^\mu \psi)_x = \frac{U_x^{\mu\dagger} \psi_{x+\hat{\mu}} - U_{x-\hat{\mu}}^\mu \psi_{x-\hat{\mu}}}{2a} \quad , \tag{4.2}$$

where the U's are the SU(3) matrices mentioned above and a is the spacing of the lattice.

From Eqns. (4.1) and (4.2) we can deduce the magnitude of the computational problem. There are $16^3 \times 32$ unknown quantities $\psi(x_\mu)$, each with 3 complex values (i.e., 6 numbers), so that the number of unknowns is 786,432. Eqn. (4.2) tells us that $4 \times 16^3 \times 32 = 524,288$ SU(3) matrices are contained in the coefficients matrix M . In fact, this is all that is needed, if, for the sake of economy in storage, one is willing to reorder data elements instead of duplicating them in a different order. Looking at Eqns. (4.1) and (4.2) it is clear that U_x is to be aligned with $\psi(x)$, as well as $\psi(x+\hat{\mu})$ and $\psi(x-\hat{\mu})$, where μ ranges over the 4 possible directions. This is not simply an offset of one unit, since we are dealing with four-dimensional space with periodic boundary conditions, and is best handled by creating index-lists pointing to the nearest neighbors in each direction. These index-lists can be utilized with the CDC CYBER 205's "gather" instruction for a fast reordering of data elements.

The number of non-zero coefficients in M does not have to be quite as high as 9.4 million. QCD is a gauge theory, which means that one can perform gauge transformations on a configuration, without changing its physical content. Since it is convenient to slice the problem by time-axis values, a natural gauge transformation will be the one that converts the configuration to the "temporal" gauge. After this transformation the SU(3) matrices associated with the time direction (between time slices) become the identity matrix for all but the last time-slice (which is treated as a special case). The computations thus simplify and the number of non-trivial real coefficients in the matrix is reduced to about 0.75 of what it would have been otherwise, i.e. just a little over 7 million values.

Without writing out explicitly all the terms in the discretized operator, it is apparent from Eqns. (4.1) and (4.2) that three time-slices are needed for applying the operator on a vector. The values of slices "t-1" and "t+1" are used even though they are connected via the identity matrix (which requires no matrix multiplication). The same dependence on neighbors exists for the spatial coordinates, but all their values are

contained in time-slice "t" anyway. In fact, we need in memory 4, and not
3 time-slices, at any one time. This we see from Eqn. (2.8): the computation of the vector $q^{(n)}$ requires slices t-1 , t and t+1 of r ;
but to compute the t+1 slice of r one also needs slice t+2 of the
vector p . This means that we need real memory sufficient for 5 time-
slices, the space for the fifth one used to initiate swapping slice t-2
which is not needed anymore with slice t+3 to be used at the next
iteration. (As a matter of fact, less than 5 time slices need to be kept
in fast memory for some of the other variables. In the temporal gauge only
3 slices of M are required.)

The vectorization of the $r^{(n)} = Mp^{(n-1)}$ and $s^{(n)} = M^{\dagger}u^{(n)}$ equations
is not achieved by vectorizing the product of the SU(3) matrix by a
3-component vector, but rather by doing the component multiplications over
many lattice sites simultaneously. Indeed, we manipulate all of the spatial
coordinates of a given time-slice in parallel.

With the slicing as described above the code executes comfortably in
less than 8 million bytes of memory (using 32-bit operands). The elapsed
time for one iteration improved substantially in the conversion to the
single pass algorithm. Timing data are about 50 secs elapsed time with the
original two-pass procedure, as follows: with 32-bit SU(3) matrices and
64-bit updated vectors; 24-25 secs when the one-pass method was adopted;
13-14 secs when the I/O channels were used concurrently for the SU(3)
matrices and the vectors, respectively; and, finally, 7-8 secs per iteration when 32-bit operands are used for the updated values. This last
change also doubles the arithmetic rate on the CDC CYBER 205. The CPU time
per iteration is still considerably smaller because the arithmetic is done
very efficiently. Utilizing 32-bit arithmetic one iteration lasts 1.32
CPU-secs (1.91 secs for 64-bit arithmetic), with an execution rate of 131
Mflops (78 Mflops for 64-bit). This includes the time spent in the "gather"
instruction (about 30% of the total, 17% for 64-bit precision) and corresponds to a rate of 193 Mflops for the 32-bit arithmetic alone. All these
figures pertain to a two-pipe CDC CYBER 205.

There is enough idle time while waiting for the I/O to complete to
test for convergence every iteration (if needed) without adding any elapsed
time to the execution. The test amounts to applying the operator a third
time to the "solution" vector and comparing it to the right-hand-side. We
then measure the L_2 , L_1 and L_{∞} norms of the residual vector. It turns
out that the convergence rate is, as expected, very sensitive to the size
of the main-diagonal elements determined by the parameter m in Eqn. (4.1).
The bigger m is the faster the convergence is. It changes from an order
of magnitude reduction in error size every 3 iterations when m=1 , to
similar decrease every 100 iterations when m = 0.025 . The evaluation of
the physical observables requires that, for a given configuration, the code
be executed with the point-like source positioned at various sites in the
lattice with a fixed value for the parameter m . Then the process is
to be repeated for different smaller values of m. When this is all
done a new configuration is formed (using Monte Carlo simulations) and the
whole procedure is repeated. A number of configurations are to be examined
this way until the statistical error of the observables becomes sufficiently
small. The analysis of the solution vector for the observables is not
part of the CG solver described here. The necessity of using the solver
so many times is what prompted us to search for an efficient algorithm for
very large systems.

ACKNOWLEDGEMENTS

We would like to thank Control Data Corporation for awarding time on the CDC CYBER 205 at the Institute for Computational Studies at Colorado State University where the code described in the the text was developed. One of the authors (C.R.) would like to thank the Directorate of CERN for the award of a Research Associateship.

REFERENCES

[1] M.R. Hestenes and E.L. Stiefel, Nat. Bur. Std. J. Res. <u>49</u>, 409(1952).

[2] L.A. Hageman and D.M. Young, <u>Applied Iterative Methods</u> (Academic Press, New York, 1981), p. 138.

[3] T. Ginsburg, in <u>Handbook for Automatic Computation</u>, Vol. II, ed. J.H. Wilkinson and C. Reinsch (Springer-Verlag, Berlin, 1971), p. 57.

[4] J. Stoer and R. Bulirsch, <u>Introduction to Numerical Analysis</u> (Springer-Verlag, Berlin, 1980), p. 572.

[5] D. Barkai, K.J.M. Moriarty and C. Rebbi, Comp. Phys. Comm. <u>32</u>, 1(1984).

IMPLEMENTATION OF THE VECTOR C LANGUAGE ON THE CYBER 205

Kuo-Cheng Li and Herb Schwetman[†]

Department of Computer Sciences
Purdue University
West Lafayette, IN 47907

ABSTRACT

Vector C, a superset of the conventional programming language C for vector processing, has been designed and implemented on the Cyber 205. In this paper, the extended features of Vector C are described briefly and then some implementation issues are presented. The performance of the code generated by the Vector C compiler is illustrated by timings for several benchmark programs. Finally, three performance models for vector instructions and their evaluations are presented. Readers are assumed to have some knowledge of the C language.

1. INTRODUCTION

There are several approaches to utilizing vector processing facilities from the viewpoint of language design and optimizing compiler development. The two obvious approaches are, (1) *Vectorization:* a vectorizer, embedded in the compiler [1] or as a preprocessor (e.g., VAST [2] and KAP [3]) to the compiler, is used to extract inherent parallelism in a sequential program (usually in the DO loop of a Fortran program) and to generate vector instructions, and (2) *Vector syntax oriented language:* a language which has *explicit* syntax for vector processing.

Various surveys [4,5,6] have shown that vectorization alone is usually less satisfactory and less efficient for designing and constructing programs; many users of vector or parallel computers prefer languages with vector or parallel syntax. Figure 1 shows two different approaches for solving a problem. By using the first approach, a user needs to convert vector operations and objects in his (her) algorithm into loops and scalar objects, and then a sophisticated compiler is required to convert these scalar objects and loops back to their original vector representations and then generates vector instructions for them. This means that there is a redundancy in the programming process, (although conventional language programmers may already be used to it), and there is the increased complexity and inefficiency in the compiling process required to extract the obscured vector code embedded in loops. Furthermore, an optimal scalar algorithm/program may not be an optimal vector (or parallel) algorithm/program. Software designers need to "think parallel" when developing software in a parallel environment. As a result, relying on an automatic vectorizer to discover inherent parallelism is inefficient, both in the use of the vectorizer and the code generated by the vectorizer. (However, in order to reuse

---

[†] The current address of Herb Schwetman is MCC, 9430 Research Blvd., Austin, TX 78759.

the existing sequential programs, mostly Fortran programs, a sophisticated vectorizer is still a must).

In [7], we listed some other reasons for developing a particular vector language, called Vector C [8], which is a superset of the conventional C language [9]. Vector C has been designed and implemented on the Cyber 205. In this paper, Section 2 describes briefly the extended features of the Vector C language. Section 3 presents some implementation issues. Section 4 presents some performance measurements to demonstrate the efficiency of the code generated by the Vector C compiler, and finally, some regression models for vector instructions and their evaluations are presented; these provide more accurate models which can be used for developing timing estimates.

2. THE VECTOR C LANGUAGE

Vector C is designed to allow users to specify vector operations directly and naturally. In this paper, only the extended features of Vector C are (briefly) presented. Basically, they include (1) a means for specifying vectors and (2) operators and data/control structures for manipulating vectors.

A vector is an aggregated data structure. In Vector C, a vector is recognized by its appearance (reference) rather than by its declaration. We define seven ways of specifying subscript ranges for referencing vectors [8], namely, (the pair of symbols [ ] below stands for optional),

(1) *initial : final* [*: increment*] - For examples, 0:4 specifies 0,1,2,3,4, and 1:10:2 specifies 1,3,5,7,9.

(2) *initial # length* [*: increment*] - For examples, 0#4 specifies 0,1,2,3, and 2#5:2 specifies 2,4,6,8,10.

(3) * - Whole-dimension selection. For example, if array V is declared as "int V[100]", then V[*] means the entire vector V[0], V[1], ..., V[99].

(4) *initial : * [*: increment*] - For example, if array V is declared as "int V[100]", then V[3:*:2] is the same as V[3:99:2].

(5) *index vector* - For example, V[iv[*]], where iv is an index vector.

(6) *bit vector* - For example, V[bv[*]], where bv is a bit vector.

(7) *descriptor* - For example, V[vd], where vd is a vector descriptor.

New data objects include the bit data type and the vector descriptor. A bit object is declared by the keyword **bit**. In the example,

        bit bv[100];

'bv' is a bit vector of length 100. A vector descriptor is used to represent a vector. In Vector C, a descriptor is specified via a special character @ (similar to '*' for pointers in C), and each descriptor has a specified type associated with it; in the example

        float @id;
        int @vd[3][2];

'id' is a descriptor which represents a floating point array, and 'vd' is an array of descriptors which represent integer arrays. A vector descriptor contains (at least) the address of the vector and the length of that vector.

Initial values for vectors can be specified at compile time using a syntax similar to that used in subscript expressions, with the addition of a dyadic repeat operator, '!'. As in C, only global (external/static) vectors of any type can be initialized in this manner. As an example,

    float a[] = {1.2, 3.6, 1.3!3, 2.1:10.0:3.3, 3.5#3}

is the same as

    float a[] = {1.2, 3.6, 1.3,1.3,1.3, 2.1,5.4,8.7, 3.5,4.5,5.5}

A vector expression is an expression which contains at least one vector reference. Mixed-data types are allowed in expressions wherever appropriate. The data type conversion rules and operator precedence rules for conventional C statements apply to vector statements. A vector relational expression is evaluated to a bit vector, called a control vector, which is used in the vector control statements. Vector control statements include

1) CONTROL expression: e.g.,
   int a[5], b[5], c[5];
   bit z[5];

   a[*] = z[*] ? b[*];
   if z[*] = 1,0,0,1,1, a[*] = 1,2,3,4,5, and b[*] = 6,7,8,9,0, then a[*] becomes
   6,2,3,9,0.

2) MASK expression: e.g.,
   a[*] = z[*] ? b[*] : c[*];
   if z[*] = 1,0,0,1,1, b[*] = 6,7,8,9,0, and c[*] = 5,4,3,2,1, then a[*] becomes
   6,4,3,9,0.

3) MERGE expression: e.g.,
   a[*] = z[*] ? b[*] <> c[*];
   if z[*] = 1,0,0,1,1, b[*] = 6,7,8,9,0, and c[*] = 5,4,3,2,1, then a[*] becomes
   6,5,4,7,8.

4) IF statement: e.g.,
   if (a[*] != 5)
           a[*]++;
   else
           a[*] = 0;
   if a[*] is 2,5,2,4,5, then the result a[*] becomes 3,0,3,5,0.

5) WHILE statement: e.g.,
   while (V[0#5] > 0) {
           ...
           V[0#5]--;
   }
   The mask (V[0#5] > 0) is updated every iteration until all the bits in the mask
   are zero bits.

6) FOR statement: e.g.,
   for (i= 0; V[0#5] <= 5 ; i++,V[0#5]++)
           B[0#5][i] = V[0#5];
   If V[*] = {1,2,3,4,5}, then the final B is
           1 2 3 4 5
           2 3 4 5 x
           3 4 5 x x
           4 5 x x x
           5 x x x x
   where x means "don't care", i.e, the original value of B is not updated. Note
   that the mask (V[0#5] <= 5) is applied to B[0#5][i] = V[0#5] and V[0#5]++.

7) SWITCH statement: e.g.,
   switch (V[0#10]) {
           case 0: break;
           case 1:
           case 2: A[2#10]++;
                   break;
           default: A[2#10] = 0;
                   break;
   }
   If V = {1,0,2,2,1,3,4,2,0,0} and A = {x,x,1,2,3,4,5,6,7,8,9,0}, then the result is A =
   {x,x,2,2,4,5,6,0,0,9,9,0}, where x means "don't care".

Nested control structures are allowed and the outer control vector affects the inner
control vector. The control vector is stacked when entering a new control structure,
and it is unstacked on exiting.

   The operators in conventional C are extended to perform element-by-element
operations. Because operators (as compared with keywords or function calls) have
the advantages of conciseness and efficient lexical analysis, several new operators
are defined in Vector C. These new operators include reduction operators (@., @+,
etc.), the vector descriptor operator (@), vector descriptor related operators (@#

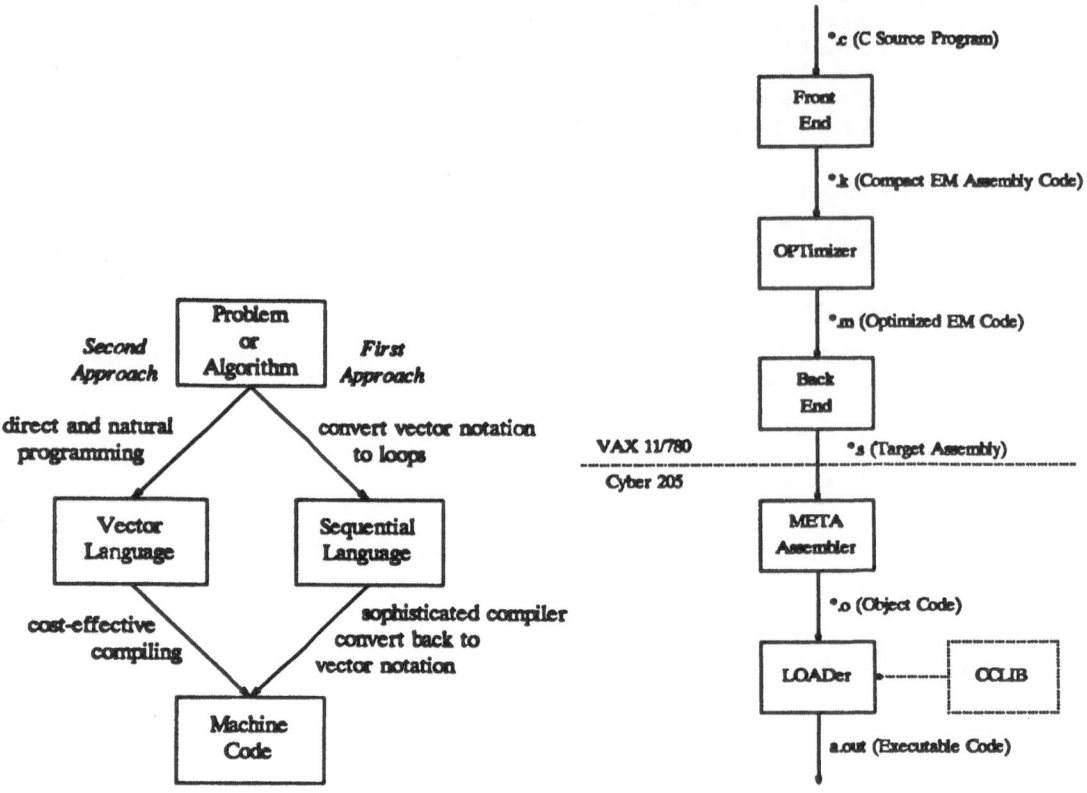

Figure 1: Problem-solving Model in Vector/Parallel Environment

Figure 2: Basic Building Blocks of VC205

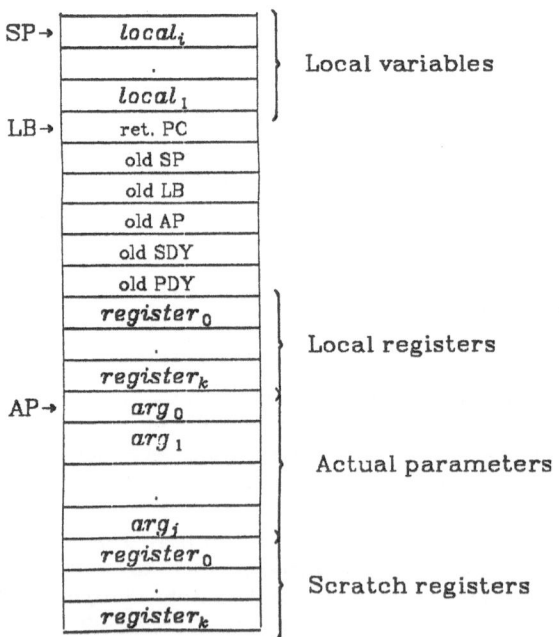

Figure 3: Activation Record for a Procedure Call (Version 0.2)

and @&), other unary vector operators (@~, @|, etc.), binary vector operators (@/, etc.), and ternary operators (@! #, etc.). For examples:

1. S = @+ VA[*];   -- vector sum over VA[*]
2. VA[*] = @| VB[*];   -- VA[i] = | VB[i] |, for all i,
3. VA[*] = @\VB[*];   -- reverse vector VB,
4. V[*] = VA[*] @/ VB[*];   -- V[i] = (VA[i] + VB[i])/2, for all i,
5. S = @|| BV[*];   -- count the number of '1' bits in a bit vector,
6. VA[*] = i @! j # n;   -- VA[*] = i, i+j, ... , i+(n-1)j,
7. BV[*] = a @1 b # n;   -- a '1' bit followed by (b-a) '0' bit until length n,

where, BV[*] is a bit vector, S is a scalar value, hence @|| is a reduction operator.

A vector descriptor is, in fact, a pointer, hence the C pointer arithmetic scheme [9] can be applied to descriptors; in other words, Vector C provides a *descriptor arithmetic* capability.

When a vector is passed as an argument to a function (or procedure), only the descriptor, either user specified (explicit descriptor) or compiler generated (implicit descriptor), is passed. A vector-valued function is declared by '@', e.g.,

    float @vfunc();

By using a dynamic space (another stack which is not in the C run-time stack [7]), vector-valued functions can be implemented.

In Vector C, some constructs have a large semantic gap from the hardware of Cyber 205, such as multiple dimensional vector subscripts, e.g., M[0#10:2][v[*]], and "selection" on the vector reduction operators [8]. This makes the implementation of Vector C on the 205 a bit complicated.

3. IMPLEMENTATION ISSUES

In order to implement the Vector C compiler in a short period of time, a compiler development tool was used. This compiler tool is called Amsterdam Compiler Kit (ACK) [10]. By using ACK, the first version of the scalar C compiler was installed on the Cyber 205 to achieve upward compatibility [7].

The implementation of Vector C on the Cyber 205 (VC205) is an enhancement on the scalar C compiler (CC205). This compiler is composed of three major parts (as shown in Figure 2), namely, Front-End (FE), Optimizer (OPT) and Back-End (BE). The FE is a modified version of PCC [11] using YACC [12] as a tool to build the parser. The FE accepts Vector C program and generates an intermediate code which then goes through OPT to perform some local optimization, and finally the BE accepts the optimized intermediate code and generates the Cyber 205 assembly code (called META).

The major implementation issues for the scalar C compiler were described in [7], and several enhancements for better performance were also presented. The implementation of Vector C compiler involves almost all the compiler development processes [13], namely, lexical analysis, syntax analysis, code optimization and code generation phases. In the following sections, the third enhancement [7] on utilizing the 205 register file is presented, and then some major issues in implementing Vector C compiler are presented.

3.1. Register Class Enhancement

The 256 registers available on the 205 were not being utilized by the generated code from CC205. Making better use of these was an "obvious" improvement. The implementation of this enhancement includes (1) modifying the front-end of CC205 to accept a **register** storage class [9] and to perform register allocation, (2) extending EM intermediate code [2] to allow the use of registers, and (3) modifying the prologue and epilogue in procedure calls to perform register swapping, and hence the procedure calls are more expensive than before [7]. The extended activation record is shown in Figure 3, where AP is the argument pointer, and SDY and PDY are the pointers for the book-keeping of the Dynamic Stack (see Section 3.2).

Table 1 shows the instructions generated by this version of CC205 (defined as Version 0.2 of CC205). The timing analysis shows that the computation is done in 7

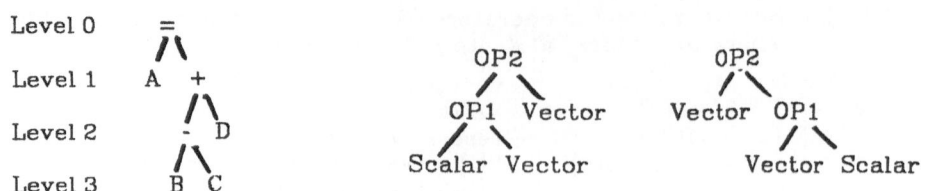

Level 0	=
Level 1	A +
Level 2	- D
Level 3	B C

Figure 4: Syntax Tree for A = B - C + D   **Figure 5:** Syntax Trees (No Linked Triadic Effect)

Table 1: Version 0.2 Instruction Mappings for a = b*c + 3

EM	CC205₂	Comments
LOR 59		load register 59 (0x3b)
LOR 60		load register 60 (0x3c)
MLI 64	mpyx r_3b,r_3c,t0	t0 = b*c
LOC 3		load constant 3
ADI 122	addx t0,c_3,r_3a	r_3a = t0 + 3

Table 2: State Transition Table

	S	V	P	G
S	S	V	P	G
V	P	V/L	L	L
P	P	L	L	L
G	G	L	L	L
L	L	L	L	L

Table 3: Execution Times (in seconds)

Benchmarks	VAX C	CC205₂
sieve.c	1.767	0.273
sort.c	10.633	2.473
fibo.c	26.617	4.379
float.c	1.650	0.086
pmva.c	12.617	1.410

Table 4: Execution Times (in seconds) and Speeds (in MFLOPS)
The numbers in parentheses are speeds

Benchmarks	CC205₂	205 FORTRAN
Multiply and Divide	0.086 (1.63)	0.125 (1.12)
Multiply only	0.022 (6.36)	0.056 (2.50)

Table 5: Execution Speeds for Matrix Multiplication (in MFLOPS)

Size	VC	Fortran	MXMPYR
10	6.86	6.98	14.80
50	31.77	31.76	43.91
100	55.12	54.30	73.16
150	72.25	72.19	91.87
200	82.53	83.07	101.54
250	93.31	93.34	112.01
300	101.77	101.43	119.85

minor cycles. The instructions generated for the same case by Versions 0.0 and 0.1 can be found in [7]; these required 120 and 32 minor cycles respectively.

3.2. Dynamic Stack

In the Vector C compiler, two run-time stacks are employed; one is the conventional C run-time stack, the other one is the dynamic stack [7]. The dynamic stack is used for storing the intermediate results of vector operations and vector-valued functions.

In this version of VC205, temporary storage is needed only when the operation is performed below the first level of a syntax tree, e.g., for the vector statement, A = B - C + D, the syntax tree is shown in Figure 4, where the "-" operation requires temporary storage, while the operation "+" does not need a temporary storage, the result can directly go to A, i.e., T = B - C and A = T + D. There are many possible choices in implementing this kind of expression, e.g., if the vector lengths are all the same and the result operand is not the right(-tree) descendant of the "=" node, then no temporary storage is needed, since the result operand can be used for storing intermediate results.

The management of the dynamic stack is done by the following principles: (1) the dynamic stack is allocated when needed by a vector instruction (or a vector-valued function) and deallocated when exiting the vector statement, and (2) when using dynamic descriptor [8] (a descriptor pointing to the vector located in the dynamic stack), the allocated dynamic stack is not de-allocated until the exit from the procedure.

3.3. Linked Triadic Operations

A special feature of the Cyber 205 hardware is the *linked triadic* operation [15], akin to *chaining* in the Cray-1 computer [16]. By using this feature, two vector operations can be combined into nearly one vector operation for certain sequences of vector instructions (e.g., +, -, *, &, |, ^, ~, and some logical compare operations). The requirements are (1) both operations must be in different functional units, (2) there must be at least one scalar and at least one vector operand, (3) logical compare operators can apply only to the second evaluated operation, and (4) other restrictions, e.g. an expression having the syntax trees shown in Figure 5 cannot have a linked triadic effect. (However, the compiler can easily solve this problem by switching the nodes of the tree, since at least one of the operators (OP1 and OP2) is associative.)

3.4. Vector Control Structures

VC205 allows nested vector control structures, and outer control vectors can affect inner control vectors; e.g.

```
if ( a[*] > 0 ) {
    ...
    if ( b[*] < 0 )
    ...
```

In this example, the first if-statement generates a control vector, say CV0 = a[*] > 0, and the second if-statement generates another control vector CV1 = CV0 & (b[*] < 0).

A control vector on the Cyber 205 only applies to integer or floating point data types and to the following operations:

1)   arithmetic operations, e.g., +, -, etc.

2)   assignment operation, i.e., =,

3)   bit-wise logical operations, such as &, |, ^, ~.

4)   The other unary, binary, ternary and reduction vector operations, such as @~, @_, @|, @', @%, @\, @^, @/, @//, @!, @., @+, @*, @< and @>.

Table 6: Execution Speeds for Different Matrix Multiplications (in MFLOPS)
† values in parentheses do not include setup time for index vectors

Size	Inner	Middle	*Outer*†	n^3
10	5.13	6.86	3.44 (3.70)	3.42
40	18.61	25.81	9.92 (10.07)	8.18

Table 7: Execution Times for Conjugate Gradient Algorithm (in μs)

Size	VC	Fortran	Speed-up(%)
8	3161	3871	18.34
16	5968	6889	13.37
32	17145	18759	8.6
64	62928	66729	5.7

Table 8: Timing for Vector Copy

Vector Length	Time (in μs)
1	0.970
11	1.050
21	1.130
31	1.210
41	1.370
51	1.450
61	1.530
71	1.610
81	1.770
91	1.850
101	1.930
111	2.010
121	2.170
131	2.250
141	2.330
151	2.410
161	2.570
171	2.650
181	2.730
191	2.810
201	2.970

3.5. Multiple Dimensional Parallelism

A multiple dimensional parallelism expression is a useful means when designing and implementing multi-dimensional (e.g. matrix) algorithms. For example, M[1:N-2][1:N-2] means the interior of an N by N matrix, M. M[iv[*]][*], where iv[*] = {0,0,1,..,N-2}, means that shifting the entire matrix M down one row. Because of the large "semantic gap" between this construct and the Cyber 205 hardware, the implementation of this type of construct is inefficient as compared to using bit vectors and thinking in "vector" way.

In a multiple dimensional parallelism expression, only the inner-most vector is treated as a real vector. The rest of the vector subscripts are implemented as loops which iterate on this vector. Some multiple dimensional parallelism expressions are still vectors; e.g., M[*][*] is the same as M[0][0#N*N] which is a vector, or M[0#5][*] is the same as M[0][0#5*N] which is again a vector. However, M[*][0#5] is a loop iterated on a vector of length five for N times. A multiple dimensional parallelism expression is actually gathered into a vector which is located in the dynamic stack before any operation is performed.

The current implementation of this construct allows all of the vector subscripts defined in Section 2, except bit vector subscripts. Table 2 shows the state transitions for handling multiple dimensional parallelism expressions, where S means Scalar, V means Vector, P means Periodic gather/scatter, G means Gather/scatter and L means Loop. When in state V with input V, the new state is V or L depending on the type of input V which could be a entire vector or a section of a vector (see the examples in the previous paragraph).

Another approach is to construct a bit vector for this multiple dimensional parallelism construct. However, the construction of a bit vector is possible only for some subscript expressions (the subscript expressions 1, 2, 3, and 4 in Section 2); for other subscript expressions, they turn out to be very difficult and inefficient to implement.

4. PERFORMANCE MEASUREMENTS AND MODELS

In the following sections, the same benchmarks [7] are used to measure the performance of the enhanced compiler (Version 0.2 of CC205), then two examples are used for comparing the performance of the code of generated by the Vector C compiler and the Fortran 200 compiler. Finally, three performance models for vector instructions and their evaluations are presented.

4.1. Performance Enhancement of CC205 Version 0.2

After installing the register storage class, the benchmarks in [7] were used again (with modifications on the local variable declarations) to compare the performance of C programs running on the VAX and the Cyber 205. Also, a *timer* was developed to measure the performance more accurately. Table 3 shows the execution times of these five benchmarks, where $CC205_2$ is Version 0.2 of CC205 which accepts register storage class and makes more efficient use of registers. Among them, the performance of *fibo.c* is degraded (as compared to Version 0.1); this is because *fibo.c* is highly recursive and the procedure call overhead (Version 0.2 has more expensive prologues and epilogues than Version 0.1) dominates the execution time of the procedure body.

In Table 3, the *float.c* has an execution time of 0.086 seconds on the Cyber 205, which means 1.63 MFLOPS, (on VAX 11/780, this was measured to be 0.085 MFLOPS). As compared to the optimal scalar floating point processing rate (50 MFLOPS), the measured rate is very small. The reasons are obvious; first, the program is about half divide operations, since the divide functional unit is not pipelined, (in comparison, for a fullword floating point multiplication, the result is available in 5 cycles by shotstop and 8 cycles at register file, while for a fullword divide operation, the result is available in 53 cycles by shortstop and 57 cycles at register file [17]). Second, there are register conflicts between statements, (the flow analysis shows that there are some flow-dependences and output-dependences [3] in this program).

The benchmark *float.c* was converted to 205 Fortran and compiled without further optimization, (since this program contains many redundant statements, additional optimizations will make the benchmark meaningless). The processing time is 0.125 second (or 1.12 MFLOPS), which is slower than the code generated by CC205 (see Table 4). This benchmark was further modified to perform all the operations using multiplication (rather than intermixing multiplication and division). These results are also shown in the Table 4.

4.2. Matrix Multiplications

There are several ways to perform matrix multiplication on a vector processor [18] (see Appendix 8.1). One of these, the middle product method, was implemented in both Vector C and 205 Fortran. Table 5 shows the execution speeds of this matrix multiplication method for different sizes and versions. In Table 5, *Size* is the matrix size (all are square matrices of *Size* by *Size*), MXMPYR is a MAGEV library routine implemented in META [19], which is called from a Vector C program, and the execution speed (in MFLOPS) is calculated as $(Size*Size*(2*Size-1))$ / *Time*.

Table 6 shows the execution speeds using the four different methods. Obviously, the *middle product method* is the best among these methods under the current vector hardware configuration.

4.3. Conjugate Gradient Algorithm

One example of problems which can exploit the vector capabilities of the 205 is solving a large system of equations, $A \cdot \vec{x} = \vec{f}$, using the *conjugate gradient* algorithm. A 205 Fortran implementation of this algorithm is given in [20]. A direct conversion to C of this Fortran conjugate gradient program is in Appendix 8.2. The timing comparison between these two programs (shown in Table 7) indicates that the Vector C compiler generates faster executing code than 205 Fortran compiler for this example. The Speed-up in Table 7 is defined as (VC - Fortran)*100 / Fortran. The Speed-up decreases as the Size increases. The reason is that there are some redundant "scalar" code generated by the Fortran compiler. When the vector length is short, the discrepancy between the performance of those two programs (VC and Fortran) is large. However, when the vector length is long, the time in executing vector instructions dominates the time in executing scalar instructions; in other word, the effect due to those redundant scalar code is minimized.

Another, more natural, version of this algorithm can be implemented. That is, one can think the problem as a matrix problem (using multiple dimensional parallelism constructs) rather than as a vector problem (using a bit vector). For example, the "vector" version is, (see Appendix 8.2)

```
(&t[1][1])[bvd] = pd;    /* expand p to t */
aptd = bvd ? 4.0 * tad - tbd - tcd - tdd -ted;
apd = apt[0][bvd];     /* compress apt into ap */
```

while the "matrix" version is,

```
t[1:N][1:N] = pd;
aptd = 4.0* td - t[i[*]][*]     /* shift down */
       - t[j[*]][*]             /* shift up */
       - t[*][i[*]]             /* shift right */
       - t[*][j[*]];            /* shift left */
apd = apt[1:N][1:N];
```

where, i[*] is {0,0:N-2} and j[*] is {1:N-1, N-1}. The rest of the variables are defined in Appendix 8.2. As mentioned before, "matrix" constructs exhibit a large semantic gap from the underlying vector hardware architecture; hence the implementation is inefficient (compared to "vector" constructs). The execution time of this "matrix" version for *Size* = 16 is 23922 μs, which is four times slower than "vector" version.

4.4. Performance Models

Three performance models of vector instructions are defined. For most vector instructions, the execution time is a function of the vector length; $Time = f_1(\,length\,)$. For periodic gather/scatter instructions, it is a function of vector length and stride; i.e., $Time = f_2(\,stride,\,length\,)$; and for bit mask generation instructions (@0, @1), it is a function of vector length and pattern specifications (called initial and field), i.e., $Time = f_3(\,initial,\,field,\,length\,)$.

Several empirical experiments were performed to gather the timing information for each vector instruction with varying vector length and stride, etc. Table 8 shows an example of the output, (actual measurements containing more than 100 data points). By examining these timing data, we can find a linear relationship between the variables (e.g., length and time). A linear regression analysis based on the least squares method [21] is performed for each vector instruction, with the model-1 estimate being

$$T = S + P * L$$

for most vector instructions, where T is the execution time, S is the vector start-up time, P is the processing time, and L is the vector length. The model-2 estimate

$$T = S + O * D + P * L$$

for periodic gather/scatter instructions, where D is the stride (or distance) and O is the factor due to stride (i.e. noncontiguous vector). The model-3 estimate

$$T = S + A * I + B * F + P * L$$

for bit mask generation instructions (@0 and @1), where I, F and L are used as I @0 F # L (see Section 2), and A, B and P are their corresponding coefficients.

Tables 9, 10 and 11 show the coefficients of the regression model for each vector instruction. In these tables, S.e.e means standard error of estimate which is defined as,

$$S.e.e = \left[\frac{\sum\limits_{i=1}^{n} (T_i - T_i)^2}{n-2} \right]^{1/2}$$

where T_i is the measured timing data, and T_i is the modeled timing result.

The instructions shown in Tables 9, 10 and 11 do not include all the vector instructions generated by the VC205, but they include timing models parameters for all vector instructions. Because some instructions share the same timing model; they have the same Issue Time (INB), Vector Busy Time (VBA) and effective PLS (pipeline size) [22]. For example, vector *min* and *max*, *maskz* and *masko*, etc, have the same timing models and parameter values.

4.5. Performance Model Evaluation

Hockney [18] defined a performance model based on two parameters: r_∞ - the maximum or asymptotic performance, i.e., the maximum rate of computation in units of operations performed per second for vector elements (e.g. MFLOPS), and $n_{1/2}$ - the half performance length, i.e., the vector length required to achieve half of the maximum performance. T, the time to perform a vector operation on length n, is defined as $T = r_\infty^{-1}(\,n + n_{1/2}\,)$. The characteristic of a computer is represented by the pair $(r_\infty, n_{1/2})$. Each computer has different values of $(r_\infty, n_{1/2})$, e.g., for the Cyber 205 with 2-pipe and 64-bit vector operation, this pair is (100,100) [8]. This timing model provides *macro − analysis*, similar to the big O-notation, and assumes all vector operations performed in a computer configuration have the same timing requirements.

As an example, compute the sum of the diagonal elements of the matrix 'a', $S = \sum\limits_{\substack{i,j=0 \\ i=j}}^{N-1} |a_{ij}|$. Two approaches for solving this problem are shown below:

```
        float  a[N][N],S;
        bit    bv[N][N];

1)      bv[*][*] = 1 @1 (N+1) # N*N;    /* 1 followed by N 0's until length N*N */
        S = @+ @| a[0][bv[*][*]];        /* compress */
```

Table 9: Regression Model-1 Parameter Values

Instruction	S	P	S.e.e
vector copy (vtov)	0.93167	0.01000	0.02379
integer add (addxv)	1.05629	0.01000	0.03551
integer mul	3.61329	0.02998	0.07191
integer div	2.34398	0.07248	0.08505
float add (addnv)	0.97501	0.01000	0.02658
float mul (mpysv)	0.97888	0.01000	0.02645
float div (divsv)	1.26931	0.06247	0.07598
random gather (vxtov)	1.12582	0.03492	0.14507
random scatter (vtovx)	1.04980	0.03073	0.14459
mask (maskv)	1.27714	0.01000	0.01734
merge (mrgv)	1.03321	0.01436	0.28224
compare (cmpne)	1.26668	0.00994	0.03377
count one (cnto)	1.67228	0.00125	0.02372
bit vector and (and)	1.18935	0.00118	0.03306
integer vector and (andv)	1.04767	0.01001	0.02669
absolute (absv)	0.97481	0.01000	0.02646
square root (sqrtv)	1.23241	0.06250	0.05360
adjust mean (adjmean)	1.04888	0.01002	0.04525
average (avg)	0.98421	0.00999	0.02874
interval	0.92881	0.02000	0.02046
sum	2.45360	0.02000	0.04506
product	2.49513	0.02000	0.04094
max	1.87524	0.02000	0.00943
dot-product (dotv)	2.77529	0.02000	0.04494
compress (cpsv)	1.28103	0.01000	0.01889
expand (mgrv,sb)	0.95698	0.01439	0.27680
reverse (vrev)	0.73034	0.01422	0.09081
linked triadic (linkv)	1.63252	0.01000	0.02390

Table 10: Regression Model-2 Parameter Values

Instruction	S	O	P	S.e.e
periodic gather	0.4746813	0.0022840336	0.031017857	0.27174
periodic scatter	1.1706221	0.0008781513	0.026522321	0.17330

Table 11: Regression Model-3 Parameter Values

Instruction	S	A	B	P	S.e.e
maskz (@0)	1.1538645	-0.00084046	0.00003738	0.00141	0.0811

Table 12: Timing Results and Relative Errors for the First Solution.

(a) Timing Results (in μs)

N	Measured	Li	CDC	Hockney
1	6.06	5.904	4.921	1.040
51	39.06	37.072	35.541	54.040
101	127.62	125.290	122.161	207.040
151	272.18	270.558	264.781	460.040
201	508.54	472.876	463.401	813.040

(b) Relative Errors

N	Li	CDC	Hockney
1	0.026	0.188	0.828
51	0.051	0.090	-0.384
101	0.018	0.043	-0.622
151	0.006	0.027	-0.690
201	0.070	0.089	-0.599

Table 13: Timing Results and Relative Errors for the Second Solution.

(a) Timing Results (in μs)

N	Measured	Li	CDC	Hockney
1	4.47	3.969	3.775	1.030
51	7.48	7.134	6.525	2.530
101	10.57	10.299	9.275	4.030
151	13.56	13.464	12.025	5.530
201	46.36	16.629	14.775	7.030

(b) Relative Errors

N	Li	CDC	Hockney
1	0.112	0.155	0.770
51	0.046	0.128	0.662
101	0.026	0.123	0.619
151	0.007	0.113	0.592
201	0.641	0.681	0.848

or

2) $S = @+ @| a[0][0 \# N : N+1]$;   /* periodic gather */

Applying Hockney's model to this example, we have, $(1.0 + 0.02N + 0.02N^2)$ μs for the first solution, and $(1.0+0.03N)$ μs for the second solution.

The timing information provided in [22] is estimated and needs further verification. These timing data are used to perform analytic timing analysis for this example, and the results are $(4.88 + 0.03N + 0.0112N^2)$ μs for the first solution, and $(3.72 + 0.055N)$ μs for the second solution.

By applying the empirical timing models developed in Section 4.4 to this example, we have $(5.8625 + 0.03004N + 0.01141N^2)$ μs for the first solution, and $(3.9054 + 0.0633N)$ μs for the second solution.

Finally, this example was coded in Vector C and executed with different array sizes (N). Tables 12.a and 13.a show the measured and modeled timing results in μs for $N = 1, 51, 101, 151, 201$. The columns under Li, CDC and $Hockney$ are the estimates using the models developed in Section 4.4, CDC engineering specification and Hockney's model respectively. Tables 12.b and 13.b show the relative errors of each estimate compared with the actual measured value. The relative error is defined as,

$$Rerr = \frac{Measured - Modeled}{Measured}$$

From these tables, we can see that the performance models developed in this section are more accurate than the other models. These models could be embedded in a compiler to determine, for a vector statement, whether to generate vector code or scalar code [23].

5. SUMMARY

The programming language C is a powerful conventional language. One of the design goals of C is flexibility. This flexibility allows us to utilize the underlying hardware efficiently. The Vector C programming language is designed (extended) with similar goals [8]. The preliminary version of VC205 has demonstrated that Vector C programs can be executed at speeds that meet or exceed those of equivalent statements from 205 Vector Fortran.

Vector C allows the implementation of vector algorithms in a natural and direct manner. Vector C solves some of the old problems in Fortran, e.g., allowing aggregate data structures, complex control flow, and a flexible argument passing mechanism. Vector C exhibits a more concise and modern syntax than 205 Vector Fortran, e.g., gather/scatter, compress/expand, vector function calls, descriptor arithmetic, and flexible nested vector/scalar control structures. Also, Vector C provides users in the C community with a means of utilizing vector processing capabilities and of accessing highly optimized software libraries (e.g., MAGEV). Hence, we believe that Vector C will be of great use to programmers in devising new algorithms for vector computers.

There are several future projects mentioned in [8], namely, the evaluation of the constructs in Vector C, and the portability, extensibility and usability of the Vector C language. We hope these problems will be studied.

6. ACKNOWLEDGMENTS

We like to thank many people in PUCC and Department of Computer Sciences who have contributed to this project. Specially, we note the contributions of Professors Saul Rosen, John Rice and Dennis Gannon. This work was supported in part by the Purdue University Computing Center (PUCC).

7. REFERENCES

1.   "CDC CYBER 200 Fortran Version 2 Reference Manual," *Control Data Corporation*, 1983.

2.   B. Brode, "Precompilation of FORTRAN Programs to Facilitate Array Processing," *IEEE Computer*, pp 46-52, September 1981.

3. D. Kuck, R. Kuhn, D. Padua, B. Leasure, and M. Wolfe, "Dependence Graphs and Compiler Optimizations," *Proc. of the 8th ACM Symposium on Principles of Programming Languages*, pp 207-218, January 1981.

4. C. Wetherell, "Design Consideration for Array Processing Languages," *Software-Practice and Experience*, pp 265-272, April 1980.

5. R.H. Perrott and D.K. Stevenson, "Users' Experiences with the ILLIAC IV System and its Programming Languages," *SIGPLAN Notices*, pp 75-88, July 1981.

6. R.H. Perrott and D.K. Stevenson, "Consideration for the Design of Array Processing Languages," *Software-Practice and Experience*, pp 683-688, July 1981.

7. K.C. Li and H.D. Schwetman, "Implementing a Scalar C Compiler on the Cyber 205," *Software-Practice and Experience*, (to appear).

8. K.C. Li and H.D. Schwetman, "Vector C - A Vector Processing Language," *Journal of Parallel and Distributed Computing*, (to appear).

9. B.W. Kernighan and D.M. Ritchie, "The C Programming Language," *Prentice-Hall*, 1978.

10. A.S. Tanenbaum, H. Van Staveren, E.G. Keizer, and J.W. Stevenson, "A Practical Tool Kit for Making Portable Compilers," *Comm. ACM, Vol 26*, pp 654-660, September 1983.

11. S.C. Johnson, "A Tour through the Portable C Compiler," *UNIX Programming Manual*, Bell Labs, 1980.

12. S.C. Johnson, "YACC - Yet Another Compiler Compiler," *UNIX Programming Manual*, Bell Labs, 1975.

13. A.V. Aho and J.D. Ullman, *Principles of Compiler Design*, Addison-Wesley, 1977.

14. A.S. Tanenbaum, H. Van Staveren, and J.W. Stevenson, "Description of an Experimental Machine Architecture for Use with Block Structureed Languages," *Informatica Rapport 54, Vrije University, Amsterdam*, 1980.

15. "CDC CYBER 205 Hardware Reference Manual," *Control Data Corporation*, 1981.

16. R.M. Russell, "The CRAY-1 Computer System," *Comm. ACM*, pp 63-72, January 1978.

17. "CDC CYBER 200 Model 205 Technical Description," *Control Data Corporation*, November 1980.

18. R.W. Hockney and C.R. Jesshope, *Parallel Computers*, Adam Higer Ltd. Bristol. 1981.

19. "MAGEV - Mathematical/geophysical vector library," *Purdue University, Computer Center*, V3.1 (3/83).

20. D. Gannon, J. Rice and S. Rosen, "CS590V - Vector and Parallel Computing," *Class Notes*, Dept. of Computer Sciences, Purdue University, 1983.

21. J. Neter and W. Wasserman, *Applied Linear Statistical Models*, Richard D. Irwin, Inc., 1974.

22. "Engineering Specification," *Control Data Corporation*, No. 10358026, September 1982.

23. C.N. Arnold, "Vector Optimization on the Cyber 205," *Proc. of the 1983 International Conference on Parallel Processing*, August 1983.

8. APPENDIX

8.1. Matrix Multiplications

Matrix multiplication, $C_{nzm} = A_{nzl} * B_{lzm}$. There are several ways to implement matrix multiplication [18], e.g.,

```
(1) inner product method
    float @bd;
    for (j = 0; j < m; j ++) {
       @bd = B[*][j];     /* gather B[*][j] into a stride-1 vector*/
       for (i = 0; i < n; i ++)
```

```
                    C[i][j] = A[i][*] @. bd;
                }
    (2) middle product method
        for (i = 0; i < n; i++)
            for (k = 0; k < l; k++)
                C[i][*] += A[i][k] * B[k][*];    /* Linked-triadic operation*/
    (3) outer product method
        for (i = 0; i < l; i++)
            C[*][*] += spread(A[*][i],3,m) * spread(B[i][*],1,n);
```

where, the first spread function expands vector A[*][i] to the right for m times, and the second one expands vector B[i][*] upwards n times.

or,
```
        for (i = 0; i < l; i++) {
            iv1[*] = i @! 0 # m;
            iv2[*] = i @! 0 # n;
            C[*][*] += A[*][iv1[*]] * B[iv2[*]][*];
            }
    (4) n³ -parallelism method
        C[*][*] = @+[2] (spread(A[*][*],3,m) * spread(B[*][*],1,n));
    or,
        float   p[n][l][m], q[n][l][m];
        for ( k = 0; k < m; k++)
            p[*][*][k] = A[*][*];
        for ( i = 0; i < n; i++)
            q[i][*][*] = B[*][*];
        p[*][*][*] *= q[*][*][*];
        for ( i = 0; i < n; i++)
            for ( k = 0; k < m; k++)
                C[i][k] = @+ p[i][*][k];
```

8.2. Conjugate Gradient Scheme

```
/*
 *      Solve a large system of equations, $A \cdot \vec{x} = \vec{f}$,
 *      using Conjugate Gradient Scheme
 */
#define N 16
#define NSQ N*N
#define M  N+2
#define NM  N*M
#define K 100


main()
{
        float x[NSQ], r[NSQ], p[NSQ], f[NSQ], ap[NSQ], t[M][M], apt[M][M];
        register float   rr, a, b, oldrr;
        register float   @tad, @tbd, @tcd, @tdd, @ted, @aptd, @pd, @fd, @rd, @xd, @apd;
        bit    bv[NM];
        register bit @bvd;
        register int i;

        @tad = t[1][1#NM];  @tbd = t[0][1#NM];  @tcd = t[2][1#NM];
        @tdd = t[1][2#NM];  @ted = t[1][0#NM];  @aptd = apt[0][0#NM];
        @pd = p[*];  @fd = f[*];  @rd = r[*];
        @xd = x[*];  @apd = ap[*];  @bvd = bv[*];


        bvd = N @1 M # NM;              /* build bit vector N one's followed by 2 zeros */

        fd = 1.0 @! 1.0 #NSQ;          /* setup f vector -- 1.0,2.0,3.0... */
        rd = fd = pd = 2.47 * fd * fd;  /* 2.47 is broadcast */
```

83

```
oldrr = rd @. rd;                    /* dot-product */
t[*][*] = xd = 0.0;

for (i = 0; i < K; i++) {
     (&t[1][1])[bvd] = pd;           /* expand p to t */

     aptd = bvd ? 4.0 * tad - tbd - tcd - tdd -ted;
     apd = apt[0][bvd];     /* compress apt into ap */

     a = oldrr / (apd @. pd);
     xd += a * pd;
     rd -= a * apd;
     rr = rd @. rd;
     b = rr / oldrr;
     pd = rd + b * pd;
     oldrr = rr;
     }
```

CYBER 205 APPLICATIONS AT CHRYSLER

G.H. Bibbins and W.W. McVinnie

Technical Computer and Instrumentation Center
Chrysler Corporation
Detroit, Michigan

INTRODUCTION

The intent of this paper is to describe Chrysler's integration and planned usage of the Control Data Corporation's CYBER 205 Vector Processor Computer installed at Chrysler's Technical Computer and Instrumentation Center. The first section covers the integration of this machine into Chrysler's technical computer system that serves a user community of 2,000 engineers and designers. The remaining sections cover specific applications that can be effectively performed on this class of machine. Some of these applications are currently being done on other computers; e.g., structural analysis and imaging. Other applications, e.g., aerodynamics and crash simulations, require the computer resources that the CYBER 205 provides to be done effectively, and, as such, are only in their infancy at Chrysler Corporation.

CHRYSLER'S TECHNICAL COMPUTER CENTER

Chrysler's Technical Computer Center consists mainly of Control Data Corporation mainframe computers broken up into clusters of two to four computers with each cluster accessed via a separate front end computer and serving a specific user area such as Power Train or Body Engineering. Data created on any computer within a cluster is directly accessible to any other computer in that cluster. Under this arrangement, if one computer in a cluster goes down, the user simply accesses one of the remaining computers with minimal loss of productivity. All of the computer clusters are in turn interconnected by means of a high speed data transfer link, thus enabling data from any cluster to be readily accessible to any other cluster. As a result, a user on the Power Train Engineering cluster can easily access data created by Body Engineering and vice versa.

The CYBER 205 represents one of the few exceptions to the cluster arrangement. It is effectively front-ended by any of the other CYBERs. The user creates a batch job which is routed to the CYBER 205 for processing. When completed, the results are returned to the originating CYBER. Depending on the application this output might consist of a picture, analysis results to be post-processed, or an output file to be printed.

With the exception of the CYBER 205, which runs only batch jobs, all of the computers support mainly interactive users with most operating in a graphics oriented environment. Over 700 terminals, primarily TEKTRONIX 40XX and 41XX, are connected into the system and communicate at a 9,600 and 19,200 baud rate. Most of Chrysler's **CAD/CAM** work is performed using the Chrysler developed programs **CONCEPT**, **QUICKCON**, and **CIPPS**. Pre- and post-processing for finite element analysis programs such as **NASTRAN** and **DYCAST** is done using another Chrysler developed program **GCSNAST**. A common data base allows all of these programs to interchange data. Thus an engineer can develop a finite element model of a component using **GCSNAST** from data that was created by a designer using **QUICKCON**.

SYNTHETIC IMAGE GENERATION

One application for Chrysler's CYBER 205 is synthetic image generation. This is the process of creating shaded color images that realistically reflect light. The results are realistic images which aid in the assessment of visual aspects of a design and in the communication of the intended design well before a costly prototype is fabricated. Two programs are used on the CYBER 205 to generate synthetic images: **MOVIE·BYU** from Brigham Young University and **SYNTHAVISION** from the MAGI Corporation.

SYNTHAVISION is a solid modeling program that "builds" its images from primitives such as cubes, cylinders, surfaces of revolution, et cetera, which are added or subtracted to make up the image. These primitives are defined interactivly using Chrysler's **CAD** system. After processing by **SYNTHAVISION**, edge boundary and cross section data can be returned back to the **CAD** data base for use in creating engineering drawings, **N/C** tapes, and structural and aerodynamic models. It can also be used to calculate mass properties of the object. This capability has been used to get weights of various components; balance camshafts, alternator fans, and distributor rotors; and to determine volumes for combustion chambers and fuel tanks.

MOVIE·BYU is a widely used program that creates its images from polygonal surface data. To date its main application has been to provide shaded images of finite element models used for structural analysis as well as color contour displays of stress and deformation distributions. It has also been used to make very realistic pictures of future car designs for Chrysler's Styling Department. As a matter of fact, this may turn out to be the biggest single use for **MOVIE·BYU** at Chrysler in the future.

Together **MOVIE·BYU** and **SYNTHAVISION** have many design and engineering applications. Both of these programs were originally run on Chrysler's CYBER 760's. The calculation speed and memory capacity of the CYBER 205 has greatly expanded their capability and increased user productivity by decreasing turn-around times to just a few minutes, even for complex pictures, many of which could not even be done on the CYBER 760.

STRUCTURAL ANALYSIS

Chrysler has been using computers to do structural analysis on automotive structures for over 20 years and this application is the only significant "number crunching" application that will be a direct transfer to the CYBER 205. The types of analysis that are currently being performed can be categorized as static, normal modes, frequency response, and transient. The major analysis programs now being used make use of the finite element method, which models complex geometries as an assemblage of elements having simple geometries connected to grid points. The type of element used depends on the structure being modeled. Typically, frame type structures are modeled with beam elements, stampings are modeled with shell elements, and castings are modeled with solid elements.

The most widely used of the structural analysis programs in the automotive industry is **NASTRAN**, which has been in use at Chrysler since 1972. Chrysler's current production version of **NASTRAN** is the MacNeal-Schwendler Corporation version and is available only on the CYBER 760's. Since **MSC/NASTRAN** is not supported on the CYBER 205, Control Data Corporation has contracted with Universal Analytics, Incorporated, to upgrade **UAI/NASTRAN** to be user compatible with **MSC/NASTRAN** and to fully optimize it for the CYBER 205. This work is not yet completed, but initial evaluations of **UAI/NASTRAN** performance on the CYBER 205 are very encouraging. Table I indicates the order of performance improvements in **CPU** time that are being observed. Large models show about an 8:1 reduction in **CPU** time with even larger reductions in real or clock time.

TABLE I: COMPARISON OF **NASTRAN** PERFORMANCE

CPU TIME, SECS

CLASS OF PROBLEM	CYBER 760	CYBER 205
Small	709	370
Medium	3,309	939
Large	20,121	2,440

The necessity to perform structural analysis on the CYBER 205 stems from the facts that more analysis is being done and models are getting much more detailed and consequently larger. In 1972, the largest models were on the order of 200 grid points and 200 elements. It was difficult to imagine a model bigger than 500 grid points. Today, the larger models are approaching 10,000 grid points and 5,000 elements (and it is difficult to imagine a 50,000 grid point model). Computer time goes up linearly with the number of grid points and quadratic with the band-width of the stiffness equations on a scalar machine and, to be efficient, equation solvers require memory proportional to the product of the number of grid points and the bandwidth. The CYBER 205 allows for faster solutions due to the vectorization and significantly more memory.

More detailed models have become possible due to the development of interactive, graphics oriented, pre- and post-processing programs. Chrysler uses an in-house developed program called **GCSNAST**. This program links directly to Chrysler's **CAD** data base and has significant surface and solid element generation capabilities. **GCSNAST** runs on any of the front end CYBERs. The data is then routed to the CYBER 205 for analysis. After the analysis, the output is returned to the front end CYBER for postprocessing and viewing of the results. A direct interface to the **MOVIE·BYU** program that runs on the CYBER 205 provides the user with a wide range of picture generation capability including shaded images and color contour plots of stresses or deformations.

A program that uses the boundary element method of structural analysis will soon be implemented for the static analysis of components such as crankshafts and bearing caps that need to be modeled using solid finite elements. This method results in a large number of simultaneous equations to solve, but unlike the finite element method, the equations are neither banded nor symmetric, making it an ideal application for the CYBER 205. The biggest advantage of the method is that only the surface of the structure needs to be modeled, resulting in modeling times estimated to be one-fifth of the time needed for an equivalent finite element model. It will also be possible to include more detail such as holes drilled through a part or small radius fillets which, from a practical point of view, can't be included in a finite element model.

CRASH SIMULATION

Historically, Chrysler has depended on component and full vehicle testing to satisfy crash performance criteria. Computer analysis was limited to a multiple degree of freedom spring-mass model in which the spring stiffness characteristics were determined by static testing of the components that they represent. This model, when used by an experienced person, yields considerable insight into the crash response, but suggested modifications must be verified by additional testing.

In an attempt to reduce the amount of testing needed to develop a crash-worthy vehicle, Chrysler has contracted with Grumman Aerospace Corporation to use the DYCAST crash simulation program. This program makes use of the finite element method to model a vehicle, as well as allowing experimental data to be incorporated into the model when appropriate. Since crash response is highly nonlinear, several thousand iterations are required for a single crash. Each iteration is essentially equivalent to one linear analysis; consequently, one solution is equivalent to several thousand linear solutions. Speed of computation is essential in making this type of analysis feasible and the CYBER 205 with its large amount of real memory and vector processing capability provides this speed. Jobs that required over a week to complete on a CYBER 760 using re-starts can now be completed in a single overnight run on the CYBER 205. The additional memory now available on Chrysler's CYBER 205 allows models in excess of 2,000 degrees of freedom to be solved, whereas the available memory of a CYBER 760 limited models to about 500 degrees of freedom. This increase in model size has helped to eliminate some of the "art" needed to develop reliable models.

AERODYNAMIC ANALYSIS

Areodynamics, as applied to the study of cars and car-like shapes, is as a numerical science, in its infancy. There is no program that can determine the "best" shape for a car. The so-called "state-of-the-art" is, however, at a point where numerical calculations can yield significant information to the engineer. While this does not remove the need for wind tunnel tests, or the engineer's expertise, it does allow for a much greater range of variables to be studied within the design cycle of a vehicle.

The flow modeling problem presented by an automobile is different in both a qualitative and quantitative sense from the problems found in the aircraft industry. After a preliminary investigation of the non-proprietary aerodynamics programs available, the McAERO design program from NASA was selected as being the most suitable for car shapes and installed on both the CYBER 760 and CYBER 205 computers. The original intent was to use this program as an evaluation tool to see how well the theory could handle car-like shapes and at the same time to see how well the CYBER 205 could handle the theory.

A car-like shape that had received extensive wind tunnel testing was modeled and analyzed. The results of this analysis exceeded expectations with calculated pressure values comparing very closely to the test values. Based on these results, consideration is being given to expanding this program to be able to handle at least some types of production problems.

The original **McAERO** program was written for a scalar type computer and has a limitation of 1,000 panels or 1,400 points in a model. To achieve the necessary accuracy, an actual car model will require between 3,000 and 10,000 panels. The time required to solve a problem can be broken down into two main divisions: problem set-up and problem solution. As the models get larger, the problem solution takes the major portion of the total problem time; consequently, the primary (and simplest) candidate for vectorization was the solution phase. This vectorization of the solution phase has been completed and has resulted in significant time reductions as expected. A 354 panel model that took 158 CPU seconds on the CYBER 760 took only four CPU seconds on the CYBER 205. Since the problem size limitations have not yet been increased, large problems have not yet been run. However, an extrapolation of results indicates that a 5,000 panel problem would require in excess of 40 hours of CPU time on a CYBER 760 compared to less than three hours on a CYBER 205. In actual fact, the CYBER 760 would probably require significantly more time than the estimate because of limited memory. In any case, it is obvious that scalar computers are not feasible for handling automotive problems. Even Chrysler's CYBER 205 will need to be expanded to at least four million words of memory to be able to effectively handle the size problems that are anticipated.

The work to date has proven that numerical aerodynamics can be a useful engineering tool at Chrysler and that the CYBER 205 is an essential piece of hardware for this type of analysis. To make **McAERO** into a truly useful tool will require the addition of new capabilities such as vorticity and separation modeling, vectorization of the problem set-up phase, and significant postprocessing to be able to visualize the results easily. Before undertaking these improvements, other commercially available programs are being investigated as a more cost effective way of meeting the needs of aereodynamic analysis.

SUMMARY

To briefly summarize Chrysler's CYBER 205 experience: It is a back-end processor accessible to all users through a large network of other CYBERs. Synthetic image generation applications running on the CYBER 205 are **MOVIE·BYU** and **SYNTHAVISION**. In the near future, Chrysler will use **UAI/NASTRAN** (by Universal Analytics, Incorporated) for structural analysis with pre- and post-processing with **GCSNAST** on the front end CYBERs. Crash simulation is accomplished on the CYBER 205 with **DYCAST** (from Grumman Aerospace Corporation). Aerodynamic analysis with **McAERO** from **NASA** is in early stages of usage. The calculation speed and memory capacity of the CYBER 205 has greatly expanded the capabilities of the above applications and has increased user productivity by reducing turn-around times.

A VECTORIZED MATRIX-VECTOR MULTIPLY AND

OVERLAPPING BLOCK ITERATIVE METHOD

Linda J. Hayes

Texas Institute for Computational Mechanics
Department of Aerospace Engineering and Engineering Mechanics
The University of Texas at Austin
Austin, Texas 78712

ABSTRACT

     An overlapping block iterative method is presented for solving large
systems of equations which result from finite element discretizations.
This algorithm is very much in the spirit of the element-by-element
techniques described by Hughes et al. and by Carey and Jiang; however, it
differs from these in that the algorithm here is a block iterative method
in the classical sense where the blocks or groups are the nodes in the
individual elements. Since any given node may appear in several elements,
this technique has been called an "overlapping" block iterative method.

     This algorithm is particularly attractive for vector processing. It
uses a special vectorized form of a matrix-vector multiply in which all
computations are done on an element level. The individual element stiffness
matrices were stacked together to create long vectors. It has been implemented on the CYBER 205 and has been compared in terms of execution time
and storage requirements against standard equation solvers such as (1) band
solver, (2) sparse matrix solver, (3) conjugate gradient, (4) SOR and (5)
symmetric SOR with conjugate gradient acceleration.

     Comparisons were made on a variety of regular grids which ranged in
size from several hundred unknowns up to 11,000 unknowns, and on several
extremely irregular grids which occur in biomedical applications. The
element-by-element overlapping block algorithm and vectorized matrix-vector
multiply were found to be very efficient, and it was particularly advantageous for problems with irregular finite element grids.

INTRODUCTION

     Numerical modeling of virtually all physical applications eventually
involves the solution of systems of equations. Finite element modeling
techniques have become very popular because they can easily treat complicated geometries, a variety of boundary conditions and multiple materials.
However, if the finite element model is very large, the associated system
of equations that must be solved is also very large. In practice, the
storage requirement and computational requirements for solving the resulting
linear system of equations often limits the model size. The purpose of this

paper is to present a new overlapping block iterative method for solving linear systems of equations that arise from finite element technique. Conceptually, the linearized, global system of equations is assembled and is then partitioned into groups or blocks which correspond to the individual elements in the finite element grid. These element size blocks are then inverted just as is done in standard block methods. In practice, all computations are done on an element level. Several researchers have proposed element-by-element iterative algorithms. They all resemble each other in that certain computations are done using individual element data structures. However, apart from that feature, the algorithms are quite different. Carey and Jiang[1] have implemented a point Jacobi method with conjugate gradient acceleration, where each of the matrix vector multiplies is done at the element level, thus economizing on storage. Hughes et al. have proposed a family of iterative methods for solving a general linear system of equations, which are based on replacing the algebraic problem $Ax = b$ by a first-order, ordinary differential equation whose steady state solution is the same as that of the original system. This has been termed parabolic regularization. They propose several different incomplete factorizations or preconditionings for the regularized problem. One of the approximate factorizations which they have proposed is based on the element matrices[2-4]. The method proposed here is a block iterative method[5-7]. The block iterative method is applied directly to the original system of equations $Ax = b$ and this iterative technique can be accelerated using conjugate gradient methods.

An essential feature of the implementation of this algorithm on the CYBER 205 was a special data structure and a fast vectorized method for performing a sparse matrix-vector multiply. This technique can be applied when a matrix arises from the finite element method and the global matrix, A, can be written as a sum of individual matrices.

$$A = \sum_{e=1}^{E} A_e. \tag{1}$$

The data structure is based on the individual matrices, A_e, and can be applied to both symmetric and to non-symmetric matrices. The number of vector operations depends on the type of elements used in the finite element modeling and is independent of the number of elements in the grid. The matrix-vector multiply and the overlapping block algorithms have been implemented and timed on the CYBER 205 and have been compared in terms of computation time and storage to other standard techniques.

ELEMENT-BY-ELEMENT MATRIX-VECTOR MULTIPLY

First, a vectorized version of a matrix-vector multiply will be described. This technique will be used in the right-hand side calculations for the iterative method which is described later. If one wants to perform the matrix-vector multiply Ax, and if the matrix A can be written in the form (1), then the multiplication can be done in the following manner

$$Ax = \sum_{e=1}^{E} A_e x_e = \sum_{e=1}^{E} F_e = F. \tag{2}$$

The subscript e denotes the contributions from a single finite element. The data structure and multiplication technique will be illustrated for the case of four noded quadrilateral finite elements. In this case, the element matrix is four by four. If one orders the element matrix in the following order

$$
\begin{bmatrix} a_1 & a_5 & a_9 & a_{13} \\ a_{14} & a_2 & a_6 & a_{10} \\ a_{11} & a_{15} & a_3 & a_7 \\ a_8 & a_{12} & a_{16} & a_4 \end{bmatrix} \begin{bmatrix} x_1 \\ x_2 \\ x_3 \\ x_4 \end{bmatrix}_e = \begin{bmatrix} F_1 \\ F_2 \\ F_3 \\ F_4 \end{bmatrix}_e
\tag{3}
$$

then one will observe that the four right-hand side contributions can be
calculated using four vector multiply instructions as

$$
\begin{bmatrix} F_1 \\ F_2 \\ F_3 \\ F_4 \end{bmatrix}_e = \begin{bmatrix} a_1 \\ a_2 \\ a_3 \\ a_4 \end{bmatrix} \begin{bmatrix} x_1 \\ x_2 \\ x_3 \\ x_4 \end{bmatrix}_e + \begin{bmatrix} a_5 \\ a_6 \\ a_7 \\ a_8 \end{bmatrix} \begin{bmatrix} x_2 \\ x_3 \\ x_4 \\ x_1 \end{bmatrix}_e + \begin{bmatrix} a_9 \\ a_{10} \\ a_{11} \\ a_{12} \end{bmatrix} \begin{bmatrix} x_3 \\ x_4 \\ x_1 \\ x_2 \end{bmatrix}_e + \begin{bmatrix} a_{13} \\ a_{14} \\ a_{15} \\ a_{16} \end{bmatrix} \begin{bmatrix} x_4 \\ x_1 \\ x_2 \\ x_3 \end{bmatrix}_e
\tag{4}
$$

Vectors of length four are not attractive, so in order to increase the
length of the vectors, the calculations of each element in the grid are
stacked one behind the other. This is shown in Figure 1. The data
structure that was used here was to place the a_1 entry from each of the
E elements in consecutive storage location. This was followed by the a_2
entries from each of the E elements, etc. The complete matrix-vector
multiply was achieved by interpreting Equation (4) where each item is an
array of length E. After the right-hand side F_e is calculated on each
element, the contributions must be assembled back into the global form.
Computation time for this entire procedure will be given in a later section.

   The number of vector operations in (4) is independent of the number of
elements in the finite element grid. It is also independent of the nodal
point ordering in the grid. Increasing the number of elements simply
increases the length of the vector operations which adds to their efficiency.
Equation (4) can be calculated in several ways. It can be calculated in
four multiply instructions and two add instructions, or the multiplies can
be stacked so that only one multiply instruction is issued and two add
instructions are required. For reasonable sized finite element grids with
fifty or more elements, the vectors are long enough that the start up time
for the extra three vector multiples can be ignored. If (4) is computed
with four multiplies, execution time is saved because the full x vector
does not have to be created.

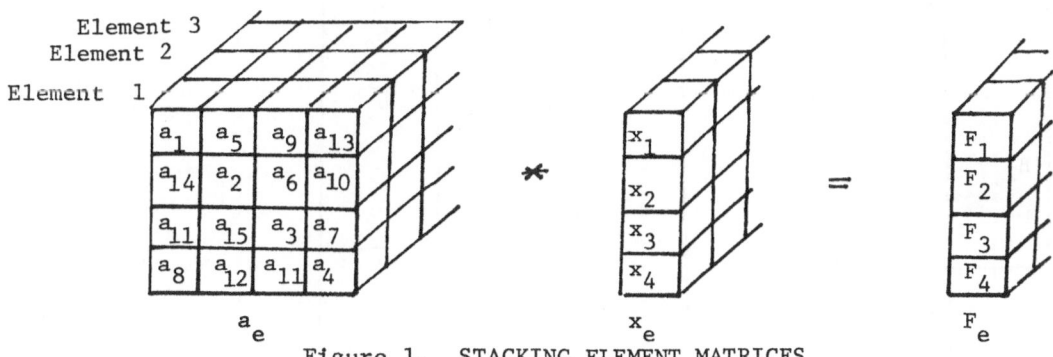

Figure 1.   STACKING ELEMENT MATRICES

If there had been eight nodes in each element in the grid, the element matrix would have been eight by eight and would have been constructed as it was in (3). If several element types are used, all element matrices are simply filled with zeros. Table 1 contains a list of various two dimensional element types and the numbers of vector multiplies and adds that are required to perform operation (3). In addition, Table 1 gives the length of the vector instructions that are involved in the multiplies and in the divide and conquer scheme for the additions.

Table 1. Vector Instructions Required For Various Element Types (E = # Elements)

Element Type (#Nodes)	#Vector Multiplies	#Vector Adds	Length of Multiplies	Length of 1st Add	Length of 2nd Add	Length of 3rd Add	Length of 4th Add
3	3	2	3E	3E	3E	–	–
4	4	2	4E	8E	4E	–	–
6	6	3	6E	12E	12E	6E	–
8	8	3	8E	32E	16E	8E	–
9	9	4	9E	36E	18E	9E	9E

OVERLAPPING BLOCK ITERATION

In this section the notation will be changed somewhat. We will consider the linear system of equations.

$$Ax = b \tag{5}$$

where A is an NxN matrix, b is a vector of length N and x is the vector of unknowns of length N. Block methods were proposed many years ago for solving large linear systems of equations (5-7). In a block method, the unknowns in the problem are decomposed into groups or blocks and the matrix (1) is partitioned according to the groups of unknowns. This leads to the structure

$$
\begin{bmatrix}
B_{1,1} & B_{1,2} & & \\
& & & \\
B_{2,1} & B_{2,2} & & \\
& & & B_{E,E}
\end{bmatrix}
\begin{bmatrix}
x_{1_1} \\
x_{1_2} \\
\vdots \\
x_{1_p} \\
\hline
x_{2_1} \\
\vdots \\
x_{2_p} \\
\vdots \\
\hline
x_E
\end{bmatrix}
=
\begin{bmatrix}
b_{1_1} \\
b_{2_2} \\
\vdots \\
b_{2_1} \\
\hline
b_{2_1} \\
\vdots \\
b_{2_p} \\
\vdots \\
\hline
b_E
\end{bmatrix}
\tag{6}
$$

The diagonal blocks $B_{j,j}$ are $N_j \times N_j$ matrices. x_{e_i} is the i-th unknown in the e-th group. Given an initial estimate for the solution vector $x^{(0)}$, one can define a block iterative method as

$$x^{(n+1)} = C^{-1} B \, x^{(n)} + C^{-1} b \qquad (7)$$

where superscripts are used for iteration numbers. The matrix B is a block matrix of the form

$$(8)$$

$$B = \begin{array}{|c|c|c|c|}
\hline
0 & B_{1,2} & B_{1,3} & \cdots \\
\hline
B_{2,1} & 0 & \ddots & \cdots \\
\hline
\ddots & 0 & B_{E-1,E} \\
\hline
& B_{E,E-1} & 0 \\
\hline
\end{array}$$

and C is the block diagonal matrix containing the block diagonal entries of A

$$C = \begin{array}{|c|c|c|}
\hline
B_{1,1} & 0 & 0 \\
\hline
0 & B_{2,2} & 0 \\
\hline
0 & & B_{E,E} \\
\hline
\end{array} \qquad (9)$$

In order to carry out the iterative procedure (7), it is necessary to be able to solve equations corresponding to the diagonal of A. These equations have the form

$$A_{j,j} \, x_j = b_j \qquad (10)$$

where x_j and b_j are the unknowns and right-hand side from the j-th block. If each block contains only one unknown, then this would reduce to a standard point method. However, the advantage of using block methods is that they will converge faster than point methods. The increase in the rate of convergence is directly proportional to the size of the matrix $A_{j,j}$ that must be inverted[5]. If there were only one block which contained all of the unknowns, C would coincide with the matrix A and the solution would be obtained directly in one step. Ostrowski[6] has shown that the convergence proofs are valid even if a node is in several groups, and he termed this an overlapping block iterative method.

In our application the unknowns in the problem will be grouped according to nodes on each element in a finite element grid. Since a given node will be in several elements, it will appear in several of the groups or blocks of unknowns. Therefore, this technique will be an overlapping block iterative method.[7] Conceptionally, one can envision the overlapping block iterative as corresponding to a larger matrix problem in which the equation for specific nodes appears several times. It will appear in each group that the node belongs to. This expanded matrix can then be partitioned as was done in(6) and an overlapping iterative block method is defined by (7). Note that if an unknown appears in more than one block at the end of each iteration, each block will have a new value for that unknown and $x_i^{(n+1)}$ is obtained by averaging the values obtained on the n

plus first iteration from each of the blocks that contain node i.

The basic overlapping block iterative method (7) can be accelerated using the conjugate gradient technique[8]. The conjugate gradient acceleration of this overlapping block method can be written

$$x^{n+1} = x^n + \lambda_n p_n \tag{11}$$

$$r^n = b - Ax^n \qquad \text{(residual)} \tag{12}$$

$$\delta^n = x^{n+1} - x^n = [D]^{-1} \sum_{e=1}^{E} C_e^{-1} (F_e - (Ax^n)_e) \quad \begin{array}{l} \text{pseudo-} \\ \text{residual} \end{array} \tag{13}$$

$$([D]_{ii} = NE_i \delta_{ij}) \qquad\qquad NE_i = \text{\# elements which contain node i}$$

$$p^n = \begin{cases} \delta^0 & n = 0 \\ \delta^n + \alpha_n p^{n-1} & n = 1, 2, \dots \end{cases} \tag{14}$$

$$\alpha_n = \frac{(\delta^n, r^n)}{(\delta^{n-1}, r^{n-1})} \qquad n = 1, 2, \dots \tag{15}$$

$$\lambda_n = \frac{(\delta^n, r^n)}{(p^n, Ap^n)} \qquad n = 0, 1, 2, \dots \tag{16}$$

The matrix-vector multiplies which appear in equations (13) and (16) were done using the elementwise data structure presented previously.

NUMERICAL RESULTS

Timing comparisons were made on the CYBER 205 at Colorado State University. Comparisons were made for both regular grids and irregular grids. The regular grids had a uniform mesh of tensor product quadratic elements on a rectangular region. The irregular grids resulted from three biomedical applications.

The element-by-element matrix-vector multiply presented here was timed and compared to the matrix-vector multiply which is found in the vectorized version of the ITPACK routines. The ITPACK routines use the Purdue Sparse Storage Pattern. For the regular grids a speed up of between 55 to 63% was observed when using the element-by-element multiply. On the irregular grids a speed up of approximately 65% was observed with the element-by-element technique for all of the irregular grids. Timing comparisons for the overlapping block iterative method are presented in Tables 2 and 3.

The timing comparisons for the regular grid are shown in Table 2 and the timing comparisons for the irregular grids are shown in Table 3. The band solver was a vectorized algorithm available in the MAJEV Library of the CYBER 205. 'J-CG ITPACK' is the vectorized version of the point Jacobi conjugate gradient which used the vectorized element-by-element matrix multiplication technique which we developed for right-hand side calculations. 'Overlap Block' is the overlapping block iterative method described here. Several observations can be made. The band solver which was used in this study was a very efficiently vectorized version of the algorithm. It was particularly fast in solving tri-diagonal systems. However, once the problem size exceeds a few hundred unknowns, iterative methods solve the linear system of equations faster than the band solver. The vectorized element-by-element matrix-vector multiply improved the performance of the point Jacobi conjugate gradient method. For this reason, the overlapping block method should be compared to the Jacobi conjugate gradient method, both of which use the vectorized element-by-element right-hand side calculations. In the case of the irregular grids, even for a few number of unknowns (N = 148) the overlapping block iterative method has definite advantages over standard equation solving techniques. For the very regular type grids, the overlapping block method is competitive with standard iterative methods but shows no special advantages. However, other special algorithms offer savings for regular grids over any of the techniques that were tried here. These techniques include generalized red/black iterative methods, and alternating direction techniques that have been extended to curved grids. The storage requirements for several standard equations solving techniques are given in Table 4. The ITPACK vectorized version of the point Jacobi conjugate gradient uses the Purdue Sparce Matrix Storage Pattern, and this is less than the storage required by the vectorized band solver which was used. Of these algorithms, the overlapping block method requires far less storage. In fact, it is comparable to the storage required by the non-vectorizable Yale Sparse Matrix routines, which are considered to have very low storage requirements for sparse matrices.

CONCLUSIONS

A matrix-vector multiply and an overlapping block iterative method for problems using finite element methods has been presented. In this method, the iteration as well as the matrix-vector multiply for the right-hand side is done in an element-by-element procedure. This iterative technique has storage savings over vectorized band solvers and standard iterative techniques even for small problems. On irregular grids, the method is faster than standard iterative methods. On regular rectangular grids. it is comparable. Execution times and storage requirements for these two procedures would not increase if the matrix were nonsymmetric.

Table 2

CYBER 205 Timing Comparisons for Regular Grids

Method	Time/iteration (Sec)	# Iterations	Total Time (Sec)
CASE A 200 elem 441 unknowns			
Bandsolver	——	——	0.053
J-CG Itpack	0.00136	33	0.045
J-CG elem	0.00044	33	0.015
Overlap. Block	0.0007	36	0.025
CASE B 800 elem 1681 unknowns			
Bandsolver			0.433
J-CG Itpack	0.0031	68	0.203
J-CG elem	0.0015	68	0.102
Overlap. Block	0.0024	60	0.144
CASE C 5408 elem 11,025 unknowns			
Bandsolver			--
J-CG Itpack	0.015	176	2.60
J-CG elem	0.0096	176	1.70
Overlap. Block	0.015	114	1.71

Table 3

CYBER 205   Timing Comparisons for Irregular Grids

Method	Time/iteration (Sec)	# Iterations	Total Time (Sec)
Case I - Arm Cross-Section		148 Unknowns	
Bandsolver			0.021
J-CG Itpack	0.00075	77	0.058
J-CG elem*	0.000225	77	0.017
Overlap. Block	0.000374	36	0.013
Case II - Heart		223 Unknowns	
Bandsolver			0.046
J-CG Itpack	0.00091	128	0.117
J-CG elem*	0.00029	128	0.037
Overlap. Block	0.00048	52	0.025
Case III - Monkey Head		351 Unknowns	
Bandsolver			0.089
J-CG Itpack	0.0011	249	0.280
J-CG elem*	0.00041	249	0.100
Overlap. Block	0.00067	95	0.064

Table 4

Storage Requirements for Matrix Solution Methods

Method	Words of Storage*
Overlapping Block	8,100
Non-Vectorizable Yale Sparse Matrix	7,444
Vectorized Purdue Sparse Matrix (Itpack)	11,466
Vectorized Bandsolver (MAJEV)	19,845

*Results for 10 by 10 QUAD9 elements, 441 unknowns.

REFERENCES

1. B.N. Jiang and G.F. Carey, Subcritical flow computation using element-by-element conjugate gradient method, Proceedings: Fifth International Symposium on Finite Elements and Flow Problems, Austin, Texas (January 1984).
2. T.J.R. Hughes, M. Levit, and J. Winget, Element-by-element implicit algorithms for heat conduction, Journal of Engineering Mechanics 109(2) (April 1983).
3. T.J.R. Hughes, M. Levit, and J. Winget, Element-by-element implicit solution algorithm for problems of structural and solid mechanics, Computer Methods in Applied Mechanics and Engineering, 36 (1983).
4. T.J.R. Hughes, M. Levit, and E. Tezduyar, New alternating direction procedures in finite element analysis based upon EBE approximate factorizations, presented at the Symposium on Recent Developments in Computer Methods for Nonlinear Solid and Structural Mechanics, ASME Joint Meeting of Fluids Engineering, Applied Mechanics and Bioengineering Divisions, University of Houston, Houston, Texas (June 20-22, 1983).
5. D.M. Young, "Iterative Solutions of Large Linear Systems," Academic Press, New York (1971).
6. A.M. Ostrowski, On the linear iteration procedures for symmetric matrices, National Bureau of Standards Report No. 1844 68:23 (August 1952).
7. A.M. Ostrowski, Iterative Solution of linear systems of functional equations, Jl. of Math. Anal. and Appl. 2:351-369 (1961).
8. D.M. Young, L.J. Hayes, and K.C. Jea, Generalized conjugate gradient acceleration of iterative methods: Part I: the symmetrizeable case, Center for Numerical Analysis Report #162, University of Texas at Austin (September, 1981).

A VECTOR ELASTIC MODEL FOR THE

CYBER 205

Nancy Schmidt Adams and
Olin G. Johnson

Research Computation Laboratory (RCL)
University of Houston

INTRODUCTION

The research project reported in this paper is the result of a joint grant from Texaco and Control Data Corporation. The focus of this project was the development of a vectorized version of the two dimensional elastic forward modeling algorithm developed by Kosloff and Reshef (1982). The program was designed for the CDC CYBER 205, and implemented on Texaco's installation. Timings obtained for Kosloff's implementaion on the RCL VAX 11/780 with an attached FPS-100 array processor indicate that the vectorized version is approximately 900 times faster on a CYBER 205.

Highlights of the project include the amount of experimentation and optimization, and the thorough testing of the code, made possible by the immense computational speed of the CYBER 205. This allowed extensive testing for a variety of source wavelets and boundary conditions, and resulted in the construction of highly efficient absorbing boundaries.

The following sections discuss various parts of the design stage and present the results. Several smapshots of the stresses, pressure, and curl are presented and compared to published results obtained by Kosloff and Reshef (1982).

ADVANTAGES OF CYBER 205 FOR LARGE SCALE MODELING

The advent of vector computers has made it feasible to solve large problems that could not be solved in a reasonable time on scalar machines. The potential for reaching the maximum execution speed on the CYBER 205 depends largely on the degree to which it is possible to 'vectorize' the scalar code and on the size of the problem. In general, larger problems can be handled more efficiently than smaller problems because longer vector operations utilize the CYBER 205 resources more efficiently. The computational power of the CYBER 205 makes it a very powerful development tool, on which both experimentation and optimization are desirable. A significant portion of the time spent on the implementation of this model was devoted to experimentation. The most obvious advantage of the CYBER 205 is the increased processing

Table 1. Values Tested for Absorbing Boundary Scheme During Experimentation.

NPOINTS	ABSORB-RATE	RESULTS AND COMMENTS
15	0.90	Strong side reflections and wrap-around.
15	0.92	Slight reflections and wrap-around noticed.
20	0.92	No reflections or wrap-around; wide strip.
15	0.94	Very slight reflections.
15	0.9445	No relfections or wrap-around.
10	0.96	Some wrap-around still evident.
10	0.94	Side relfections noticeable.

speed. There are other advantages implicit to vector programming:

- vector code is more natural to scientific problem solutions than is scalar code.

- vector code is concise, readable, and understandable.

- vector code tailors the application to the hardware.

In the optimizing stage an attempt was made to 'vectorize' as much of the scalar code as possible and to remove many redundant operations. A storage scheme that minimized large page faults and maximized vector lengths was designed for maximum efficiency. Also, many of the optimized routines in the CYBER 205 math and geophysical library (MAGEV) were utilized (Control Data Corporation, 1981 (b)).

PROGRAM DESIGN FEATURES

A variation of the transparent boundary scheme developed by Kosloff and Reshef (1983) is implemented in this program. The wave amplitudes of the stresses and the time derivatives of the stresses are gradually eliminated in a number of grid-points (NPOINTS) along the boundaries by multiplying with values that taper smoothly from 1.0 to ABSORB-RATE (Schmidt, 1984). Various values for NPOINTS and ABSORB-RATE were tested during experimentation. The results are given in TABLE 1. The values NPOINT=15 and ABSORB-RATE=0.9445 performed best experimentally.

Table 2 shows the data structutes used in the program. The maximum size model that can be run has grid dimensions of 256 x 256. A large page on the CYBER 205 consists of 65,536 64-bit words (Control Data Corporation, 1981 (a). A 256 x 256 array stored in 32-bit, half precision words, occupies half of a large part of storage. Since the arrays are of fixed dimensions, 11 large pages are needed to run the program, regardless of the model size.

Table 2. Summary of the Data Structures Used in the Program.

DATA STRUCTURE	DIMENSIONS	USEAGE	LARGE PAGES
SXX	(256,256)	TO STORE	1/2
SYY	(256,256)	THE STRESSES	1/2
SXY	(256,256)	AND	1/2
DSXX	(256,256)	THEIR	1/2
DSYY	(256,256)	TIME DERIVATIVES	1/2
DSXY	(256,256)		1/2
DENS	(256,256)	TO STORE THE	1/2
LAMBDA	(256,256)	MATERIAL	1/2
MIU	(256,256)	DESCRIPTIONS	1/2
LAM2MIU	(256,256)		1/2
FX	(256,256)	TO STORE THE FORCE	1/2
FY	(256,256)	HISTORIES	1/2
FXY	(256,256)		1/2
WAVE	(200)	NORMAL DISCRETE FUNCTION	-
TEMPXX	(256,256)	TO STORE THE	1/2
TEMPXX	(256,256)	INTERMEDIATE RESULTS	1/2
TEMPXX	(256,256)	IN THE COMPUTATION	1/2
TSAVE	(256,256)	THE STRESSES	1/2
TPART	(256,256)	TEMPORARY STORAGE	1
TSAVE	(256,256)	USED TO COMPUTE THE PARTIAL DERIVATIVES	1

Several comments are in order before describing the program:

A) $(\lambda + 2\mu)$ is computed once and stored in the array LAM2MIU, saving two vector multiplications and two vector additions very time-step.

B) 1/p is computed once and stored in the array DENS, thus a vector multiplication is performed every time-step instead of a less efficient vector division.

C) The terms $ACC_X = (1/\rho \left[\frac{\partial \delta_{xx}}{\partial x} + \frac{\partial \delta_{xy}}{\partial y} \right])$ and $ACC_Y = (1/\rho \left[\frac{\partial \delta_{xy}}{\partial x} + \frac{\partial S_{yy}}{\partial y} \right])$ are computed once per time step and saved in temporary storage to avoid recomputation.

D) The following parameters are input by the user:

Grid dimensions (NX & NY)
Distance between grid-points (DX & DY)
Time steps (NT)
Increment between time steps (DT)
Maximum frequency (FMAX)
Interval between snapshots (ISNAP)
Source Coordinate (ISX, ISY)
Type of source (ITYP)
The structure of the earth and the material composition.

The following program description is divided into routines. The data structure names are in Table 2. Although the operations are all vectorized, for simplicity of presentation the vectorization was ignored. All references to FFT's denote complex FFT's. In the following, the terms K_x and K_y denote the spatial wavenumbers; a $\sim$ above a symbol represents a 1-Dimensional (1D) spatial Fourier transformation; and i $= \sqrt{-1}$. The functions PARTIAL-X and PARTIAL-Y have as input a descrete function f, with NY rows and NX columns.

INPUT: Read in the (above) variables and material descriptions.

SETUP: Compute λ and μ and store in LAM and MIU respectively;
Compute ($\lambda + 2\mu$) and store in LAM2MIU;
Compute $1/\rho$ and store in DENS;
Compute the normal function WAVE using FMAX;
Compute $\partial F_x/\partial x$, $\partial F_y/\partial y$, and $\partial F_x/\partial y + \partial F_y/\partial x$ using ISX, ISY and ITYP - then store in FX, FY, and FXY respectively.

PARTIAL-X (f): (Approximates $\frac{\partial f}{\partial x}$)

        Take forward FFT of f;
Multiply $\tilde{f}$ by iK_x;
Take inverse FFT of $iK_x\tilde{f}$ to obtain result.

PARTIAL-Y (f): (Approximates $\frac{\partial f}{\partial y}$)

        Transpose the matrix f   (f^T);
Take forward FFT of f^T;
Multiply $\tilde{f}^T$ by iK_y;
Take inverse FFT of $iK_y\tilde{f}^T$ (∂fy^T);
Transpose ∂fy^T to obtain result.

STEP: Initialize SXX, SXY, SYY, DSXX, DSXY, and DSYY to zeroes.

        Repeat the following for ISTEP = 1 to NT:

            TEMPXX = SXX
            TEMPXY = SXY
            TEMPYY = SYY

    1)   (Compute ACC_x and store in TEMPXX and TEMPXY)

            TEMPXX = DENS * [PARTIAL-X(TEMPXX) + PARTIAL-Y(TEMPXY)]
            TEMPXY = TEMPXX

2)   (Compute ACC$_Y$ and store in TEMPYY and TSAVE)

        TEMPYY = DENS * [PARTIAL-X(TEMPXY) + PARTIAL-Y(TEMPYY)]
        TSAVE = TEMPYY

3)   (Compute $\ddot{\delta}_{xy}$ and store in TEMPXY)

        TEMPXY = PARTIAL-Y(TEMPXY) + PARTIAL-X(TSAVE)
        TEMPXY = TEMPXY + WAVE(ISTEP) * FXY
        TEMPXY = MIU * TEMPXY

4)   (Compute $\ddot{\delta}_{xx}$ and store in TEMPXX)

        TEMPXX = PARTIAL-X(TEMPXX)
        TEMPXX = TEMPXX + WAVE(ISTEP) * FX
        TEMPXX = LAM2MIU * TEMPXX + LAMBDA * TEMPYY

5)   (Compute $\ddot{\delta}_{yy}$ and store in TEMPYY)

        TEMPYY = PARTIAL-Y(TEMPYY)
        TEMPYY = TEMPYY + WAVE(ISTEP) * FY
        TEMPYY = LAMBDA * TEMPXX + LAM2MIU * TEMPYY

6)   (Compute $\dot{\delta}$xx, $\dot{\delta}$yy, $\dot{\delta}$xy, δxx, δyy, and δxy and store in DSXX, DSYY, DSXY, SXX, SYY, and SXY, respectively)

        DSXX = DSXX + DT * TEMPXX
        DSYY = DSYY + DT * TEMPYY
        DSXY = DSXY + DT * TEMPXY
        SXX = SXX + DT * DSXX
        SYY = SYY + DT * DSYY
        SXY = SXY + DT * DSXY

RESULTS AND COMPARISONS

The vectorized elastic forward model was tested on three different problems that have known analytical solutions. We now analyze the results for these problems and compare them to the published results obtained by Kosloff and Reshef (1982). In the following discussion, a figure referred to by a number without the qualifiers (a) or (b) refers to both figures. The terms P-wave and S-wave refer to compressional waves and shear waves, respectively.

Lamb's Problem

This problem deals with the propagation of waves in a two-dimensional homogeneous elastic halfspace. The model has dimensions (NX x NY) of 256 x 128 with 20 feet between grid-points. The material has P and S velocities of 3460 and 2000 feet/second, respectively. Figures 1 and 2 are snapshots where the force applied is a vertical force. Figures 3 and 4 are shapshots where the waves are initiated by a pressure source.

Figure 1(a) is a snapshot of pressure after 500 milliseconds. The corresponding figure 1(b) is a comparable plot obtained by Kosloff and Reshef (1982). These figures only contain P-waves and reflected P-waves (Pp and Sp) as expected. Figure 2(a) is a snapshot of the curl of the acceleration fields after 500 milliseconds. Figure 2(b) shows the corresponding result from Kosloff and Reshef (1982). These figures display only the S-waves and reflected S-waves. The difference in the velocities of P-waves and S-waves can be seen by comparing Figures 1 and 2.

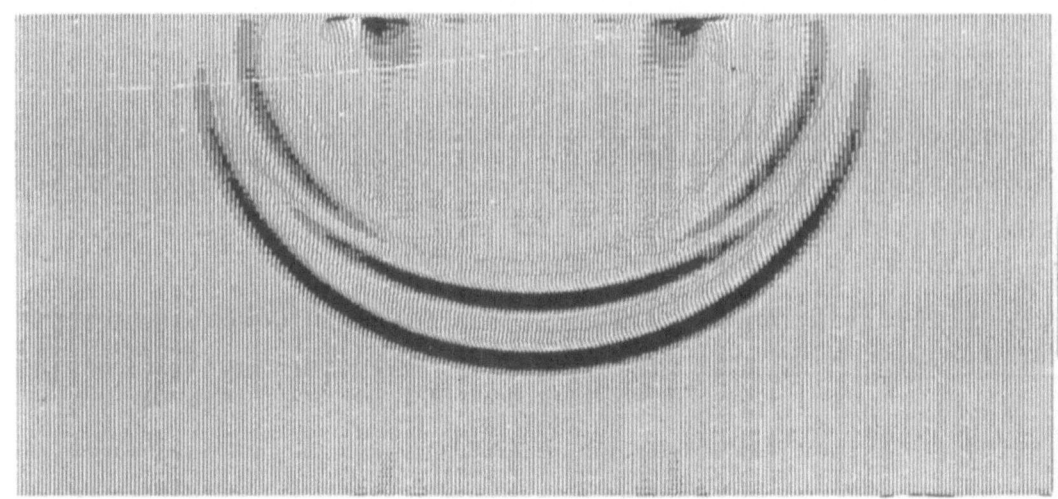

Fig. 1 (a)   Pressure at T=500 milliseconds.

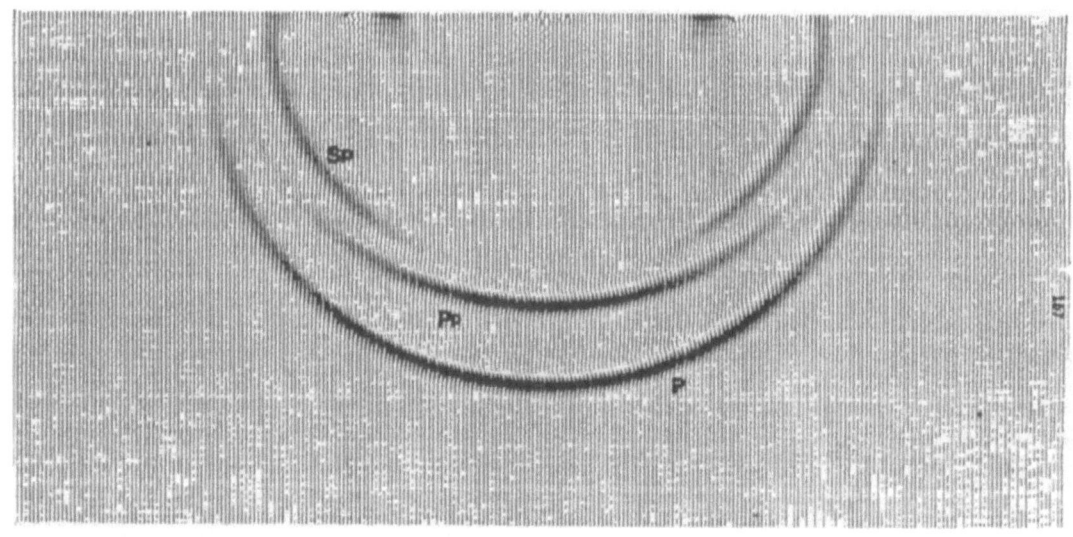

Fig. 1 (b)   Pressure (Kosloff and Reshef).

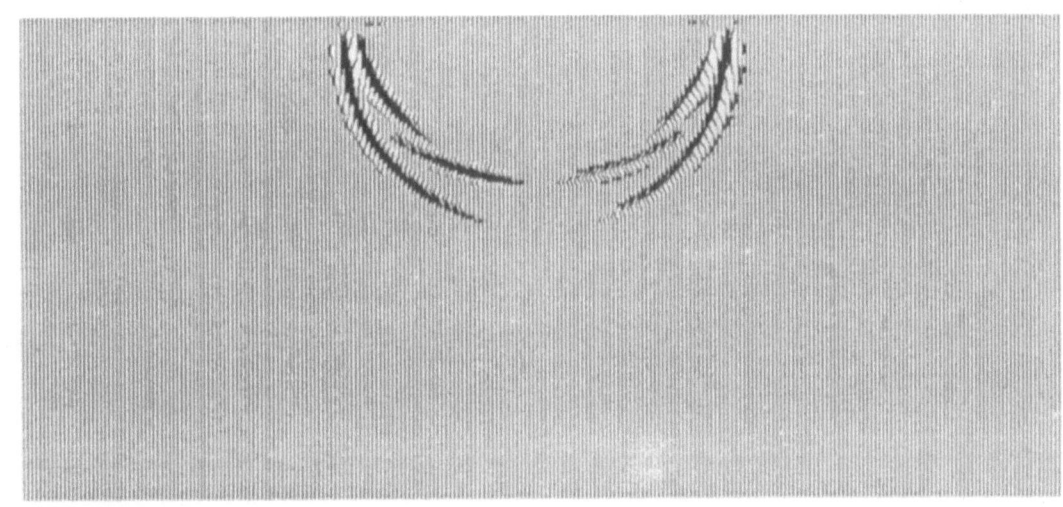

Fig. 2 (a) Curl of acceleration at T=500 milliseconds.

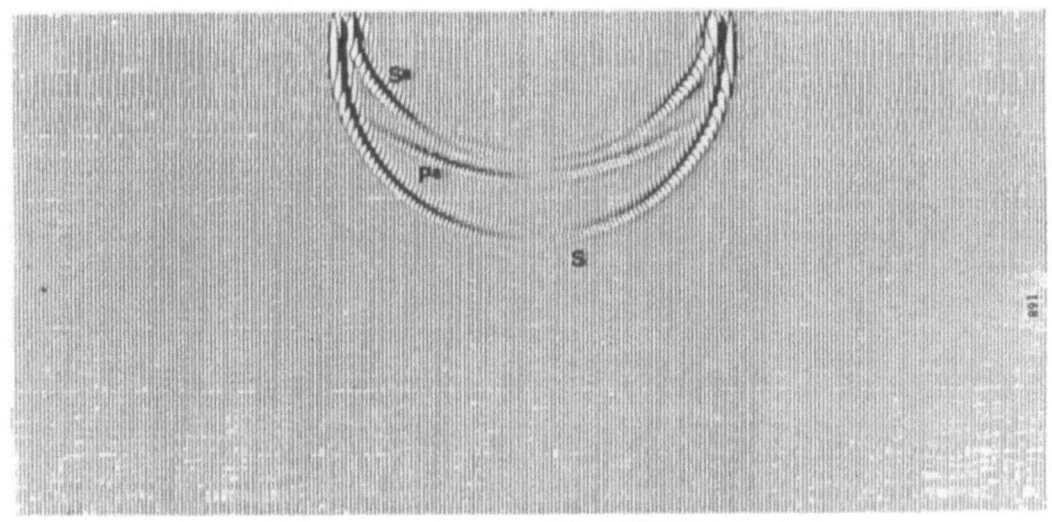

Fig. 2 (b) Curl of acceleration (Kosloff and Reshef).

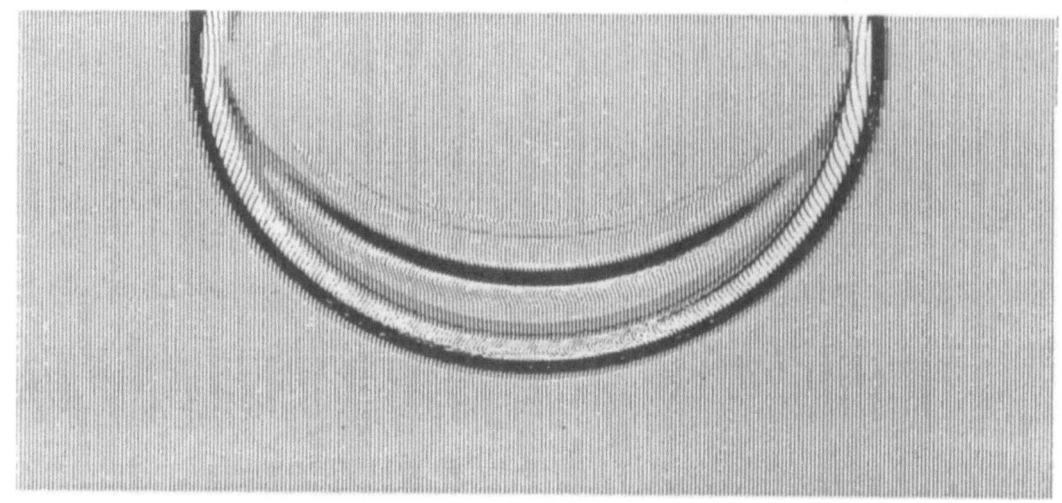

Fig. 3 (a) Pressure at T=500 milliseconds.

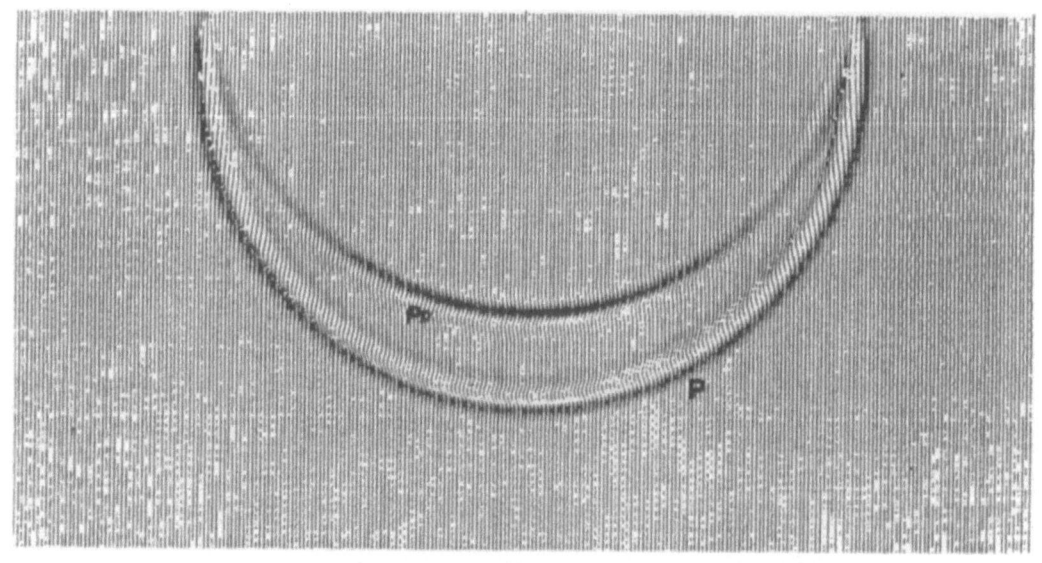

Fig. 3 (b) Pressure (Kosloff and Reshef).

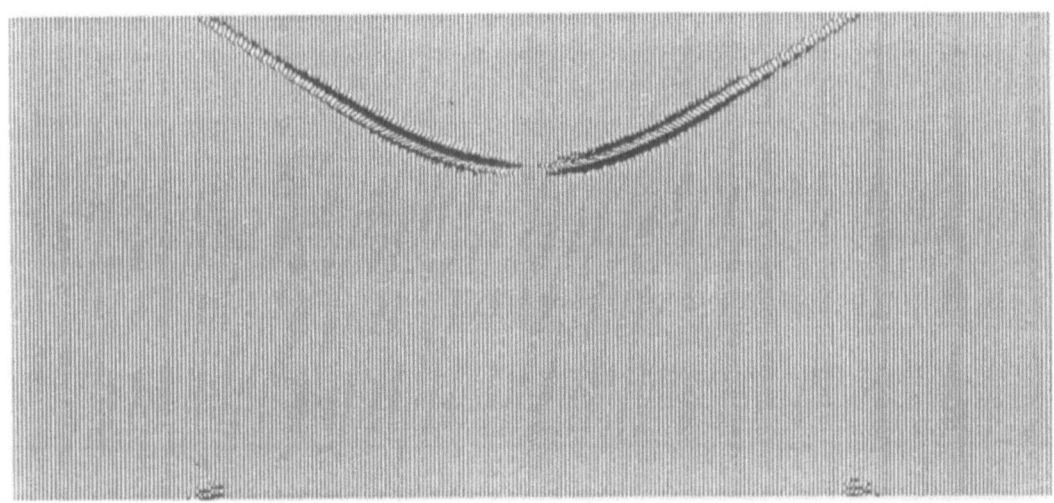

Fig. 4 (a) Curl of the accelerations at T=500 milliseconds.

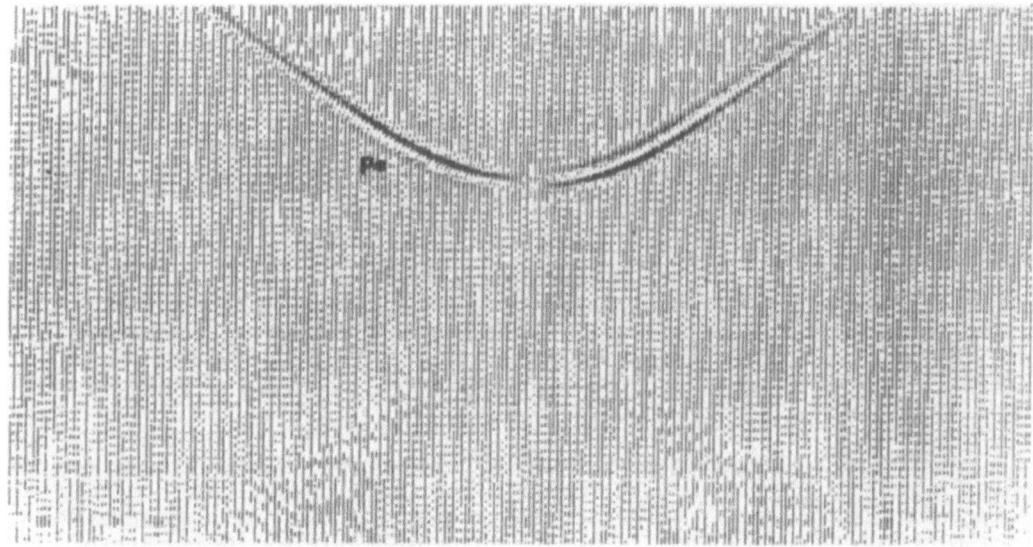

Fig. 4 (b) Curl of the accelerations (Kosloff and Reshef).

Table 3. Timings on the CYBER 205

NX x NY x NT	ABSORBING STEPS	SNAPSHOT SETS	TIME-LINE POINTS	205 TIME (s)
128 x 256 x 1000	800	0	0	143.55
256 x 256 x 1000	500	0	0	273.47
256 x 256 x 1000	500	1	0	282.16
256 X 256 X 1000	500	0	256	293.89
256 X 256 X 1000	1000	0	128	294.07
256 X 256 X 1000	1000	0	256	303.82
256 X 256 X 1000	500	10	256	377.31

Figure 3(a) and 4(a) are snapshots of pressure and curl, respectively, after 500 milliseconds. The corresponding figures 3(b) and 4(b) show the comparable plots obtained by Kosloff and Reshef (1982). Figure 3 contains only P-waves and reflected Pp-waves. Figure 4 of the curl of the acceleration field contains only the converted PS waves.

Solid Over Fluid

This problem considers the propagation of waves in a two-dimensional medium which is an elastic halfspace over a fluid halfspace. It tests the accuracy of our program on a model with a fluid/solid interface. The fluid layer (P velocity~2000 ft/second) is the bottom half of the grid and the solid layer (P and S velocities ~ 3460 and 2000 ft/second) is the top half. The source is introduced in the fluid medium, 8 grid-points below the fluid/solid interface. The model has dimensions (NX x NY) of 128 x 128 with 20 foot spacing between grid points. The waves are initiated by a pressure source.

Figures 5(a), 6(a), and 7(a) are snapshots of curl, δxy, and pressure respectively, after 400 milliseconds have elapsed. The corresponding figures, 5(b), 6(b) and 7(b) are the comparable snapshots obtained by Kosloff and Reshef (1982). In Figures 5 and 6 there are no waves in the fluid, since S-waves do not propagate in a fluid.

TIMINGS

The timings obtained on several test runs on the CYBER 205 are given in Table 3. 'Snapshot sets' is a set of 5 different snapshots - δxx, δyy, δxy, pressure, and curl. The column 'ABSORBING STEPS' is a rough approximation of the number of time-steps absorption took place.

Table 4 shows the detail of the time (in seconds) taken to perform certain tasks. Notice that as the model size increases, the computational time per point decreases; absorbing time also decreases as the model size increases. This is understandable as longer vector operations are more efficient. However, the time to output a set of snapshots remains constant since I/O is performed with scalar code.

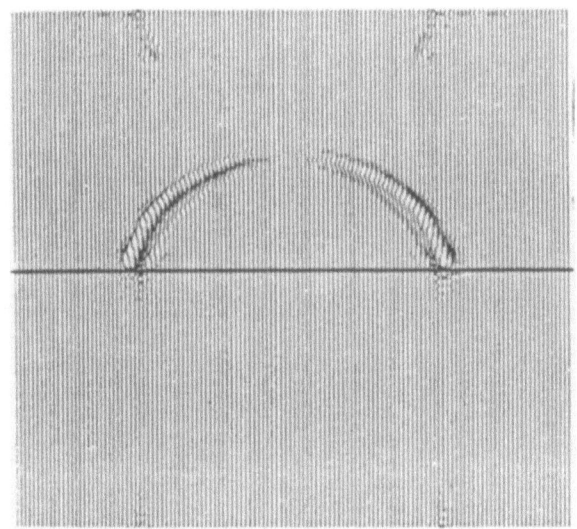

Fig. 5 (a)   Curl of accelerations at T=400 milliseconds.

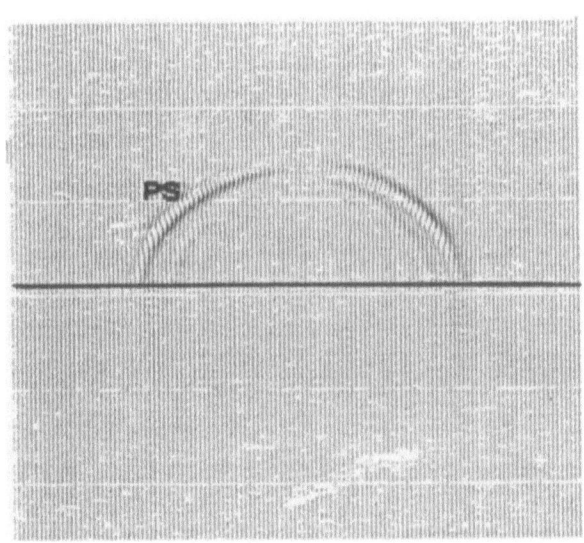

Fig. 5 (b)   Curl of accelerations (Kosloff and Reshef).

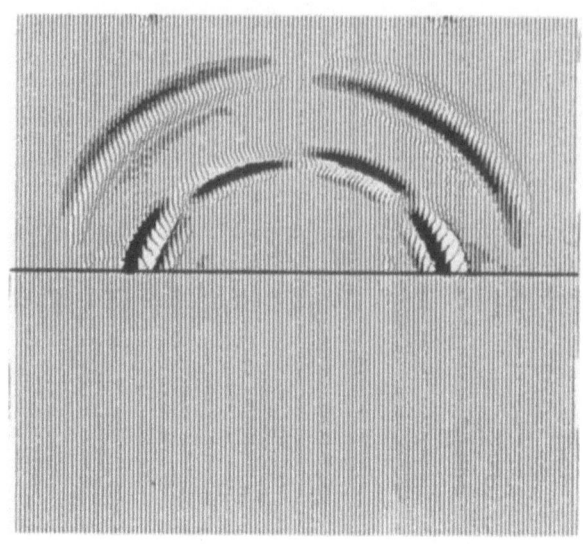

Fig. 6 (a) δ_{xy} at T=400 milliseconds.

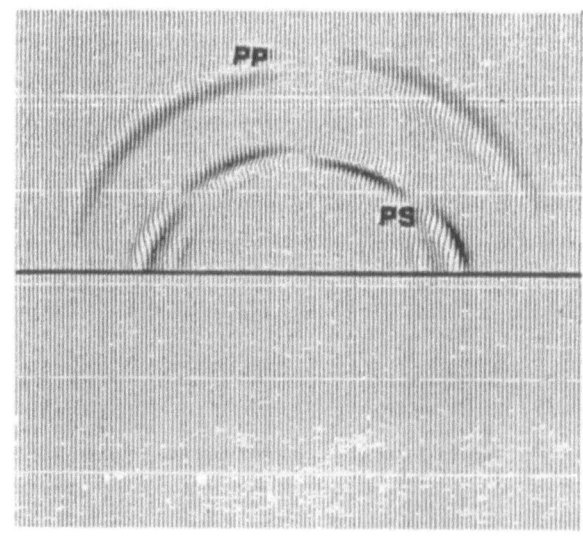

Fig. 6 (b) δ_{xy} (Kosloff and Reshef).

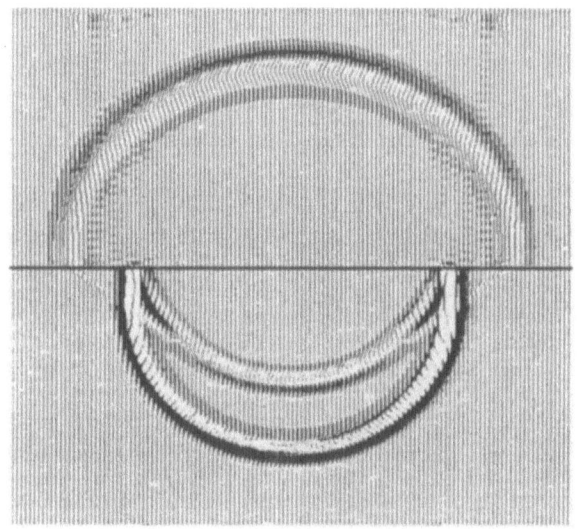

Fig. 7 (a)   Pressure at T=400 milliseconds.

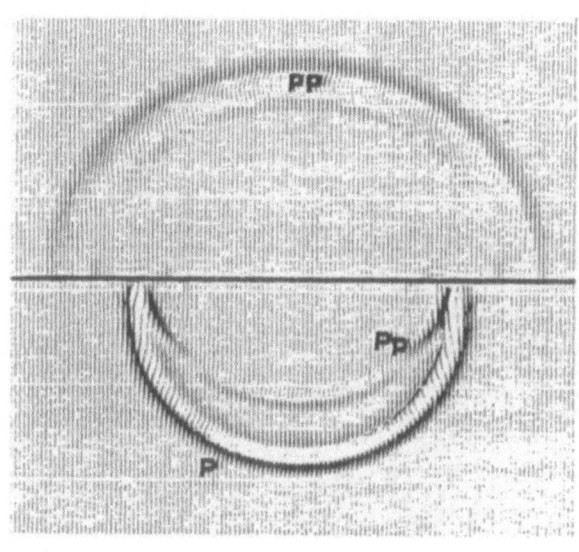

Fig. 7 (b)   Pressure (Kosloff and Reshef).

Table 4 Time detail for different models on the CYBER 205.

MODEL SIZE (NX x NY x NT)	COMPUTATION (Sec)	COMPUTATION per point	SNAPS per set	ABSORB per 100 steps
128 x 128 x 1000	70	.00000427	2.2	1.1
256 x 128 x 1000	134	.00000409	4.4	1.8
256 x 256 x 1000	263	.00000401	8.7	2.1

A 64 x 64 model was timed for 200 time-steps on the VAX-11 780 with the FPS-100 AP using the program written by Kosloff and Reshef. This model took 2193 seconds of CPU time and 5940 seconds of AP time, for a total of 8133 seconds. The equivalent model took a little less than 9 seconds on the CYBER 205. This is approximately 900 times faster than the VAX-AP time. Since longer vectors are more efficiently manipulated on the CYBER 205, we estimate that the amount of time saved will increase as the model size increases.

The ratio obtained here is much higher than other researcher's results (Cheng and Johnson 1982). Other researchers usually report an improvement ratio of approximately 60:1 for the CYBER 205 to the VAX/11 780 with the FPS-100 AP. The great improvement in the timing ratios we obtained is probably due to both the complexity and the size of the model. The elastic forward model manipulates a large amount of data, which implies much data motion on the FPS-100. Since data motion is an expensive operation, it accounts for a substantial amount of the processing time.

REFERENCES

Cheng, T., and Johnson, O. J., 1982, 3D Vector Forward Modeling: Seismic Acoustics Laboratory Fifth Year Progress Review, 10:210-228.

Control Data Corp., 1981(a), "CDC Cyber 200 Fortran Version 2", Control Data Corporation, St. Paul.

Control Data Corp., 1981(b), "MAGEV Library Utility", Control Data Corporation, St. Paul.

Kosloff, D., and Reshef, M., 1982, Elastic Forward Modeling, Seismic Acoustics Laboratory Fifth Year Semi-Annual Progress Review, 10:115-185.

Kosloff, D., and Reshef, M., 1983, A General Non-reflecting Boundary Condition for Discrete Wave Propagation Algorithms, Seismic Acoustics Laboratory Sixth Semi-Annual Progress Review, 12:155-164.

Schmidt, N., 1984, "Fourier Techniques for an Elastic Forward Model Implemented on the CDC CYBER 205", M.S. Thesis - University of Houston.

AN ENHANCED VECTOR ELASTIC MODEL FOR THE

CYBER 205

Nancy Schmidt Adams and
Bjorn Mossberg

Petroleum Technology Center
Control Data Corporation
Houston, Texas   77027

INTRODUCTION

FORM2DE is a two-dimensional elastic forward modeling program designed and developed especially for the CYBER 205. The first development effort was pursued at Texaco in Houston, using their CYBER 205, and resulted in a kernel that solved the mathematical problem, as documented in "A Vector Elastic Model for the CYBER 205" [1]. The research was supported by a joint research grant from Texaco and Control Data Corporation.

The second development effort, performed at the Petroleum Technology Center of Control Data, transformed the kernel to the application program FORM2DE. The many lessons learned the first time through made it possible to further optimize and improve the kernel.

The following sections discuss the various improvements, and the resulting gains in performance and ease-of-use.

IMPROVEMENTS TO THE ORIGINAL IMPLEMENTATION

The original program used complex FFT's to approximate derivatives. The most significant change made to the kernel was the introduction of algorithms where redundant work was eliminated, i.e., real FFT's. Further, the grid dimensions are no longer restricted to be powers of 2, but can instead be composite numbers of the form $2^p \cdot 3^q \cdot 5^r$ ($p \geq 2$).

Additional flexibility has been achieved by means of dynamic dimensioning. In the original implementation the maximum grid size was 256 x 256, since this program had fixed array dimensions - 11 large pages were required to run a model of any size. FORM2DE needs only one large page to run a 64 x 64 model and three large pages to run a 128 x 128 model.

The performance gain resulting from other code optimization, and from the use of real FFT's varies with the model size. Reductions of the total CPU time of up to 67 percent are recorded in Table 1.

Table 1. Benchmark results for the original kernel and FORM2DE, run on a 2-pipe CYBER 205.

MODEL SIZE (NX x NY x NT)	ORIGINAL KERNEL (CPU seconds)	FORM2DE (CPU seconds)
64 x 64 x 200	9	3
128 x 128 x 1000	72	35
128 x 256 x 1000	140	64
256 x 256 x 1000	270	140
512 x 512 x 1000	-	540

TRANSFORMING THE KERNEL TO AN APPLICATION

The most challenging part of the second development effort was the design of interfaces for input and graphics output. The input data to a forward modeling program includes the physical structure of the earth. The new interface gives the user several options for the creation of this structure. The most attractive option is, perhaps, to build the model on a Landmark workstation. However, alternatives exist for those without access to one of those.

Snapshots of the stresses, pressure, and curl comprise the bulk of the output from FORM2DE. Standard SEG-Y files are generated for these snapshots. Animated movies of the snapshots can be produced on a Landmark workstation. Also, standard plots can be obtained from any plotter that can handle SEG-Y tapes.

FORM2DE BENCHMARKS

Table 1 gives the measured times from runs of several different models, using both the original kernel and the new application program, FORM2DE. Since the benchmarks were performed on different machines, the comparisons are approximate. On the average, FORM2DE is about twice as fast as the kernel. For the smallest model reported, the speed increase of FORM2DE was about 2500, as compared to a pilot implementation on a VAX 11/780 with an attached FPS-100 Array Processor.

REFERENCES

[1] Adams, N. S., Johnson, O. G.: "A Vector Elastic Model for the CYBER 205", October 31, 1984, Supercomputer Conference, Purdue University.

STABILITY AND SENSITIVITY OF CORRELATION FUNCTIONS

IN A SINGLE-COMPONENT FLUID

John Kerins

Corporate Research and Development
The Standard Oil Company (Ohio)
4440 Warrensville Center Road
Cleveland, Ohio   44128

L. E. Scriven and H. T. Davis

Department of Chemical Engineering
    and Materials Science
University of Minnesota
Minneapolis, Minnesota   55455

ABSTRACT

By computing the pair correlation function $g(r;n,T)$ and the direct correlation function $c(r;n,T)$ at a multitude of (n,T) points, we have mapped out the three solution spaces associated with the hypernetted-chain, Percus-Yevick and Born-Green-Yvon-Kirkwood approximations for a single-component Lennard-Jones fluid. For each of the HNC, PY and BGYK approximations, we find a locus of turning-point singularities in the (n,T) plane. On each locus there is a critical temperature which we identify as the liquid-vapor critical temperature of the LJ fluid. To further characterize these solution spaces, we have investigated the parametric sensitivity of $g(r)$ to $c(r')$. These calculations fully utilized supercomputer resources of large central memory and of vectorization for matrix construction and for full-matrix Gauss elimination.

BACKGROUND: FLUID STRUCTURE AND CORRELATION FUNCTIONS

Although the average density n at any point in a homogeneous single-component fluid is constant, the fluid does have a microscopic structure characterized by the pair correlation function $g(r)$. If we choose the location of any one particular particle in the fluid as a reference origin, then the local density of particles a distance r away is $ng(r)$. In a simple fluid the particles interact pairwise through a pair potential $u(r)$ which is strongly repulsive at short range, and hence $g(r)$ vanishes as $r \to 0$; typically $g(r)$ rises sharply from 0 to a maximum value greater than unity at a distance r of one particle diameter σ and then decays to its asymptotic value of unity within several multiples of σ (fig. 1b). The specific structural details of g(r) depend on the pair potential u(r), the bulk density n, and the temperature T.

In this work we have modeled the pair interaction potential u(r) between particles at $\underline{r}'$ and $\underline{r}'+\underline{r}$ by a truncated and shifted Lennard-Jones (LJ) potential (fig. 1a):

$$u(\underline{r}',\underline{r}'+\underline{r}) = u(|\underline{r}|) = \begin{cases} 4\epsilon \left[\left(\frac{\sigma}{r}\right)^{12} - \left(\frac{\sigma}{r}\right)^{6} \right] - u_o & , \quad 0 \leq r \leq l_o \\ 0 & , \quad r \geq l_o \end{cases} \qquad (1)$$

The distance l_o for truncation of the LJ potential was fixed at 6σ; the shift constant u_o is chosen such that u(r) is continuous at $r=l_o$. With the pair potential specified, we can focus on changes in the structure of g(r) as a function of the parameters n and T.

Closely related to the pair correlation function g(r) is the direct correlation function c(r) defined through the Ornstein-Zernicke (OZ) equation[1]:

$$c(r) = g(r) - 1 - n \int d\underline{s} \, [g(|\underline{s}|)-1] \, c(|\underline{s}-\underline{r}|) \quad . \qquad (2)$$

Simpler in structure than g(r), c(r) typically approaches a negative value as $r \to 0$, has one positive peak at $r \simeq \sigma$, and decays rapidly to 0 as r increases beyond σ (fig. 1c).

In addition to describing the microscopic fluid structure, the correlation functions also determine the macroscopic thermodynamics of the fluid. The isothermal bulk modulus B ($\equiv (kT)^{-1}(\partial P/\partial n)$ with P the bulk-phase pressure and k Boltzmann's constant) is calculated from the direct correlation function;

$$B(n,T) = 1 - n \int d\underline{r} \, c(r) \quad . \qquad (3)$$

In principle all other thermodynamic properties of the fluid could be generated through B by appropriate integration and differentiation, but we will restrict the thermodynamic aspect of our investigation to the parametric dependence of B on n and T.

An exact statistical-mechanical calculation of either g(r) or c(r) is not feasible, but three approximations, the hypernetted-chain[2] (HNC), the Percus-Yevick[3] (PY) and Born-Green-Yvon-Kirkwood[4,5] (BGYK), are all known to yield qualitatively correct forms for g and c. Each approximation provides a second relation between g and c which can be used with the OZ equation to give a set of two equations for the two unknown functions. With $\beta=1/kT$, the HNC approximation,

$$c(r) = g(r) - 1 - \ln[g(r)] - \beta u(r) \quad , \qquad (4)$$

and the PY approximation,

$$c(r) = g(r) \, [1 - \exp(\, \beta u(r) \,)] \quad , \qquad (5)$$

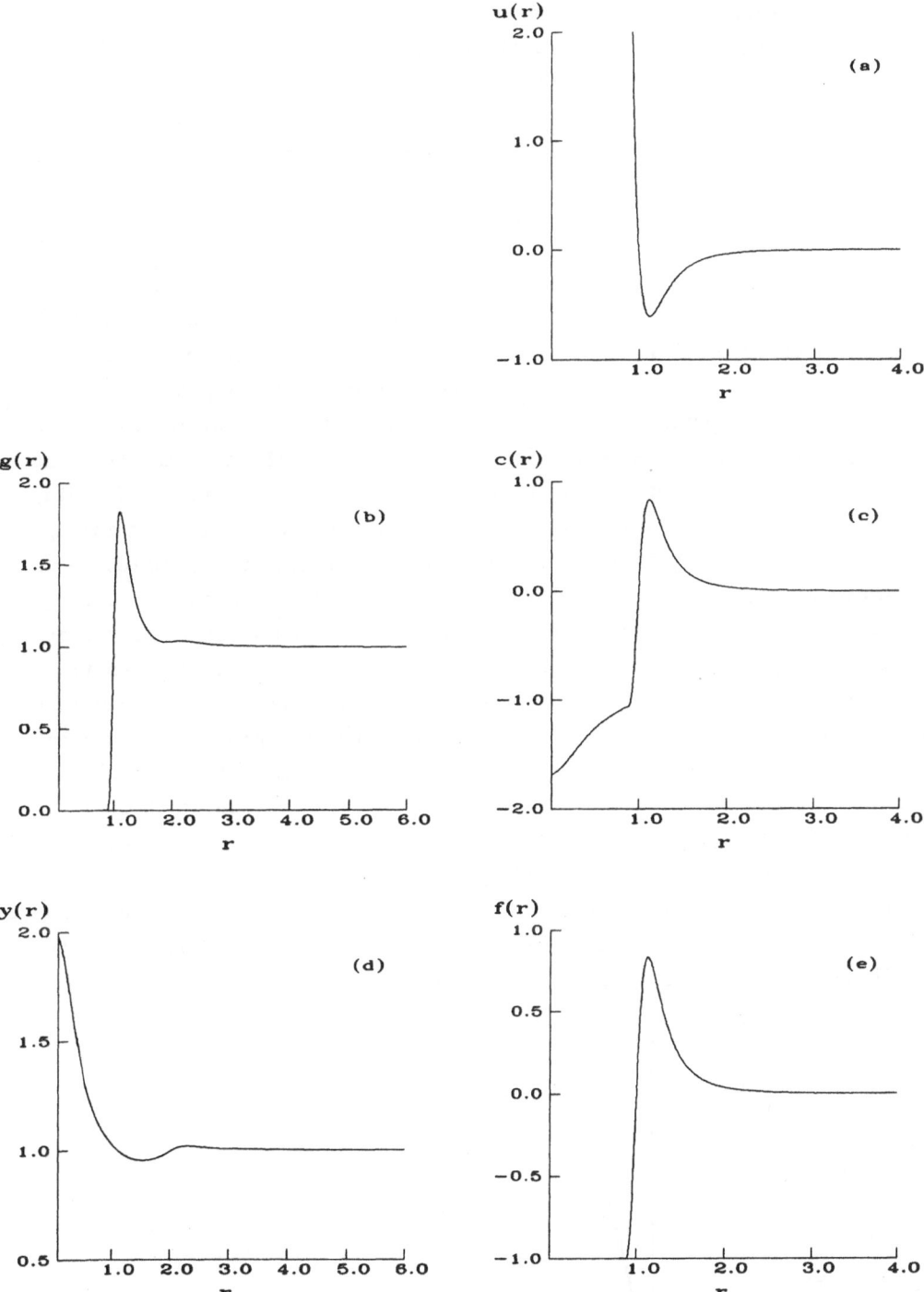

Fig. 1. At T=1.65, (a) the scaled Lennard-Jones potential (eqns. 1,7); for density n=1.30: (b) HNC pair correlation function g(r), (c) HNC direct correlation function c(r), and (d) HNC y(r) function (eqn. 8). In (e) the Mayer function f(r) at T=1.65 (eqn. 9).

both state a simple relation between c and g. The BGYK approximation, which follows from the Kirkwood superposition approximation[4] as applied to the second member of the Born-Green-Yvon correlation function hierarchy,[5] is

$$\ln[g(r)] + \beta u(r) = \pi n \int_0^\infty dt\ t[g(t)-1] \int_{|r-t|}^{r+t} ds\ g(s)\frac{du(s)}{ds}\ \frac{s^2-(r-t)^2}{r}$$

$$+ \pi n \int_0^\infty dt\ 4t^2[g(t)-1] \int_{r+t}^\infty ds\ g(s)\ \frac{du(s)}{ds}\ . \tag{6}$$

The BGYK approximation strictly involves only g(r), but the associated c(r) can be computed by using the BGYK g(r) in the OZ equation. Since the background and performance history of these approximations have been reviewed elsewhere,[6,7,8] we will not comment any further on their origin.

Given the LJ potential (1) and OZ equation (2), each of the HNC, PY and BGYK approximations allows explicit calculation of both g(r) and c(r), and hence B, as functions of density n and temperature T. Now for a LJ fluid, there is a region in the (n,T) parameter plane associated with liquid-vapor coexistence.[7] This region is enclosed by a coexistence curve, inside which thermodynamically stable liquid and vapor states could coexist (fig. 2). Outside the coexistence curve, there is one unique thermodynamically stable state at the given (n,T). Our specific goal is to determine the behavior of the correlation functions g(r;n,T) and c(r;n,T) and the thermodynamic function B(n,T) on approach to the two-phase coexistence region from the surrounding one-phase region. In the next section we summarize the central equations, and we outline our finite-element discretization of these equations in the section Numerical Approach. In the Results section, we display representative features and discuss the physical meaning of our calculations. In the last section, Extensions, we examine the functional sensitivity of the correlation functions and comment on the implications of this sensitivity for other studies on fluid microstructure.

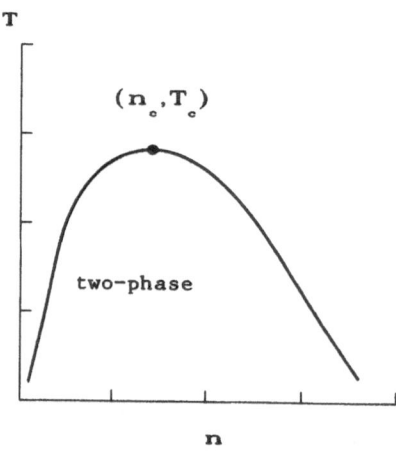

Fig. 2. In the (n,T) plane, the coexistence curve seperates a two-phase, liquid-vapor region from a surrounding one-phase region. The maximum temperature on the curve, T_c at density n_c, is known as the critical temperature.

EQUATIONS: ITERATION, CONTINUATION AND PARAMETRIC SENSITIVITY

Before proceeding to the HNC, PY and BGYK equations, we first introduce dimensionless parameters. Lengths are scaled by the LJ particle diameter σ, temperature is measured relative to the LJ well depth ϵ, and energies are scaled by the thermal energy:

$$r^* = r/\sigma \quad , \quad n^* = n\sigma^3 \quad , \quad T^* = kT/\epsilon \quad , \quad u^*(r) = u(r)/kT \ . \quad (7)$$

From this point on we will suppress the astericks with the understanding that all parameters and functions are dimensionless. Second, we define two additional functions, $y(r)$ and $f(r)$, which arise in all three approximation schemes. $y(r)$ is a correlation function (fig. 1d),

$$y(r) = g(r) \exp[\beta u(r)] \quad , \quad\quad\quad\quad\quad (8)$$

in which the explicit contribution of the pair potential $u(r)$ to $g(r)$ has been removed. In the Mayer-function $f(r)$ (fig. 1e),

$$f(r) = \exp[-\beta u(r)] - 1 \quad , \quad\quad\quad\quad\quad (9)$$

the divergent behavior of the pair potential as $r \rightarrow 0$ is suppressed.

With these definitions, the result of substitution of the HNC approximation (4), in the OZ equation (2), can be written in terms of a functional residual $R[y(r);r]$ which vanishes at all r for an equilibrium solution $y(r;n,T)$.

$$R[y(r);r] = \ln[y(r)] - \frac{2\pi n}{r} \int_0^\infty ds \ s\{[f(s)+1]y(s) - 1\} \quad\quad (10)$$
$$\times \int_{|r-s|}^{r+s} dt \ t\{[f(t)+1]y(t) - 1 - \ln[y(t)]\}$$

Given $y(r)$, $g(r)$ and $c(r)$ can be immediately calculated by equations (8) and (4), respectively. Similar residuals could be written for the PY and BGYK approximations. In the remainder of this section our derivations will be limited to the HNC equation (10), with the understanding that similar results follow in the PY and BGYK cases.

Our first step in analyzing the HNC equation (10) is to assume that $y(r)=1$ for $r \geq l_1 \geq l_0$; then the limits for the integrals in the residual are bounded by l_1. As long as $g(r)$ is not too long-ranged, so that $g(r) \simeq 1$ for $r \geq l_1$, this truncation is physically sensible. To examine in detail the effects of truncation, we will treat l_1 as an additional parameter; i.e. $y = y(r;n,T,l_1)$. The HNC integral equation for $y(r)$ is solved by Newton iteration. Given an initial guess $y_{i=0}(r)$, the next iterate $y_{i+1}(r) = y_i(r) + \delta y_i(r)$ is found from a linear expansion of $R[y_{i+1}(r)]=0$ about $R[y_i(r)]$,

$$0 = R[y_{i+1}] \simeq R[y_i] + \int d\underline{s} \ \frac{\delta R[y_i;r]}{\delta y_i(s)} \delta y_i(s) \quad . \quad\quad (11)$$

Iteration is repeated until a specified convergence criterion is satisfied, with the last iterate taken as the equilibrium solution $y(r;n,T,l_1)$. The kernel $\delta R(r)/\delta y(s)$ of the integral equation for $\delta y_i(r)$ is the functional derivative of $R[y_i(r)]$ with respect to $y_i(s)$ and is known as the Jacobian $J(r,s)$ of the Newton method.

$$J(r,s) = \frac{\delta R(r)}{\delta y(s)} = \frac{\delta(r-s)}{y(r)} - \frac{2\pi n}{r} \left\{ s[f(s)+1] \int_{b(r,s)}^{a(r,s)} dt\ t\{[f(t)+1]y(t)-1-\ln[y(t)]\} \right.$$

$$\left. + s\{ [f(s)+1] - \frac{1}{y(s)} \}\ \int_0^{l_1} dt\ t\{[f(t)+1]y(t)-1\}\ \theta(t;r,s) \right\}$$
(12)

where $\delta(r)$ is the Dirac delta function and

$$\theta(t;r,s) = \left\{ \begin{array}{ll} 1 & ,\quad a(r,s) \leq t \leq b(r,t) \\ 0 & ,\quad \text{otherwise,} \end{array} \right.$$
(13)

with $a(r,s) = \min(\ |r-s|,\ l_1)$ and $b(r,s) = \min(\ r+s,\ l_1)$.

The Jacobian also determines whether the solution $y(r;n,T,l_1)$ is locally unique with respect to the specified parameters. Given a solution $y(r)$ of the HNC equation (10) at fixed values of n, T, and l_1, suppose there is a nearby but distinct solution $\bar{y}(r)$ of (10) at the same parameter values which can be written as $\bar{y}(r) = y(r) + \epsilon p(r)$, $\epsilon<<1$. (Note that we restrict $p(r)$ to vanish for $r>l_1$ as part of the condition that $\bar{y}(r)$ is close to $y(r)$.) When terms of order ϵ^2 are neglected, $p(r)$ satisfies the linear integral equation $0 = \int ds\ J(r,s)\ p(s)$. This equation has nontrivial solutions $p(r)$ if and only if the kernal $J[y;r,s]$ is singular. When the only solution to (17) is the trivial solution, $p(r)\equiv0$, then $y(r)$ is locally unique. Thus the singularities of the integral operator J locate, for the specified set of parameters, the points where multiple solutions are to be found.

To check the assumption $y(r)=1$ for $r\geq l_1$, we calculate the residual $R(r)$ for $l_1<r\leq 2l_1$; if $R(r)$ differs appreciably from 0 in that range, then l_1 is increased. Distinct from the question of consistency of the truncation approximation, we can also examine the sensitivity of the solution $y(r)$ to the parameter l_1. An equation for the parametric sensitivity function $\partial y(r)/\partial l_1$ is generated by differentiating $R[y(r);r]$, $0\leq r\leq l_1$, with respect to l_1 at fixed n and T.

$$-\int ds\ J(r,s)\ \frac{\partial y(s)}{\partial l_1} = \frac{2\pi n}{r}\left\{ \lim_{s\to l_1^-} s\{[f(s)+1]-1\} \int_{l_1-r}^{l_1} dt\ t\{[f(t)+1]y(t)-1-\ln[y(t)]\} \right.$$

$$\left. + \lim_{s\to l_1^-} s\{[f(s)+1]y(s)-1-\ln[y(s)]\} \int_{l_1-r}^{l_1} dt\ t\{[f(t)+1]y(t)-1\} \right\}.$$
(14)

The inhomogeneous term in this sensitivity equation is fixed for the given $y(r)$ and l_1 and depends strongly on the value of $y(l_1^-)$. If the inhomogeneous term vanishes and $J(r,s)$ is nonsingular, then $\partial y(s)/\partial l_1 = 0$ and the solution $y(r)$ is independent of the truncation l_1.

By first-order continuation, a good intial guess $\bar{y}(r)$ for the solution at $n+\delta n$, T and l_1 can be generated from the solution $y(r;n,T)$ and $J(r,s)$. First-order expansion of $y(r;n+\delta n)$ in δn about $y(r;n)$ yields

$$\bar{y}(r;n + \delta n) = y(r;n) + \frac{\partial y(r)}{\partial n} \delta n \simeq y(r;n+\delta n) \quad . \tag{15}$$

Just as for the sensitivity $\partial y(r)/\partial l_1$, the functions $\partial y(r)/\partial n$ is determined from the equation

$$\int ds\ J(r,s)\ \frac{\partial y(s)}{\partial n} = - \left(\frac{\partial R}{\partial n}\right)_y \quad . \tag{16}$$

Similar manipulations are possible to calculate $\partial y/\partial T$, but in this case the resulting equation is complicated enough that the guess $\bar{y}(r;T+\delta T)=y(r;T)$ with δT small is more practical.

Suppose that $|\partial y(r)/\partial n| \gg 1$ at $r=r_0$. Then it is advantageous to parameterize $y(r)$ by T, l_1 and $y_0=y(r_0)$, with n now a dependent variable. The initial guess $\bar{n}$ and $\bar{y}(r)$, $0 \le r \le l_1$ $r \ne r_0$, for a change δy_0 in the parameter y_0 requires calculation of $\partial y(r)/\partial y_0$; these partial derivatives are found as solutions of the linear integral equation given by differentiating $R[y]$ with respect to y_0.

$$0 = \left(\frac{\partial R}{\partial n}\right) \left(\frac{\partial n}{\partial y_0}\right) + \int ds\ J(r,s) \left(\frac{\partial y(s)}{\partial y_0}\right) \quad , \tag{17}$$

subject to the normalization condition

$$\frac{\partial y(r = r_0)}{\partial y_0} = 1 \quad . \tag{18}$$

By first-order continuation we can still find, at fixed l_0, l_1 and T, an initial guess $\bar{n}$ and $\bar{y}(r)$, $0 \le r \le l_1$, for a change δy_0 in the parameter y_0.

$$\bar{y}(r;\ y_0 + \delta y_0) \simeq y(r;y_0) + \left(\frac{\partial y(r)}{\partial y_0}\right) \delta y_0 \quad , \qquad r \ne r_0 \quad , \tag{19}$$

$$\bar{n}(y_0 + \delta y_0) \simeq n(y_0) + \left(\frac{\partial n}{\partial y_0}\right) \delta y_0 \quad . \tag{20}$$

By allowing for flexibility in our choice of continuation parameter, we can insure a very good intial guess $\bar{y}(r)$. For example, at turning points in the density n (see fig. 4), $\partial y(r)/\partial n \gg 1$ for many r. If the continuation parameter is $y_0=y(r_0)$, then $\partial y(r)/\partial y_0$ and $\partial n/\partial y_0$ are less than one (due to eqn. 18), and $\bar{n}(y_0+\delta y_0)$ and $\bar{y}(r;y_0+\delta y_0)$ from (19) and (20) will be close to the equilibrium solutions $n(y_0+\delta y_0)$ and $y(r;y_0+\delta y_0)$.

NUMERICAL APPROACH: FINITE-ELEMENT DISCRETIZATION

Given the above integral equations for Newton iteration (11), parametric sensitivity (14) and first-order continuation (15-20), we adopt the finite-element method[10] for numerical solution of these equations. The intent of the finite-element approach is to approximate the function $y(r)$ locally over small subdomains. The unknown solution $y(r)$ is approximated by an expansion with coefficients y^k in a finite set of N linear functions $H^k(r)$;

$$y(r) \simeq \sum_{k=1}^{N} y^k H^k(r) = \underline{y} \cdot \underline{H}(r) \quad , \tag{21}$$

$$H^k(r) = \begin{cases} \dfrac{r - r_{k-1}}{r_k - r_{k-1}} \, , & r_{k-1} \leq r \leq r_k \, , \\[2ex] \dfrac{r_{k+1} - r}{r_{k+1} - r_k} \, , & r_k \leq r \leq r_{k+1} \, , \\[2ex] 0 \, , & \text{elsewhere} \, . \end{cases} \tag{22}$$

The uniform mesh of N points $r_k = (k-1)h$, where $h = l_1/(N-1)$, divides the domain of interest $0 \leq r \leq l_1$ into $N-1$ subdomains. We also make the Swartz-Wendroff[10] (SW) approximation in our discretization. Under this approximation, any nonlinear function of r is written as a linear expansion in $\{H^k\}_{k=1}^{N}$ with nonlinear coefficients; e.g., with $y(r)$ unknown and $f(r)$ known,

$$\ln[y(r)] = \Sigma \ln[y^k] \, H^k(r) \quad ,$$
$$f(r)y(r) = \Sigma \, f(r_k) \, y^k \, H^k(r) \quad . \tag{23}$$

In the integral equations for $y(r)$ and its derivatives, the finite-element and SW approximations amount to specifying a simple quadrature rule, with the difference between the approximate and exact solution controlled by the size h of the subdomain.

Since no choice of expansion coefficients y^k will satisfy the condition $R[y;r]=0$ for all $0 \leq r \leq l_1$, we demand that the residuals $R[y;r_k]$ vanish at each of the nodes r_k, $k=1,2,\ldots,N$. This collocation version of the weighted-residual technique in finite elements results in N coupled, nonlinear algebraic equations for the N expansion coefficients y^k.

$$0 = R[\underline{y};r_k] = \ln[y^k] - \frac{2\pi n}{r_k} \sum_{i=1}^{N} \sum_{j=1}^{N} \left\{ s_i [f(s_i)+1]y^i \right\} \tag{24}$$

$$\times \left\{ t_j [f(t_j)+1]y^j - 1 - \ln[y^j] \right\} A^{ijk} \; ;$$

$$A^{ijk} = \int_0^{l_1} ds \, H^i(s) \int_{a(s_i,r_k)}^{b(s_i,r_k)} dt \, H^j(t) \quad , \quad i,j,k = 1,2,\ldots,N \quad . \tag{25}$$

This set of N residual equations is solved through the discretized version of the Newton method embodied in equation (24); the kernel $J(r,s)$ is now a matrix $J[\underline{y};r_i,s_j]$ in which the elements J_{ij} depend on the y^k. Iteration is repeated until the Euclidean norm of the update vector $\delta\underline{y}_i$ or the Euclidean norm of the residuals for the ith iterate is less than $10^{-10}\sqrt{N}$. We monitor the truncation approximation by calculating the extra residuals $R[\underline{y};\bar{r}_k]$ at the nodes $\bar{r}_k = l_1 + r_k$, $k=1,2,\ldots,N$, outside the domain $0\leq r\leq l_1$. If each of these extra residuals is less than 10^{-6}, then we regard the assumption $y(r) = 1$ for $r>l_1$ as valid.

The discrete versions of the sensitivity (14) and continuation (15-20) equations are found by using the same scheme of finite-element expansion, SW approximation and nodal-point collocation. For the BGYK equation, a discretized form of the OZ equation (2) was used to fix the coefficients c^k in an expansion of $c(r)$, given the BGYK coefficients y^k. In our implementation of the Newton method the practical limit to N, the number of nodes, is dictated by the computer time necessary to calculate the *full N×N matrix* J_{ij}; solution by Gauss elimination of the linear equations involving J_{ij} took slightly less time. Most of our calculations were for a domain spacing $h=1/15$ with N between 90 and 361, depending on the choice of l_1, $6<l_1\leq24$. The algorithm was designed to optimize vectorization, and did implement special vector structure. For example: Q8SSUM with variable length and base position for the vector argument was used on the Cyber 205 in evaluating inner integrals; the general matrix solver QQGEL from the QQLIB Fortran subroutine library was used to solve the linear algebraic systems. Production runs were carried out either on a Cyber 205 owned by Control Data Corporation or on a Cray 1B machine at the University of Minnesota.

Values of $g(r)$ and $c(r)$ at the node points r_k, $k=1,\ldots,N$, are fixed by the expansion coefficients y^k. The integral for the thermodynamic bulk modulus B (3) was computed by first using a cubic spline interpolation on the nodal values of c to fix the values of c on a mesh of 3N uniformly spaced points in $0\leq r\leq l_1$, and then using Simpson's rule on the finer mesh. This is because the thermodynamic integral contains r^2 which becomes large.

RESULTS: DENSITY TURNING POINTS AT LOW TEMPERATURES

In this section we summarize results for the HNC, PY and BGYK cases at high and low temperatures. At high temperatures, $T\geq1.65$, a single solution branch $y(r;n,T)$, $0<n\leq0.6$, which is insensitive to the cutoff l_1, is found for each of the HNC, PY and BGYK equations. The three approximations yield very similar correlation funcitons at low densities. A variance between the BGYK and HNC or PY results is apparent at intermediate densities, $0.2\leq n\leq0.4$, especially in $c(r)$; at high densities, $n>0.4$, clear differences between the

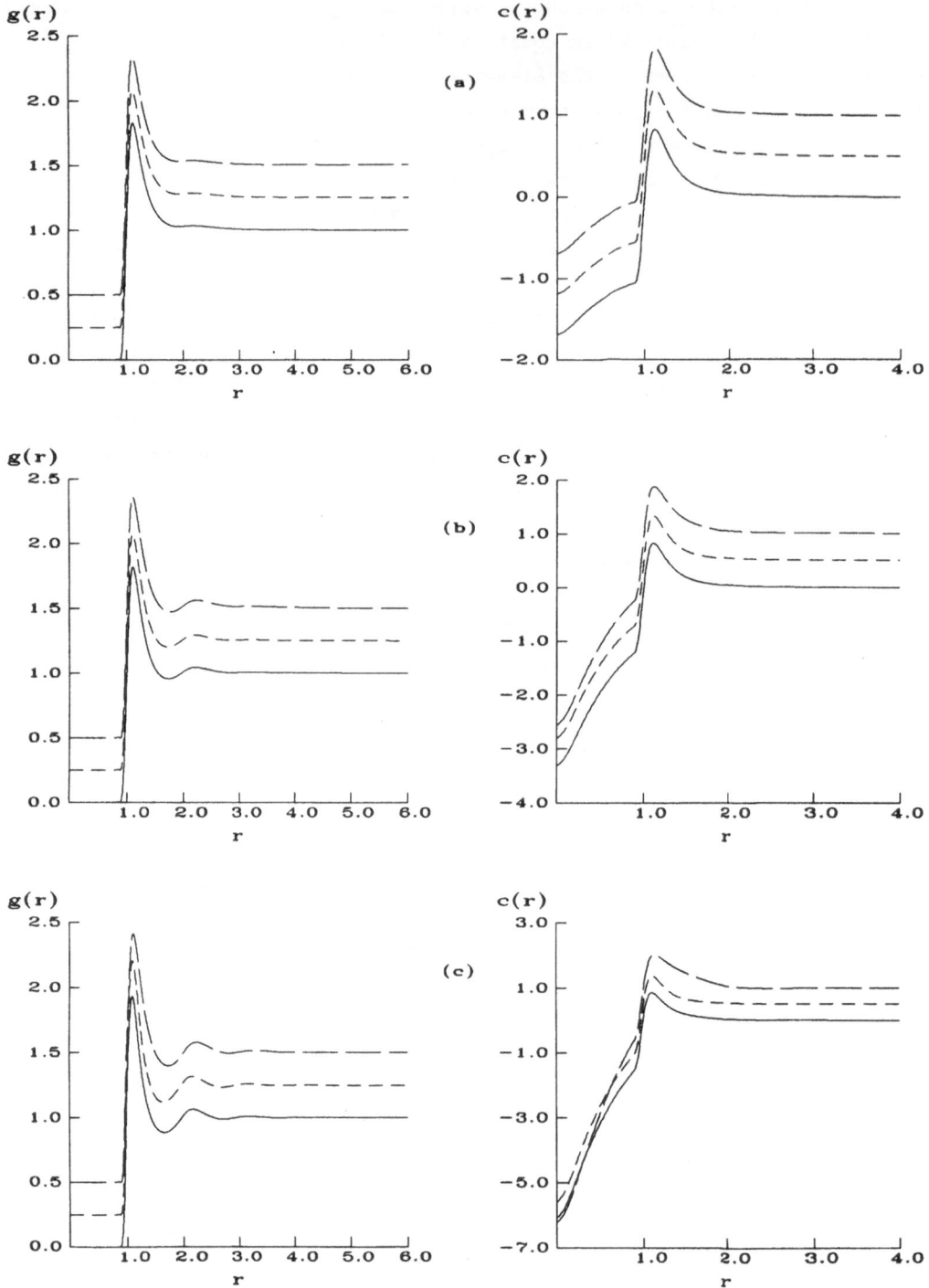

Fig. 3. g(r) and c(r) under the HNC (———), PY (– – –) and BGYK (— —) approximations at T=1.65 and density n equal: (a) 0.10, (b) 0.30, and (c) 0.50. PY and BGYK curves have been shifted up by 0.25 and 0.50 for g(r) and by 0.50 and 1.0 for c(r), respectively.

HNC, PY and BGYK cases are visible, again especially in c(r) (fig. 3). Since each approximation neglects distinct factors in the true local arrangement of fluid particles, differences are more evident at densities where more particles are packed in the neighborhood of the reference particle. Nevertheless, the correlation functions have the same qualitative form at T=1.65. Having established a common and similar high temperature baseline for the HNC, PY and BGYK correlation functions, we now examine the solution spaces at low temperatures.

At a temperature T=1.30, instead of a single HNC solution branch crossing from low to high density, there are two branches, one at low densities and the second at high densities. Both branches have a turning point in the density. In figure 4 we map the solution space as projected into the (n,-c(0)) and (n,B) planes. Tracking the lower half of a low-density solution branch past the turning point, and then moving up in temperature, we find a second high-temperature solution branch; this second high-T branch corresponds to thermodynamically unstable fluid, i.e. B(n,T)<0, and is partner to the previously found stable high-T branch. As the turning point is approached at T=1.30 from stable (B>0) vapor phase, increased sensitivity to l_1 is found just below or beyond the turning point. Solutions beyond the turning point have a very long tail, and the approximation y(r)=1 for r≥l_1 is certainly breaking down. Indeed, the exact location of the turning points is a function of l_1 (fig. 5). As T is increased from 1.30, the two turning-point branches both extend until at a unique critical temperature T_c(HNC)≃1.35 they just touch; on further T increase two disconnected branches

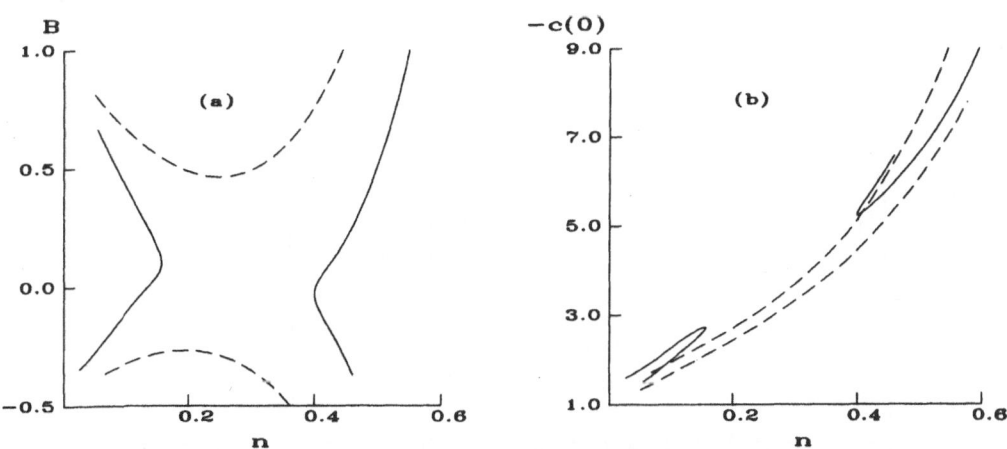

Fig. 4.   (a) Both branches of B(n) at T=1.30 (————) have positive or stable and negative or unstable portions. In contrast B(n,T=1.65) (– – –) has a positive branch and a negative branch. (b) The solution space in the (n,-c(0)) plane at T=1.30 (————) and T=1.65 (– – –).

reappear, each crossing from low to high density.

We identify T_c as the thermodynamic critical temperature for the HNC approximation (see fig. 2). Consequently at $T<T_c$(HNC) there is no solution branch for the correlation functions which extends across the two-phase co-existence region. The thermodynamic bulk modulus B(n,T) (3) is a singular function of the density n at the turning points. We believe that these turning-point singularities are associated with the nucleation/condensation singularities[12-14] which exist at the coexistence curve. For the thermodynamically stable fluid states, the single solution branch and the two disconnected solution branches reflect the single fluid state and the liquid-vapor states found at temperatures $T>T_c$ and $T<T_c$, respectively.

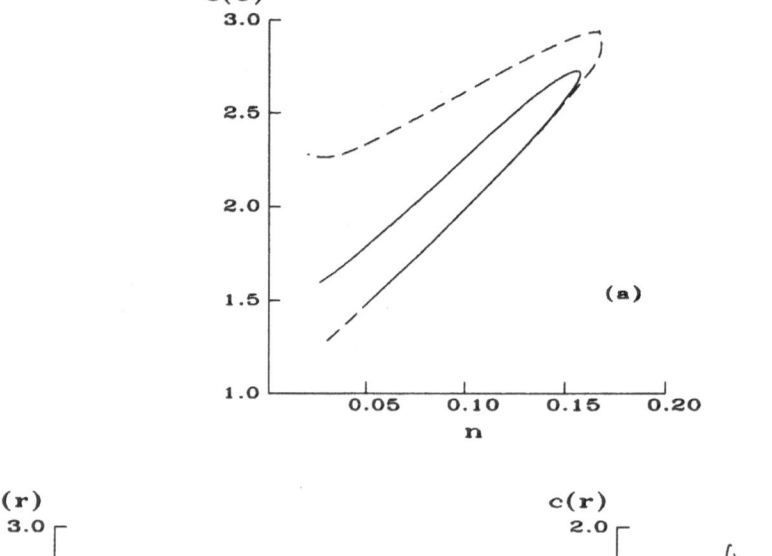

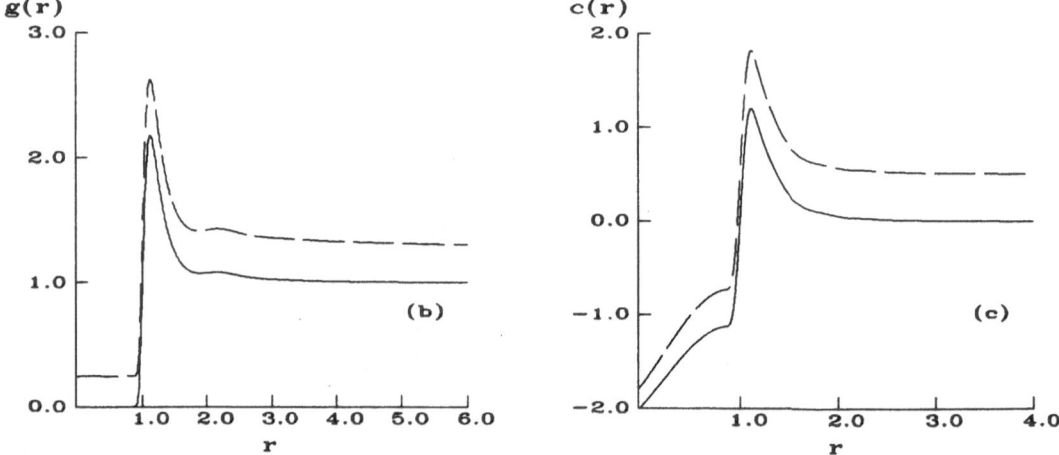

Fig. 5. At T=1.30, location of the turning point on the low-density HNC
solution branch depends on the cutoff l_1, as shown in (a) for
l_1=6.0 (– – –) and l_1=12.0 (———). For n=0.10 (b) and (c) show
the stable, vapor-phase g and c functions (———) which are
independent of $l_1 \geq 6.0$ and the unstable-phase g and c at l_1=12.0
(– –, g shifted up by 0.25, c shifted up by 0.50).

For the PY approximation, there is a different critical temperature $T_c(PY) \simeq 1.30$ which separates two classes of behavior for the correlation functions. At $T > T_c(PY)$, there is one stable solution branch which extends from low to high density and is insensitive to the choice of $l_1 \geq 9.0$. At $T < T_c$ this one solution branch develops a singular structure (fig. 6). As the density is increased from that of stable vapor phase, the solution branch first turns back to lower densities, and then turns again to higher densities. As mapped in the $(n, -c(0))$ plane, the exact location of this pair of turning points depends on l_1; the second one moves to very low density and very large $|c(0)|$ as l_1 increases until at $l_1 = 12.0$ the single $T < T_c$ solution branch appears as two distantly connected pieces.

Although there is a critical temperature $T_c(PY)$ and a low density singularity in $B_{PY}(n, T < T_c)$ due to the turning point, there is no high density singularity in B_{PY}. We still identify $T_c(PY)$ as the fluid critical temperature for the PY approximation, and still interpret the $T < T_c(PY)$ singularities in the bulk modulus as signaling liquid-vapor states, although the signal under the PY approximation is weaker than under the HNC approximation.

Finally we turn to the BGYK case, where we find results similar to the PY case. For $T > T_c(BGYK) \simeq 1.55$, a single solution branch crosses from low to high density with no turning points and no particular sensitivity to $l_1 \geq 9.0$. At $T < T_c(BGYK)$ there is still only one solution branch for $y(r;n)$, but there is a pair of turning points on this solution branch, just as in the PY case. The location of this turning-point pair shifts very slowly as l_1 increases, but even at $l_1 = 24.0$ the pair is close in both n and $|c(0)|$ (fig. 7).

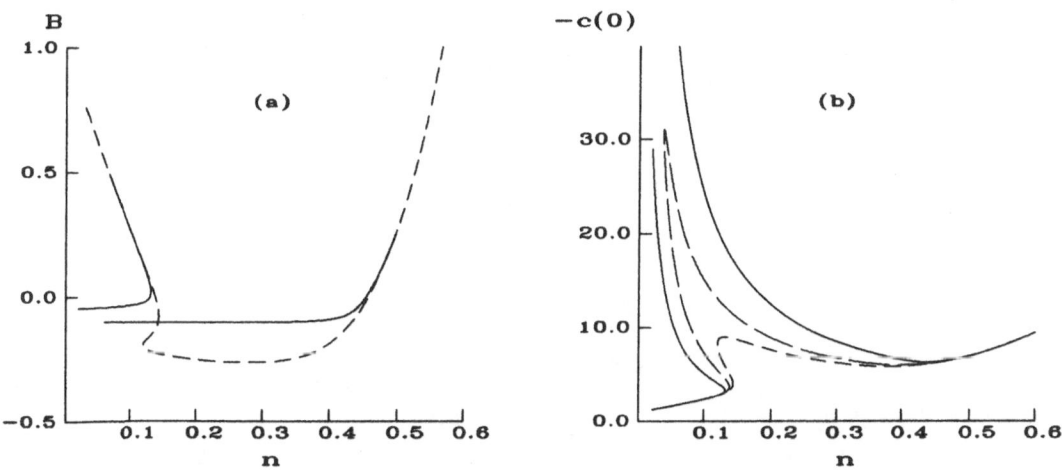

Fig. 6. (a) The bulk modulus B(n) at T=1.15 for l_1 equal 6.0 (– – –) or 12.0 (———). (b) At the subcritical temperature T=1.15, the turning point in the PY correlation-function space, as mapped into the $(n, -c(0))$ plane, depends on l_1; shown is l_1 equal 6.0 (– – –), 6.8 (— —) and 12.0 (———).

Unforturnately, unrealistic features appear in c(r) at intermediate and high densities due to the inadequacy of the Kirkwood superpositon approximation at these densities. This causes unrealistic behavior in $B_{BGYK}(n,T)$ at intermediate and high densities (viz. B at n>0.35 and T=1.65>T_c(BGYK) is negative), but these artifacts are independent of the presence or absence of low-density turning points. We again identify T_c(BGYK) as the BGYK thermodynamic critical temperature, and interpret the turning-point singularities in $B_{BGYK}(n,T)$ as a weak signal of liquid-vapor states.

We conclude that the HNC, PY and BGYK equations reflect, although as three different distorting mirrors, the existence of a liquid-vapor transition region. The location of the liquid-vapor critical temperature T_c depends on the approximation scheme. The HNC, PY and BGYK results all reveal singularities in B(n,T) at T<T_c, and hence show a limitation in extending the idea of well-behaved correlation functions into the coexistence region.

Several computational features in the background of the above conclusions deserve comment. Those solution branches with turning points were tracked in 60-100 steps using adaptive continuation at several l_1 between 6.0 and 24.0. For each approximation our conclusions for T<T_c are based on 6-8 isotherms. Although supercomputer resources were invaluable for solution of N×N full matrix systems, with typically N≃200, the real advantage of the supercomputer for this study is the quick and cost-effective calculation of solutions at many (n,T,l_1) parameter points. The full structure of the solution space for the correlation functions would not have been apparent if

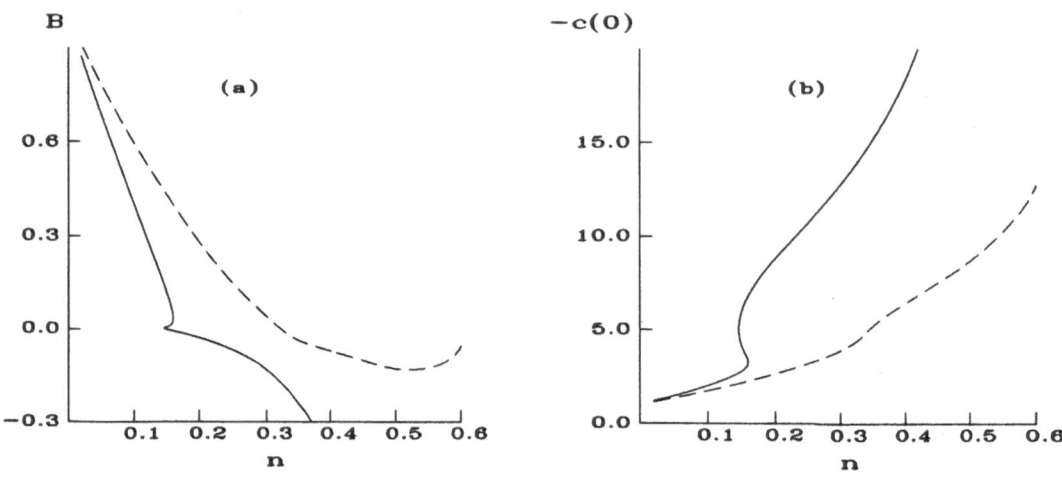

Fig. 7. The single BGYK solution branch, as mapped into (a) the (n,B) plane and (b) the (n,-c(0)) plane, is shown at the super-critical temperature T=1.65 (– – –) and the subcritical temperature T=1.30 (———).

we had been restricted to fewer parameter points. Also we have suppressed discussion on the questions of Newton iteration, stability, continuation and parametric sensitivity (although in the previous section the appropriate equations were presented to indicate the scope of the study). Supercomputer cost efficiency made it feasible to investigate these questions which probe $g(r;n,T,l_1)$ and $c(r;n,T,l_1)$ in more detail.

EXTENSIONS: FUNCTIONAL SENSITIVITY OF g(r) TO c(r)

In this last section we continue comparison of the HNC, PY and BGYK approximations with a brief discussion of a functional sensitivity, and comment on the usefulness of this functional sensitivity in other calculations of fluid structure. Since $g(r)$ is related to a probability distribution ($ng(r)$ is the probability of finding a particle at a distance r from a central reference particle), we can understand the features of $g(r)$ at a physical level: $g(r)=0$ for $r \le 1$ because of the finite size of the reference particle, the peak in $g(r)$ near $r=1$ represents a shell of neighbors, and $g(r)$ decays asymptotically to unity because the reference particle has a finite range of influence. On the other hand, $c(r)$ has a simpler form as a function of r than $g(r)$, and usually shows a more distinct variation in structure under each of the HNC, PY and BGYK approximations (see fig. 3c). By examining the functional sensitivity of the pair correlation function $g(r)$ to the direct correlation function $c(r)$, we can highlight the coupling of the simple structure of c with its physical effect in $g(r)$. If $g(r)$ is treated as a functional of $c(r)$ in the OZ equation (2), then the functional sensitivity $S(r,r')$ is given as

$$S(r,r') = \delta g(r)/\delta c(r') - \delta(r-r')$$

$$= \frac{2\pi n}{r} \int_0^\infty ds \int_{|r-s|}^{r+s} dt \; S(s,r') \; tc(t) \; + \; \frac{2\pi n r'}{r} \int_{|r-r'|}^{r+r'} ds \; s[g(s)-1+c(t)]$$

(26)

$S(r,r')$ indicates how $g(t)$ will change at $t=r$ for minor changes in $c(t)$ at $t=r'$. We may rewrite equation (26) for $S(r,r')$ as

$$S(r,r') = \int ds \int dt \; S(s,r')K(s,t) + F(r,r')$$
(27)

where the kernal $K(s,t)$ is a functional of $c(t)$ and the inhomogeneous piece $F(r,r')$ is a functional of $c(t)$ and $g(t)$. Since the kernel $K(s,t)$ is independent of r', the two-dimensional system for $S(r,r')$ effectively decouples into a set of one-dimensional systems. Using collocation for $S(r,r')$ after a finite-element discretization under the SW approximation, we find a linear set of N equations for $S(r_j,r_k')$ to be solved for N right-hand sides $F(r_j,r_k')$, $j,k = 1,\ldots,N$, where N is the number of finite element nodes.

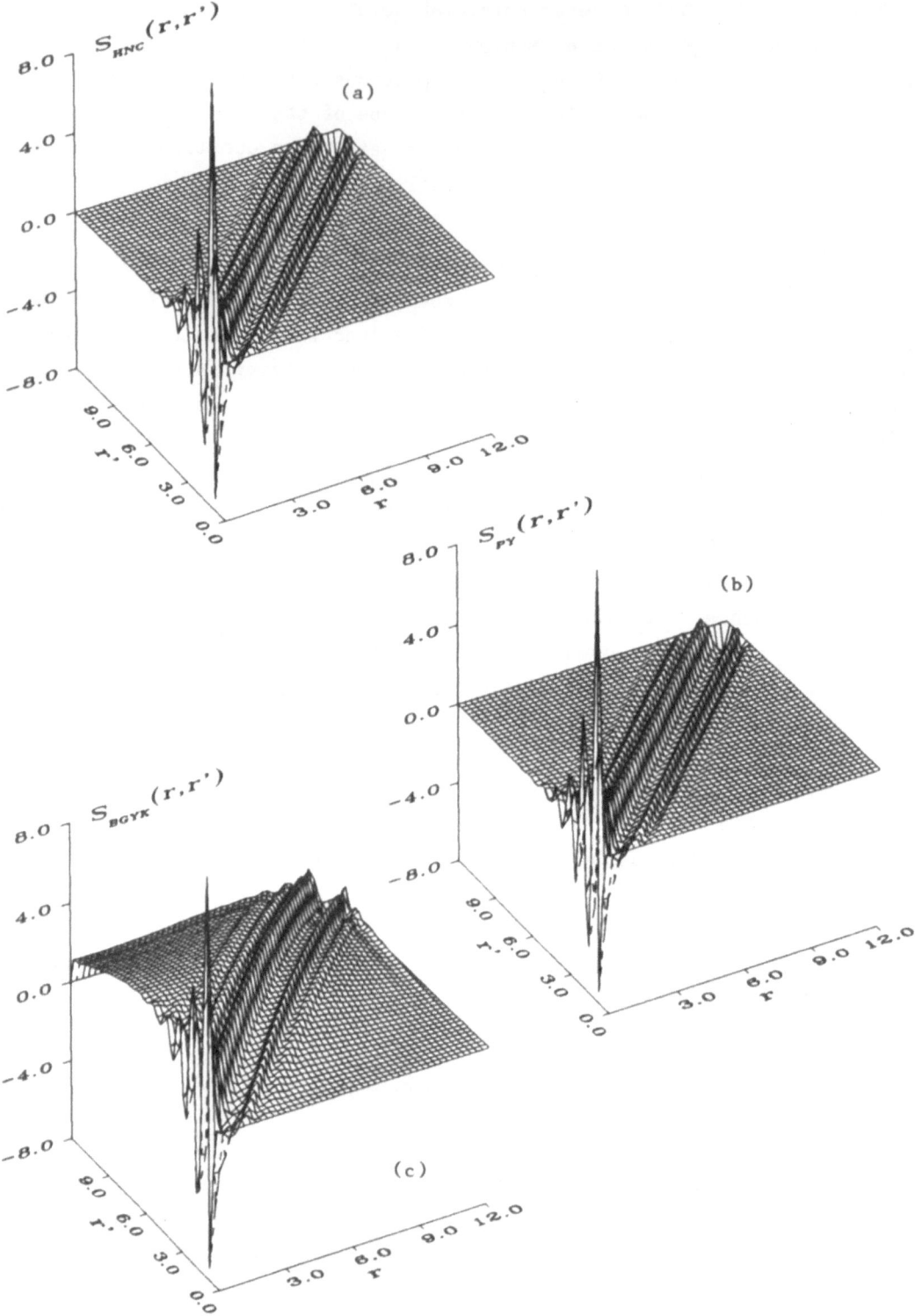

Fig. 8. The sensitivity S(r,r') for the (a) HNC, (b) PY, and (c) BGYK
cases at T=1.65 and n=0.50 (cf. fig. 3). Visible portions of
the surface underside are dashed.

In figure 8 S(r,r') is plotted for the HNC, PY and BGYK correlation functions at T=1.65 and n=0.50 (cf. fig. 3). The qualitative structure of the surface S(r,r') is the same in all three cases. Strong ripples near the (r,r') origin show g(r) near r=0 to be very sensitive to the short-range structure of c(r). Apparently the HNC, PY and BGYK approximations are all accurate enough for c(r) that g(r) is well-behaved near r≃0, but this is not true for other approximations[15]. The "banded" valley-ridge features about r=r' indicate that g(r) is dependent on the values of c(r') in the neighborhood of r=r'; the OZ equation has a more subtle structure than a simple mapping of the tail of g(r) into the small-r form of c(r).

The peaks in $S_{PY}(r,r')$ for r≃0 are slightly higher than those in $S_{HNC}(r,r')$, and the valleys slightly lower. $S_{BGYK}(r,r')$ has a much different structure than S_{PY} or S_{HNC}, reflecting the pronounced differences in the BGYK correlation functions compared to the PY or HNC correlation functions. In particular, for all r, g(r) is sensitive to the tail in c(r) (i.e. $S_{BGYK}(r,r'≃12) \neq 0$). Figure 9 shows sensitivity plots for the HNC results at T=1.30 and n=0.10 for stable and unstable states (see fig. 11 and 12). The change in the unstable-state surface from either the stable-state surface or the surfaces in figure 8 implies that a drastic alteration in sensitivity accompanies the change in thermodynamic stability. Although we do not give details here, further analysis is necessary to fully untangle the physical interpretation of $S(r,r')$[15].

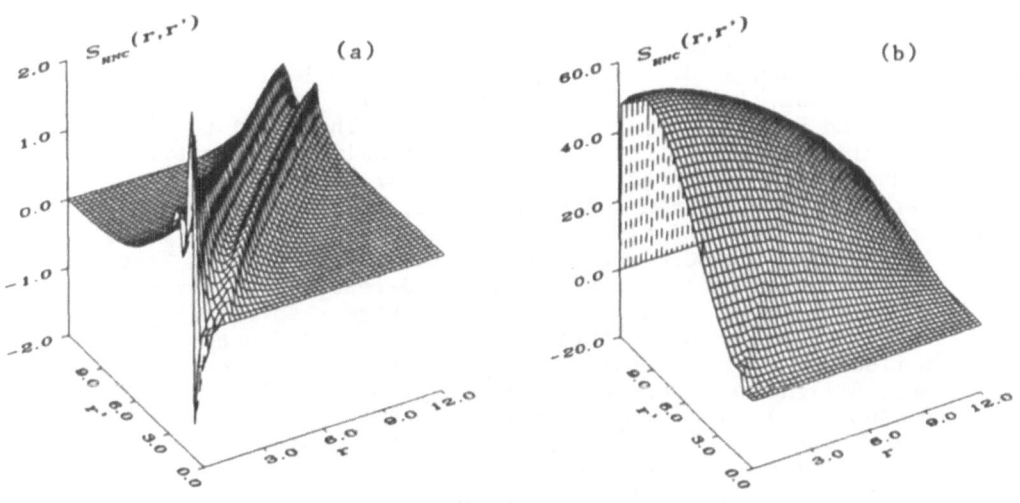

Fig. 9. The HNC sensitivity at T=1.30 and n=0.10 for the (a) stable and (b) unstable correlation functions (cf. fig. 5).

Our main purpose in introducing S(r,r') is to illustrate a novel analysis in the study of fluid microstructure. For example, given an approximation for the direct correlation function $c(\underline{r})$ in a nonuniform fluid, $S(\underline{r},\underline{r}')$ can be used to design a quasi-Newton or banded-matrix approximation to calculate $g(\underline{r})$;[16] this is important since the number of unknowns $g(\underline{r})$ is typically of order one million in a nonuniform fluid[15]. Or, in the particular case of an simple fluid near a solid surface, a clearer understanding of the influence of the solid on the fluid can be obtained by examining the sensitivity of the density profile $n(\underline{r})$ to the solid-fluid potential $U(\underline{r})$. In this context the supercomputer plays a critical role in expanding the range of questions we can afford to ask about fluids, and hence in the scope of information we can obtain about fluids.

ACKNOWLEDGEMENTS

This work was supported in part by grants from the U.S. Department of Energy and the Petroleum Research Foundation. We would like to acknowledge computing grants from Control Data Corporation and the University of Minnesota Computing Center. Assistance with the graphics by Gary Griffin and Judi Cleary is also acknowledged.

REFERENCES

1. L. S. Ornstein and F. Zernicke, Proc. Akad. Sci. (Amsterdam) 17:793 (1914).
2. J. M. J. van Leeuwen, J. Groeneveld and J. De Boer, Physica, 25:792 (1959); E. Meeron, J. Math. Phys., 1:192 (1960); M. S. Green, J. Chem. Phys., 33:1403 (1960); T. Morita and D. Hiroike, Prog. Theor. Phys., 23:1003 (1060); L. Verlet, Nuovo Cimento, 18:77 (1960); G. S. Rushbrooke, Physica, 26:259 (1960).
3. J. K. Percus and G. J. Yevick, Phys. Rev., 110:1 (1958).
4. J. G. Kirkwood, J. Chem. Phys., 3:300 (1935).
5. M. Born and H. S. Green, "A General Kinetic Theory of Liquids," Cambridge, London (1949); J. Yvon, "Actualities Scientifiques et Industriel," Herman et Cie, Paris (1935).
6. J. Kerins, L. E. Scriven and H. T. Davis, to be published. This paper contains a detailed discussion of results and of interpretation for the PY, HNC and BGYK equations for the LJ potential. Extensive references on published algorithms and results are given.
7. J. A. Barker and D. Henderson, Rev. Mod. Phys., 48:587 (1967).
8. R. O. Watts, pp. 1-70 in: "Statistical Mechanics," a specialist periodical report of the London Chemical Society, vol. 1, K. Singer, ed., Billings and Sons Ltd., London (1973)
9. J. P. Abbott, J. Comp. Appl. Match., 4:19 (1978).
10. G. Strang and G. J. Fix, "An Analysis of the Finite Element Method," Prentice-Hall, Englewood Cliffs (1973)
11. B. Swartz and B. Wendroff, Math. Comp., 23:37 (1969).
12. M. E. Fisher, Physics (N.Y.), 3:255 (1967).
13. J. S. Langer, Ann. Phys. (N.Y.), 41:108 (1967).
14. K. Binder, Ann. Phys. (N.Y.), 98:390 (1976).
15. J. Bellare, J. Kerins, L. E. Scriven and H. T. Davis (unpublished).
16. J. Bellare (private communication).

VECTORIZATION OF WEATHER AND CLIMATE MODELS FOR THE CYBER 205

P.W. White

Meteorological Office
Bracknell, United Kingdom

INTRODUCTION

The United Kingdom Meteorological Office installed a Cyber 205 computer in August 1981. All operational numerical weather forecasting was transferred from the existing IBM 360/195 to the Cyber about one year later. The Cyber was by that time also being used extensively for research work on climate studies, on meso-scale modelling and on developing new techniques to improve operational data analysis and weather forecasting. The Meteorological Office's Cyber 205 has one megaword of main memory, two vector pipelines, seven 819 disks accessible via LCN along three paths. The Cyber is connected to two IBM front-end computers (currently a 3081D and a 370/158) via LCN. All the main peripheral devices (disk and tape units, plotters, printers, card readers, microfiche and microfilm output devices etc) are attached to the front-end computers. Consequently all output from the Cyber 205 has to be passed across the links to the front-end computers before being processed and stored.

During the operational weather forecasting time periods, and during runs of large research tasks, the Cyber is run in 'stand alone mode', that is with only one user program in the machine (for shorter periods, mainly during weekday working hours, the machine is run in 'multi-programming mode' for development jobs). By making the entire memory available to a single program the wall clock time is minimised by reducing (or in most cases eliminating) the need for paging during program execution. This is an important requirement for numerical weather forecasting; clearly forecasts that are available early are of greatest value. The minimum amount of main memory used by the operating system is 2 large pages, leaving a maximum of 14 large pages for user programs during stand alone sessions. This limit on the size of main memory available has been one of the principal constraints on the design of efficient software for many application programs, although the small number of data paths and slow transfer rates (relative to the speed that the CPU can complete floating point computations) to the Cyber disks have also imposed restrictions. In order to make more effective use of the main memory, half precision has been used for most of the programs and this has the added bonus that faster vector computation is possible for many instructions (tests have indicated that the use of half precision instead of full precision has a negligible effect on the results of weather prediction and climate models). However, because the half precision Fortran compiler was not available when program development began Q8 calls have been used for all vector arithmetic; Q8 calls have also been used for much of the scalar arithmetic to improve efficiency and Q7BUFIN and Q7BUFOUT have been used to provide asynchronous I/O. Early in 1985 the Cyber is due to be upgraded by adding an additional Mword of main memory (in fact replacing the present 1 Mword of bi-polar memory with 2 Mwords of MOS memory). In the enhanced machine it is planned to improve the effectiveness of the operating system by allocating to it a larger region of main memory, but users will be permitted to use up to 24 large pages. It is anticipated that this will allow efficient multiprogramming for a larger portion of the day, though there are several research projects and some operational tasks that will require the full extended memory. These will still be run on a stand alone basis.

Several years prior to placing the order for the Cyber 205 it was realised that the coming generation of super-computers would require programming techniques significantly different from those that had been developed for serial computers, such as the IBM 360/195. It was also evident that computer design was proceeding along several different paths but it was not clear which aspects of the new computer architectures had advantages or disadvantages for computations of the type involved in numerical models of the atmosphere. To help resolve these questions the Meteorological Office decided to run a number of tests to examine the practical problems involved in using vector computers. A complete numerical weather forecasting model was programmed and tested on the CDC Star-100 and the Cray 1A using both high and low level languages (paper studies were also done on a number of other computers under development at that time). The experience gained during this exercise, and during the 18 months of program development on a Cyber 203 at CDC Arden Hills between the order and delivery of the Cyber 205, was particularly valuable in enabling efficient vectorization techniques to be devised. This paper discusses some aspects of programs, written both for operational and research use, that required special techniques to achieve effective vectorization. Also discussed is the impact on program design of the size of main memory and of the I/O transfer rates.

ANALYSIS OF METEOROLOGICAL OBSERVATIONS

Atmospheric observations are made throughout the world from instruments on the earth's surface, on ships, on buoys, on balloons, on aircraft and on satellites. The observations are exchanged on an international basis among the member states of the World Meteorological Organisation. A global telecommunications system has been established to enable the observations to be transmitted to users within as short a time as is practicable: in general about 85% of observations are received at the Meteorological Office within three hours of being made. About 3/4 million observations are collected in the Meteorological Office data bank each day. The spacing of the observations throughout the globe is uneven, the measurements being fairly numerous over land areas such as Europe and North America, but appreciably more sparse over oceanic areas such as the Pacific.

The numerical weather prediction model needs to be provided with three dimensional fields (covering the Earth through a depth of the atmosphere from ground level to a height of about 25 km) of wind (northward and eastward components), temperature, surface pressure and humidity at specified times each day to act as starting points for predicting the evolution of weather systems for times in the future.

The Meteorological Office's model holds these quantities on a regular latitude/longitude grid at 15 levels in the atmosphere, the size of the three dimensional array comprising a total of about 350,000 grid points. The problem of analysis of observations is one of interpolation from the inhomogeneous observation array to the regular analysis array. The task is complicated by a number of additional considerations: (i) observations from different types of instrument have differing inherent accuracies and should consequently be given different weights in the interpolation, (ii) observations received at the data bank may be incorrect due to instrumental malfunction, or human or transmission errors; these need to be automatically detected and rejected from the system, (iii) in data sparse areas, observations may be too far away to provide sufficient information content to be useful in the interpolation; in general the information provided by an observation (and therefore its weight in the interpolation) falls off with distance, and (iv) for reasons related to the physics of the atmosphere the quantities interpolated onto the analysis grid are not independent of one another; there are approximate relationships between the wind, the surface pressure gradient, the temperature gradient, the vertical velocity and the humidity. Meteorological analysis cannot therefore be treated as a purely mathematical interpolation problem but must satisfy conditions of physical consistency.

Details of the analysis scheme have been described by Lyne et al (1983), but an outline will be given here to highlight those aspects involving vector programming techniques. To simplify the handling of large quantities of observational data, they are first sorted into latitude bands and only those observations made within 3 hours of the analysis time are used. Each line of latitude of the regular analysis grid is considered in turn, dealing with all the grid points (192) on that line and all the levels (15) through the depth of the atmosphere as a single entity. A first guess for the analysis is obtained from a numerical forecast predicted from a previous data time; the analysis procedure then consists of calculating corrections to this first guess as indicated by the observations. In regions of the atmosphere where observations are sparse only minor changes may be made to the first guess but in data rich areas the corrections ensure that the analysis fields fit the observations to within their expected observational errors. There are two separate stages: the calculation of the interpolation weights and the assimilation of the corrections.

If F represents one of the meteorological variables being analysed (wind, pressure, temperature or humidity), then the correction δF to the first guess is calculated from a weighted average of the corrections suggested by the observations surrounding the analysis grid point.

$$\delta F_a = \sum_{d=1}^{N} w_{a,d} (O_d - G_d) \qquad (1)$$

where G is the first guess for F, O is the observed value of F and subscripts a and d indicate positions on the analysis and observational data grids respectively. N is the total number of observations used in calculating the corrections. The parameter $w_{a,d}$ represents the weight that the observation at position d has on the calculation of the correction at position a; it is calculated from the set of simultaneous equations.

$$A\underline{W} = \underline{S} \qquad (2)$$

where $\underline{W}$ is the vector $(w_{a,1}, w_{a,2}, \ldots\ldots\ldots, w_{a,N})$ and the matrix A and the vector $\underline{S}$ depend on the error co-variances of the observations and the first guess. To avoid having to invert large order matrices at each analysis grid point, the number of observation N is limited to a maximum of 7. The criteria for selecting the most useful subset of observations is as follows: observations more than 600 km away from the gridpoint are ignored as being too far away to influence the interpolation, the 20 nearest observations are examined and 7 selected on the basis that, considered by themselves, they would reduce the expected error of the analysis by the greatest amount (this depends essentially on the diagonal elements of A and the error co-variances of the first guess).

Because of limitations on the size of memory available for working space, grid points are considered 8 at a time. The distances between the grid points and all the observations in the region of influence are calculated; this allows vector lengths of 8 x M to be used, where M is the number of observations. The 20 nearest and then the subset of 7 most influential observations are selected in a calculation sequence involving scalar outer loops and inner vectorization. A 'Gather' index array is set up to link the selected observations to the grid points as indicated in Fig 1 (since an observation may contribute to the corrections at several grid points the 'Gather' function must be used rather than 'Compress' to extract the relevant observations).

In calculating the weights from equation (2) it must be noted that, for many grid points, fewer than 7 observations may be selected because observations lie too far away. As a consequence the matrices A may be of order 0 x 0 or 1 x 1 or .... or 7 x 7. The method of vectorization is to identify those points with the same number of observations selected and to perform their matrix inversions simultaneously (vector lengths are then equal to the number of grid points with a specified number of selected observations, see Fig 2).

Assimilation of the observationally inferred corrections is done in conjunction with the forecast model so that the necessary physical consistency between the variables is maintained. If F now represents one of the variables (wind, pressure, temperature or humidity) being forecast, then the rate of change with time of F_a (the value of F at

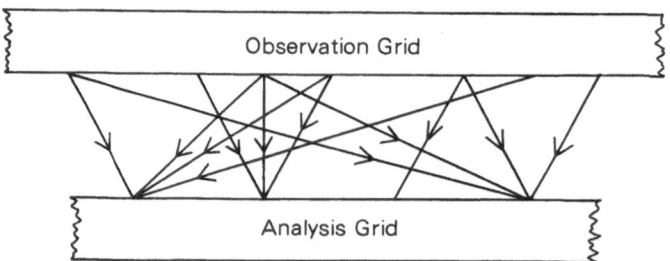

Fig. 1. Schematic diagram illustrating the Gather index linkages set up between the analysis grid and the observation grid.

grid point a) in the forecast model is governed by an equation of the form

$$\frac{\partial F_a}{\partial t} = \psi a \qquad (3)$$

whose ψ_a represents the full gamut of atmospheric forcing terms affecting F_a. Assimilation of the corrections is then achieved by replacing (3) by

$$\frac{\partial F_a}{\partial t} = \psi a + \lambda \sum_{d=1}^{N} w_{a,d} (O_d - F_d) \qquad (4)$$

The parameter λ is chosen to be small enough to ensure that the corrections are made gradually over a period of 6 hours during the integration of equation (4). The observations and their weights

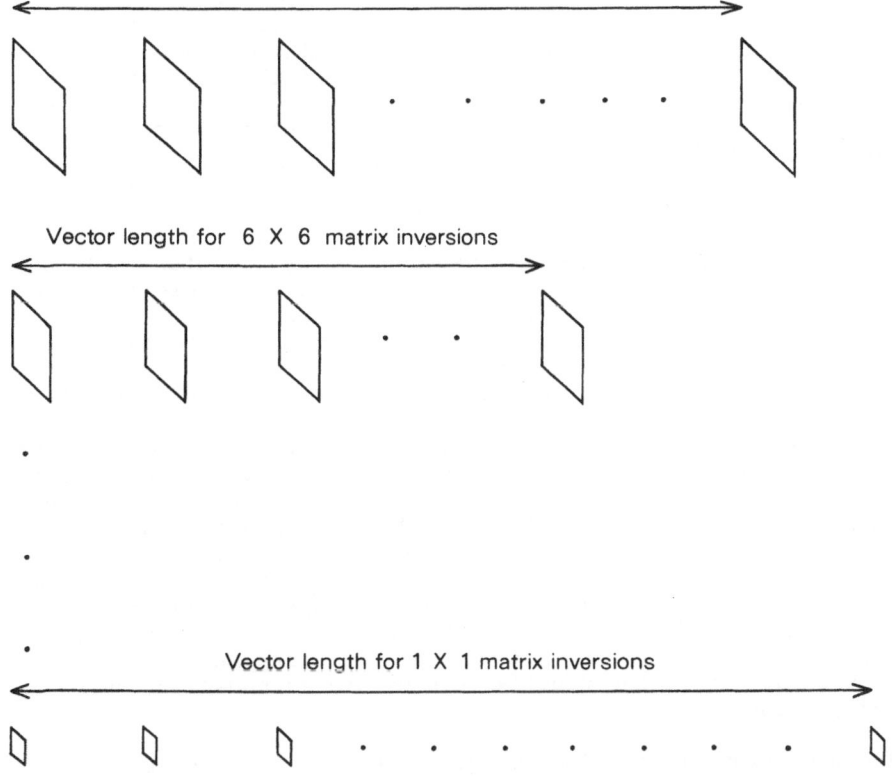

Fig. 2. Schematic diagram illustrating the rearrangement of matrices of different orders to facilitate vectorization.

(relating to each grid point) are gathered using the index array already
set up, the values of F are interpolated to the observation points d
(using bi-linear interpolation) and the sums evaluated. Vectorization
of this part of the calculation is relatively straightforward.

Although only partial vectorization has been achieved the code
showed an increase in speed by a factor of about 10 over an earlier
purely scalar code. Substantial data array sizes are involved and the
vectorization techniques involve extensive selection and reordering
procedures. Even though the data on disk are packed into low precision,
the size of data sets input to the weights calculation exceed 1 Mword,
while data sets totalling over 3 Mwords are passed from this stage to
the assimilation stage. These exceed the size of the memory by a
considerable margin and care has to be taken to arrange the storage on
disk to optimise the data transfer along the three paths available.

WEATHER FORECASTING AND CLIMATE MODELS

Although the weather forecasting and climate models have much in
common, differences arise because of the contrasting ways in which the
models are used. The forecasting model is designed for predicting the
weather for a few days ahead. The model is run operationally each day
throughout the year and consequently efficiency and reliability of the
software is a prime requirement. Modifications and improvements have to
be rigorously tested before introduction and changes are made to the
model fairly infrequently. One of the most important aspects governing
the design of the model is the need to reduce the wall clock time to a
minimum, so that the forecast is in the hands of the user with the least
possible delay. The climate model, however, is run for much longer
periods (years of simulated time rather than days) to undertake research
into possible causes of climate change. Unlike the forecasting model,
it does not have to be run to a strict schedule but, because such long
runs are involved, facilities must be provided for processing and
storing large quantities of intermediate results and for restarting the
program following cancellation by the operator or computer failure (the
forecast model runs in a sufficiently short time to enable it to be
simply resubmitted in such circumstances). Because the climate model is
a research facility, modifications are frequently made to the program to
try out new ideas. More attention has to be paid to the ease with which
this can be done rather than to absolute computational efficiency.

The basic principle underlying both the weather forecasting and
climate models is the solution by numerical techniques of equations
governing the conservation of momentum, heat, mass and water vapour for
the atmosphere in a frame of reference rotating with the Earth.
Non-conservative aspects of atmospheric motion (usually arising as a
consequence of motion on scales too small to be explicitly resolved by
the model) must also be included in the model specification: examples
are friction at the Earth's surface, turbulent diffusion, convection,
radiation, and the release of latent heat following the formation of
cloud and rain. These are usually referred to as sub-grid scale
physical processes.

The evolution of the wind, temperature, surface pressure and
humidity is predicted on a three dimensional mesh of grid points within
the atmosphere from ground level to a height of about 25km. From the
programming point of view the terms in the basic conservation equations
are relatively easy to vectorize because they are calculated in the same
way at each of the grid points of the mesh. An important decision that
has to be made before designing the program is the best way to arrange
the data within memory. The global atmosphere is first transformed onto

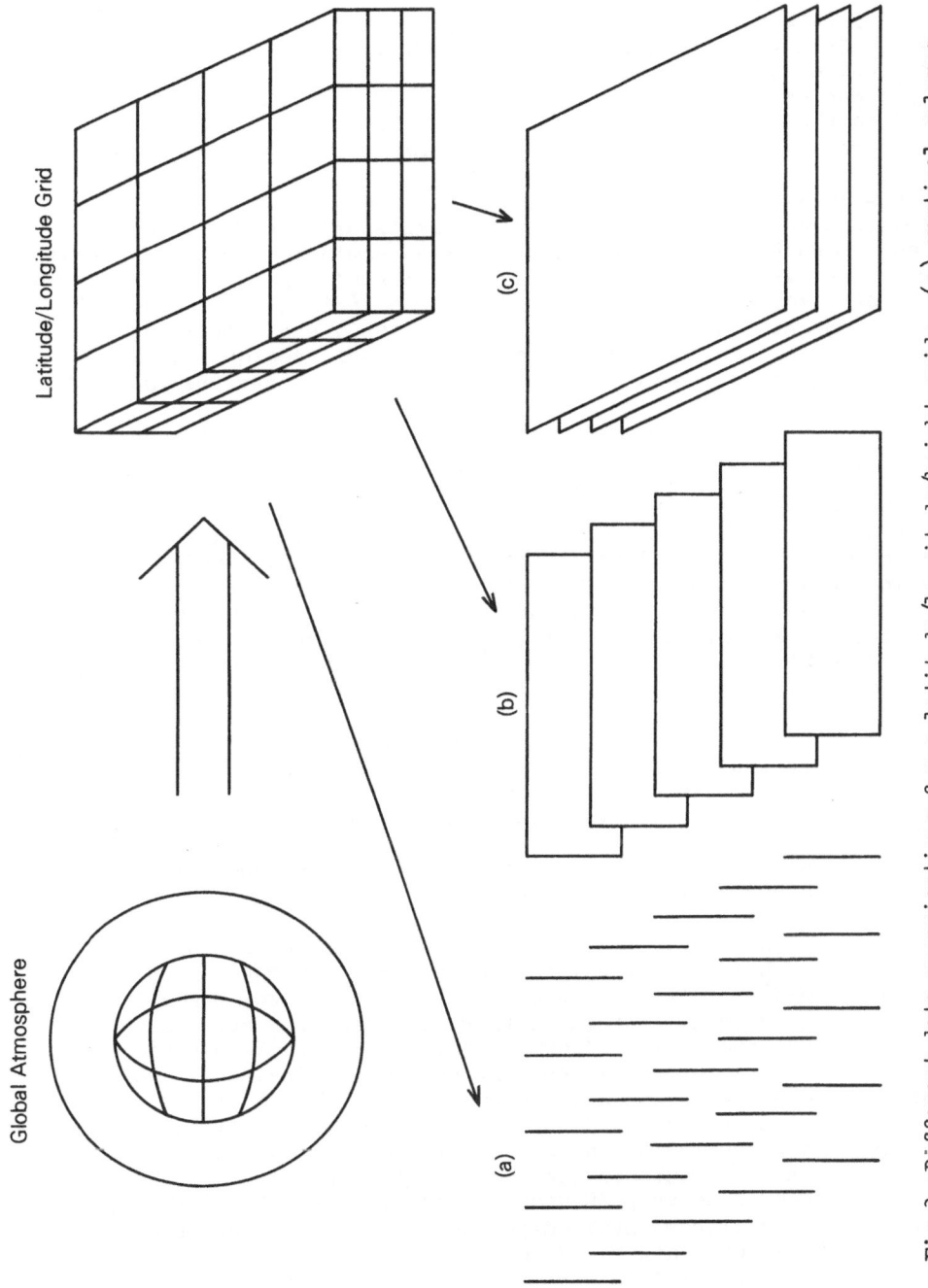

Fig 3. Different data organisations for a latitude/longitude/height grid: (a) vertical columns, (b) vertical slices, (c) horizontal slices.

a latitude/longitude/height grid as indicated in Fig 3. The total number of grid points is L x M x N where L is the number of levels, M the number of points around each line of latitude and N the number of points between the north and south poles (for the weather forecasting model L = 15, M = 192, and N = 120 while the dimensions that are usually used for the climate model are L = 11, M = 96, N = 72). On the IBM 360/195 (a serial computer) data was stored in vertical columns (Fig 3a). This was considered the most appropriate choice because most sub-grid scale physical processes involve changes in the vertical distributions of predicted quantities. In addition the amount of data involved in the calculations of a single column was sufficiently small to be entirely contained within the cache memory of the IBM 360/195.

For a vector computer, organisation of the data in columns is not appropriate because vectors of only L elements are too short for efficient computation (vector start-up times dominate the calculations). The two other ways illustrated in Fig 3 of organising the data allow the use of longer vector lengths. Vertical slices (Fig 3(b)) are used in the climate model (vector lengths 11 x 96 = 1056) while horizontal slices (Fig 3(c)) are used in the forecast model (vector lengths 192 x 120 = 23,040). Vertical slices are convenient because they allow the model to be programmed in a flexible way so that slices not being operated on can be paged out of the memory. The use of horizontal slices provides the longest vector lengths and is consequently the most efficient, although for some aspects of the calculation of sub-grid scale processes in the forecast model the data is reorganised into blocks (several vertical slices collected together). A difficulty that arises in the horizontal slice format is that there is insufficient memory in the 1 Mword memory of the Cyber 205 for the working space required. As an interim measure, until the memory is increased to 2 Mwords, the slices have been split into four segments giving vector lengths of nearly 6000. Tests have indicated that the ability to use full horizontal slices should lead to a 10% increase in speed.

Vectorization of the calculations for sub-grid scale processes is more difficult because they involve non-contiguous sub-sets of the full three dimensional data array. The spatial distribution of these sub-sets may be fixed or may vary in time, depending on the process being considered. For example, the physical nature of the surface heat balance over the sea is quite different to that over land: it is convenient to store a bit vector (zero over the sea, one over the land) to use as a mask allowing sea surface calculations to be performed separately from those over the land surfaces. A different problem arises in the calculation of rainfall production. Condensation of water vapour is assumed to arise when the air rises to a level where it becomes super-saturated. A test is therefore made for super-saturation at each time step and a bit vector constructed to single out those points where rainfall calculations must be made. Dickinson (1982) has examined the relative advantages for these calculations of Gather/Scatter, Compress/Expand and Control Store instructions. He showed that the selection of the most efficient method depended on the fraction of "active" points in the array and on the ratio of the number of Gather/Scatter or Compress/Expand instructions and the number of floating point operations involved in the subsequent calculation. He concluded that Gather/Scatter was the fastest technique if less than about 20% of the array points were active, but otherwise the choice lay between Compress/Expand and Control Store. In the coding for the forecast and climate models, Gather/Scatter has been avoided (even when Dickinson's criterion indicates that it would be the most efficient) because limitations in memory size do not allow sufficient space for the required full precision index vectors to be set up.

Table 1. Wall clock time for various tasks in a run of the climate model for one simulated year.

Task	Wall clock time
Climate model	$16\frac{1}{2}$ hours (CPU = $15\frac{1}{2}$ hours)
Subtasks	$3\frac{1}{2}$ hours (CPU = $\frac{1}{2}$ hour)
Restarting	$\frac{3}{4}$ hour
Data transfer to/from front end	3 hours

The forecast model takes 3 minutes 40 seconds to complete a one day forecast for the global atmosphere. Because it is contained entirely within main memory the wall clock time is almost the same as the CPU time. However, since daily results have to be stored for subsequent processing and each run of the model is used for forecasting for 6 days ahead, the net efficiency (as measured by the ratio of CPU time to wall clock time) of the model in operational use is less than 100% because of the time taken to write to disk. Individual routines of the model are some 30 to 50 times faster than could be achieved by equivalent Assembler Language routines on the IBM 360/195 (indicating that vectorization techniques have been particularly successful). The climate model is also contained within main memory and a similar degree of vectorization has been achieved. However, since climate model runs extend over several simulated years, it is clearly impracticable to store all the model results on a daily basis. Only a limited sub-set is stored for each simulated day, but complete dumps of the data area are retained (primarily for restarting purposes) every 10 days of the simulation. In addition 30 day and 90 day means of all model variables are calculated and stored. Programs to do this ancillary calculation and the necessary data transfers are initiated as sub-tasks of the main program. There is insufficient memory to contain both the model and the sub-tasks and consequently initiation of sub-tasks leads to paging. There is also insufficient main memory to allow data transfer to and from the front end computers to take place concurrently with program execution on the Cyber 205. Table 1 indicates times taken by various tasks during a run of one simulated year of the climate model. The net efficiency of 67% will improve substantially when the memory of the Cyber 205 is increased for 1 Mword to 2 Mwords. Plans are also under consideration to extend the climate model to include multilevel interactive ocean model. The combined atmosphere-ocean model is expected to use the full user extent (24 large pages) of the enhanced main memory.

REFERENCES

Dickinson, A., 1982, Optimizing numerical weather forecasting models for the Cray-1 and Cyber 205 computers. Comp Phys Communic., 26: 459-468.

Lyne, W.H., Little, C.T., Dumelow, R.K., and Bell, R.S., 1982, The operational data assimilation scheme. Met O 11 Tech Note no 168, unpublished Meteorological Office report.

NUMERICAL SIMULATION OF ATMOSPHERIC

MESOSCALE CONVECTION

Wen-Yih Sun and Wu-Ron Hsu

Department of Geosciences
Purdue University
West Lafayette, IN 47907

INTRODUCTION

The amount of sensible heat and moisture supplied to the atmosphere from the sea surface is tremendous over the East China Sea and the areas off the east and the south coasts of the United States during the winter. Many interesting and important meteorological disturbances on a variety of temporal and spatial scales develop or intensify in the regions. The Kuroshio flows northeastward through the East China Sea, resulting in a steep horizontal sea-surface temperature gradient (see Fig. 1). In winter, whenever a cold polar continental air mass moves out from the Asian Continent, it is modified rapidly as the cold dry air encounters the warmer sea surface. A similar situation exists off the east coast of the United States, where the Gulf Stream is a warm, north bound ocean current just like the Kuroshio.

The data of cold air outbreak during the Air Mass Transformation Experiment (AMTEX) have been analyzed by Satio (1975), Ninomiya (1975, 1976), Nitta (1976), Ninomiya and Akiyama (1976), Agee and Lomax (1978), and many others. Ninomiya and Akiyama (1976) presented the thermal structure and wind profile as in Fig. 2. They found that a relatively shallow (1-2 km in depth) layer of modified cold air mass is capped by a strong inversion layer which distinctly separates the undisturbed steady westerly layer above from the transformed northerly layer below. A prevailing downward motion exists near the cloud base or the inversion layer. The vertical mixing of momentum in the mixed layer is mainly produced by convection associated with the warm water surface. The inversion base is lower to the northwest and is generally higher over the southwestern part of the AMTEX area.

In order to study the marine-type planetary boundary layer in a strong baroclinic zone, as well as the air mass transformation during a cold air outbreak, a three dimensional mesoscale model has been developed at Purdue University. The preliminary results show that the observational wind, temperature profiles, the slope of the inversion and the prevailing subsidence are well simulated by this model due to a fine resolution in both vertical and horizontal directions, a reasonable turbulence parameterizations as well as condensation and evaporation are

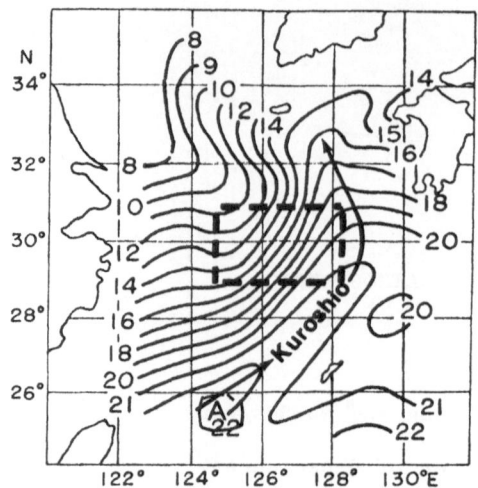

Fig. 1. The mean sea surface water temperature (°C) over the East China Sea in February from 1953 to 1957. The dashed rectangular region is the domain of the model.

calculated precisely here. It was also noted that the influence of the lateral boundary became a serious problem after 13 hours of integrations. For the time being, we are working on improving the boundary conditions in order to have better results in the mature stage of convection.

FUNDAMENTAL EQUATIONS

The normalized pressure coordinate σ is used in the vertical direction. σ is defined as

$$\sigma = (P-P_T)/[P_s(x,y,t) - P_T] = (P-P_T)/P_*(x,y,t) \tag{1}$$

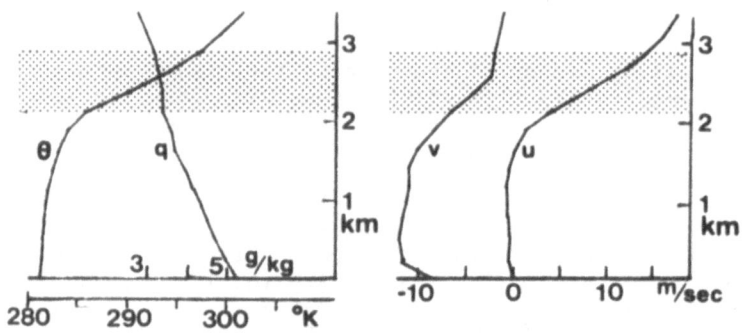

Fig. 2. Vertical distribution of Θ, q, u, and v over area A in Fig. 1 averaged for the period February 25-27, 1974. The stipled layer is the inversion layer.

where P is the pressure, P_T = 700 mb is pressure at top, P_s is pressure at bottom. The momentum equations become:

$$\frac{\partial}{\partial t}(P_*U) = -D(U) + fP_*V - \frac{\partial}{\partial x}(\emptyset P_*) + \frac{\partial(\emptyset\sigma)}{\partial\sigma}\frac{\partial P_*}{\partial x} + P_*F(U) \qquad (2)$$

$$\frac{\partial}{\partial t}(P_*V) = -D(V) - fP_*U - \frac{\partial}{\partial x}(\emptyset P_*) + \frac{\partial(\emptyset\sigma)}{\partial\sigma}\frac{\partial P_*}{\partial y} + P_*F(V) \qquad (3)$$

where the operator D is defined as

$$D( ) = \frac{\partial P_*U( )}{\partial x} + \frac{\partial P_*v( )}{\partial y} + \frac{\partial P_*( )}{\partial\sigma} \quad . \qquad (4)$$

F will be defined later; the other symbols are conventional.

The hydrostatic equation is

$$\frac{\partial\emptyset}{\partial\ln P} = -RT(1+0.61\ q-q_\ell) \qquad (5)$$

where q is the specific humidity, q_ℓ is liquid water content and T is absolute temperature. The governing equation for the total specific humidity, $q_w = q+q_\ell$ for a shallow convection without precipitation is:

$$\frac{\partial(P_*q_w)}{\partial t} = -D(q_w) + P_*F(q_w) \qquad (6)$$

The thermodynamic equation for equivalent potential temperature, $\theta_e = \theta + L/Cp\ \theta/T\ q$, becomes

$$\frac{\partial(P_*\theta_e)}{\partial t} = -D(\theta_e) + P_*F(\theta_e) \qquad (7)$$

The conservation of mass is given by

$$\frac{\partial P_*}{\partial t} = -\int_0^1 \nabla_\sigma \cdot (P_*V)\ d\sigma \qquad (8)$$

and the equation for the "vertical" velocity is

$$\dot\sigma = -\frac{1}{P_*}\int_0^\sigma \nabla_\sigma \cdot (P_*V)d\sigma + \frac{\sigma}{P_*}\int_0^1 \nabla_\sigma \cdot (P_*V)d\sigma \qquad (9)$$

The mixing length theory is applied to calculate the diffusion term F as following

$$F(\psi) = \frac{\partial}{\partial z}(K_h \frac{\partial\psi}{\partial z}) \qquad \text{for } \psi = \theta_e \text{ or } q_w$$

$$F(\psi) = \frac{\partial}{\partial z}(K_m \frac{\partial\psi}{\partial z}) \qquad \text{for } \psi = U \text{ or } V \qquad (10)$$

Following Deardorff's hypothesis (1980), the eddy coefficient is given by

$$K_M \sim L\sqrt{E} \qquad (11)$$

where E is the turbulence kinetic energy, which is represented by a prognostic equation

$$\frac{dE}{dt} = \frac{g}{\Theta_o} \, \overline{w'\Theta_v'} - \overline{u_i'w'} \, \frac{\partial U_i}{\partial z} - \frac{\partial}{\partial z} \, \overline{[w'(E'+P'/\rho_o)]} - \varepsilon \qquad (12)$$

The length scale L is defined similarly to Sun and Ogura's (1980). The detail formulation of the turbulence kinetic energy and eddy fluxes will be presented in other papers.

The similarity equations proposed by Businger et al. (1971) are applied at the surface layer to calculate the eddy fluxes of heat, moisture and momentum from the water surface.

NUMERICAL METHODS

Procedure

The split scheme (Gadd, 1978) was employed in this model. The forcing terms associated with gravity waves in the prognostic equations 2, 3, 6 and 7 are integrated with a very small time step (32 seconds) to allow the propagation of the fast moving gravity waves which can be easily excited during the simulation. The other terms are solved only once in every three small time steps (referred to as "big time step") to save computing time. The procedure for the integration will be briefly described as follows.

Small time step: The forward–backward scheme (Gadd, 1978; Sun, 1984) was used. It has saved the computing time for the fast propagating gravity waves by 50%. The vertical advection terms for temperature and moisture along with the conservation of mass (Eq. 8) were calculated before solving the hydrostatic equation (Eq. 5). The vertical advection terms, pressure gradient terms and Coriolis force terms in the momentum equations were then integrated using the new pressure and geopotential height field. Finally, vertical velocity $\dot{\sigma}$ can be obtained by solving Eq. 9.

Big time step: Horizontal advection terms are solved by the Crowley (1968) second order scheme in flux form. The diffusion term was calculated by the trapezoidal rule in time and central difference in space, such as:

$$\frac{\psi^{n+1} - \psi^n}{\Delta t} = 1/2(F^n[\psi] + F^{n+1}[\psi])$$

where ψ^n is any one of the prognostic quantities at the nth time step. Businger's (1971) similarity theory was applied to obtain the heat flux, moisture flux, and momentum fluxes at the surface layer. These fluxes appeared as the lower boundary condition for solving the implicity diffusion scheme. Note that the effect of condensation and evaporation does not appear explicitly in the fundamental equations. This is because both the equivalent potential temperature and the total specific humidity are conserved for these processes if no precipitation is allowed as we assumed in this model, but they are still required in order to calculate the specific humidity and the liquid water content in the hydrostatic equation (Eq. 5).

Grid System

The domain of interest is shown by the rectangular box in Fig. 1. The lower left corner is defined as the origin. The longer side of the rectangular region is defined as the x-axis or I-axis. The other side is defined as the y-axis or J-axis. It is a 380 km by 225 km domain

148

Fig. 3. Arakawa's type-C-
lattice.

Fig. 4. Vertically staggered
σ-coordinate.

with top at 700 mb. The vertical grid distance is about 160 m with
higher resolution (80 m) for the lowest layer, but the temperature,
humidity and wind are calculated at the height of ~40 m above the
surface due to a staggered lattice being applied. The grid distances
are 10 km and 7.5 km for x and y direction, respectively. The model
consists of 39 x 30 x 20 grids. We intend to reduce the horizontal grid
distances once the problem on the lateral boundary condition is improved.
Arakawa's C grid (1972) was adopted. It is shown in Figs. 3 and 4.

Computer Program

The program was written almost completely in "explicit vector
assignment statements" to ensure the efficiency of the calculation.
There are many boundaries in this model. Besides the lateral, lower and
upper boundaries, we have to use different formula to calculate the
physical processes along the cloud boundaries and boundaries created by
the inversion. Many "bit vectors" were devised and incorporated into the
program to enable us to vectorize these complicated calculations. "Bit
vector" and many "Q8V" functions have proven to be very useful features
of the Cyber 205 machine for our model.

An earlier version of the program written in the traditional
FORTRAN code was run on a CDC-6600 machine. It was found that the CPU
time was 31 times higher than the CPU time required by Purdue's 205
machine. However, this comparison was based on the calculation for a
smaller domain with fewer grid points than we have right now. The
efficiency of the Cyber 205 computer will probably be even higher if we
used it on a large domain. Debugging program on the 205 machine,
however, has been a frustrating experience for us.

RESULTS

Temperature, Moisture, and Heat Flux

Figure 5 shows the vertical profiles of virtual potential temperature (Θ_v) at every 270 time steps which is equivalent to about 2 hours
and 24 minutes in model time for a point near the center of the domain.
Dashed line represents the initial condition of the environment which is
very cold with a uniform temperature distribution in the vertical direction (-6°C for this particular point). As the air mass was heated by
the warm (15°C) sea water underneath, a marine type planetary boundary
layer developed. At around the 9th hour of the model time, it grew to

149

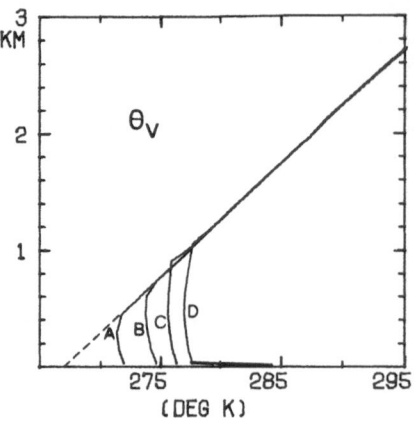

Fig. 5. Vertical profiles of Θ_V for a point near the center of domain. Lines A,B,C, and D correspond to profiles at model time 2 hr. 24 min., 4 hr. 49 min., 7 hr. 13 min., and 9 hr. 38 min., respectively. Dashed line represents the initial profile.

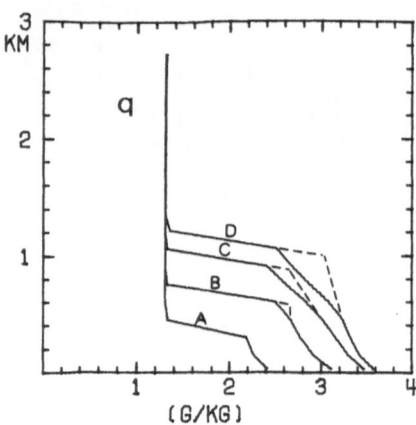

Fig. 6. As Fig. 5, but for vertical profiles of mixing ratio q (solid lines) and total water content q_w (dashed lines).

about 1000 m in depth (represented by the right most line in Fig. 5). This boundary layer consists of (1) a shallow surface layer (lowest 40 m in the model) with very unstable condition as temperature decreases rapidly with height; (2) lower mixed layer in which virtual potential temperature is neutral to slightly unstable; and (3) upper mixed layer characterized by the presence of cloud and slightly stable virtual potential temperature profile. For the air above the "capping inversion", the potential temperature remains undisturbed. This is consistent with observation shown in Fig. 2.

Water vapor was pumped up from the water surface as the boundary layer grew. It was evenly distributed within the mixed layer (Fig. 6), but was unable to penetrate through the capping inversion. Since temperature decreases with height within the mixed layer, the saturated mixing ratio is lower at the top than that at the bottom of the mixed layer. Air became saturated, and cloud developed near the top of the mixed layer starting from the 4th hour (Fig. 7). The dashed lines in Fig. 6 show the profiles of the total water content (vapor plus liquid) which was very uniform throughout the mixed layer as observed.

Fig. 8 depicts the profiles of heat flux $\overline{w'\Theta_v'}$. We can see that the source of heat came from the water surface. Due to the release of latent heat of water vapor, a second maximum of heat flux occurred within the cloud layer in later stages of the integration. These results agree very well with the results of a previous model by Deardorff (1980) (Fig. 9).

The vertical cross-section of virtual potential temperature along J=14 is presented in Fig. 10. Again, we see that the temperature lapse rate changed with height within the mixed layer due to the presence of

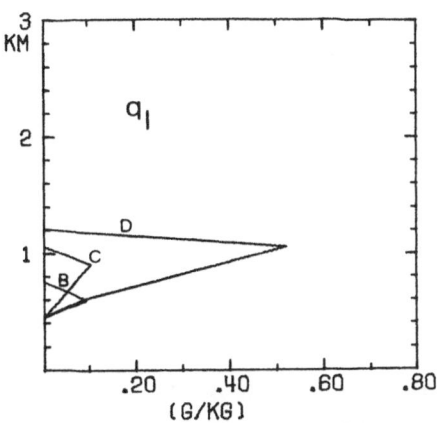

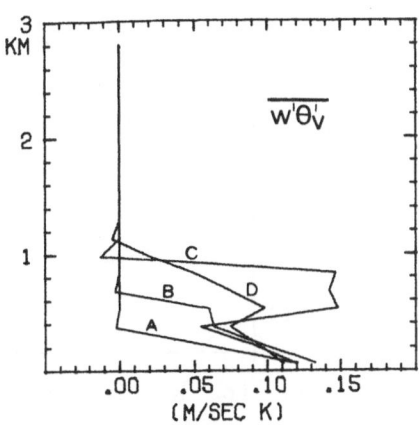

Fig. 7. As Fig. 5, but for vertical profiles of liquid water content q_ℓ.

Fig. 8. As Fig. 5, but for vertical profiles for heat flux $\overline{w'\theta_v'}$.

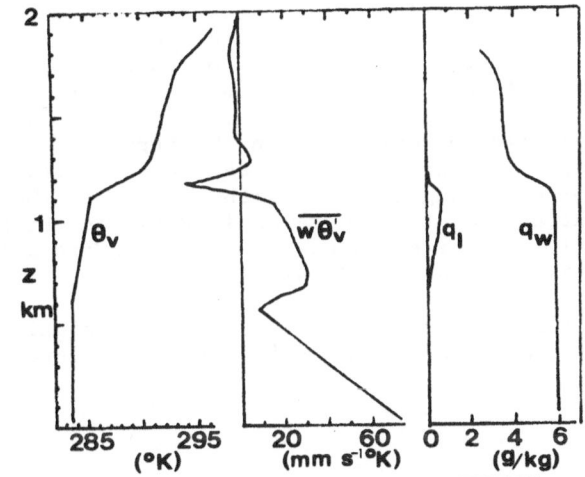

Fig. 9. Vertical profiles of θ_v, $\overline{w'\theta'}$, q_ℓ, and q_w obtained by Deardorff's (1980) model (case 6).

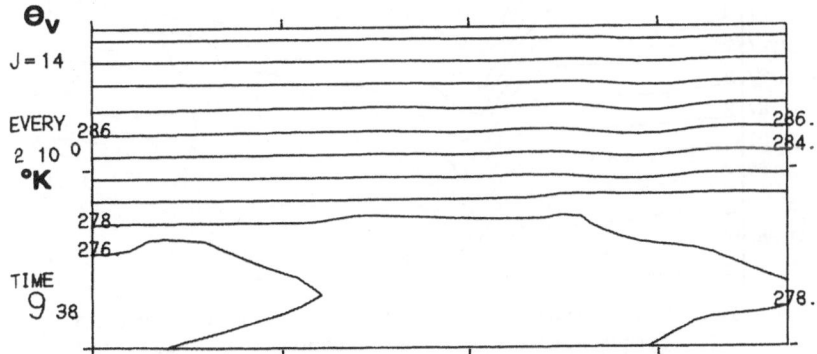

Fig. 10. Vertical cross-section of virtual potential temperature θ_v in x-z plane at the 14th grid point in y-direction, at model time 9 hr. 38 min.

151

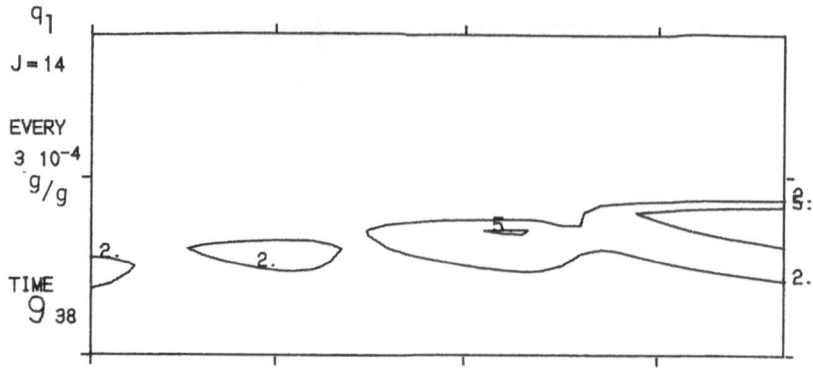

Fig. 11. As Fig. 10, but for liquid water content q_ℓ.

cloud. The depth of the mixed layer was the highest at the right hand side of the domain, since we had imposed a higher sea surface temperature there. This is illustrated even better by examining Fig. 11, the vertical cross-section of liquid water content. Even though the sea surface temperature increased linearly toward the right, the increase of the depth of the mixed layer toward the right is not perfectly uniform as shown by the jumps of cloud top in Fig. 11. This is owing to the limitation of our model in the aspect of vertical resolution. The capping inversion was lifted at every 160 m (ΔZ) intervals instead of ascending smoothly with time as in the nature. Liquid water content appears to have several breaks in the horizontal direction of Fig. 11. They are associated with downward motion in the region. This will be explained in detail later.

Fig. 12 shows the horizontal cross-section of virtual potential temperature at the level of K=12. At this particular level and particular

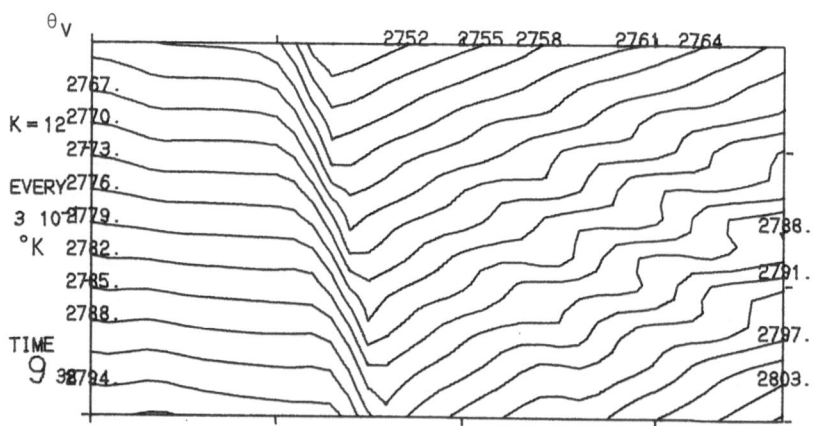

Fig. 12. Horizontal cross-section of virtual potential temperature θ_v in x-y plane at the 12th grid point in σ-direction at model time 9 hr. 38 min. The height of this plane is about 1000 m above surface.

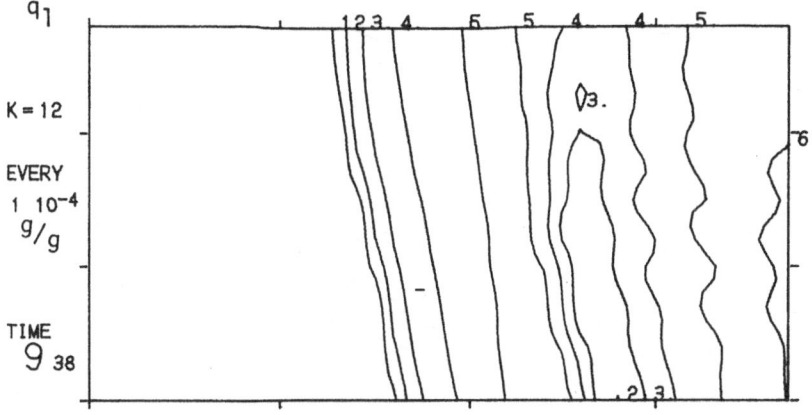

Fig. 13. As Fig. 12, but for liquid water content q_ℓ.

time, the grids at the left one-third of the domain was still above the inversion while the grids at the right was already in the mixed layer. Virtual potential temperature at the left maintained its initial values (uniform in x-direction, linearly decreasing in y-direction) while the virtual potential temperature at the right has been modified. The maximum θ_v occurred at the lower right corner where the convection was the strongest. As for the liquid water content at the same level, we had cloud-free region at the left versus cloudy region at the right (Fig. 13).

Wind Speeds and Pressure Fields

Vertical profiles of horizontal wind speeds are shown in Fig. 14 and Fig. 15. Dashed lines represent initial wind profiles. The initial

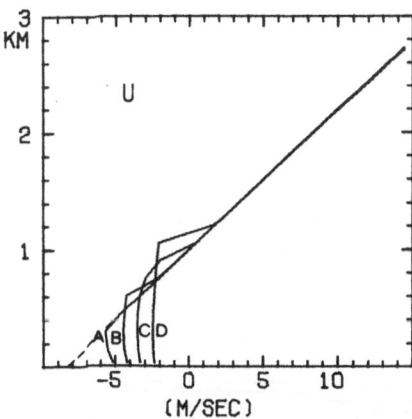

Fig. 14. As Fig. 5, but for vertical profiles of U-component of wind speed. Dashed line represents the initial profile.

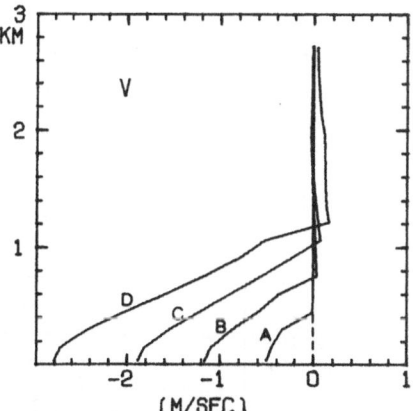

Fig. 15. As Fig. 6, but for vertical profiles of V-component of wind speed. Dashed line represents the initial profile.

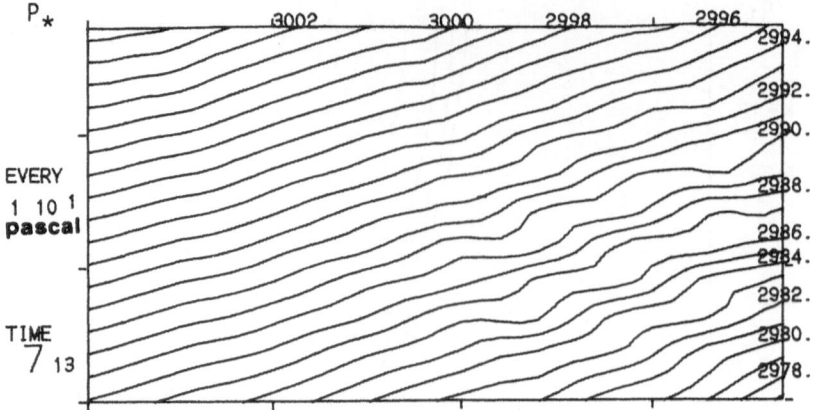

Fig. 16. Horizontal distribution of $P* = P_s - P_T$ around the 7th hour.

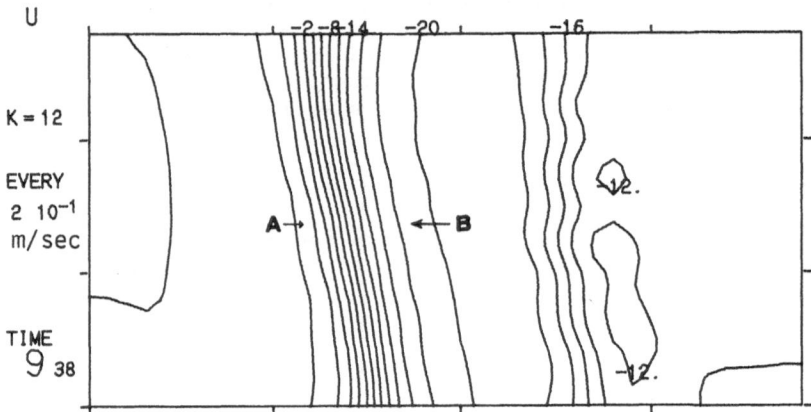

Fig. 17. As Fig. 12, but for U-component of wind speed.

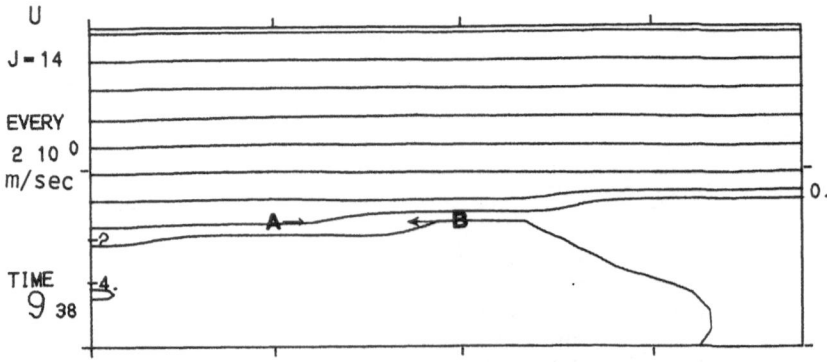

Fig. 18. As Fig. 10, but for U-component of wind speed.

strong vertical wind shear in the U-component is consistent with the
thermal wind relation, since air temperature decreased with y initially.
The U-component of wind was also well mixed in the boundary layer. But
the V-component of wind increased with height within the mixed layer.
Since the water temperature is warmer at the lower right corner, the
temperature (or density) increases (or decreases) from the northwest
toward southeast. The horizontal pressure gradient toward southeast
built up gradually with time as shown in Fig. 16. Therefore, in the
mixed layer, northerly wind increases (Fig. 15), but easterly wind
decreases with time (Fig. 14). It is also noted, a strong vertical
mixing also decreases the wind speed in x-direction. Eventually, the
U-component wind becomes very weak, but the northerly wind becomes
dominant in the mature stage as shown in Fig. 2.

Vertical Motion

Vertical motion is the result of horizontal divergence of wind as
revealed by the continuity equation. By looking at Fig. 17, the horizontal cross-section of the U-component of wind at level K=12, we found
that the wind speed is still at its initial value of about 0.3 m/sec at
the left one-third of the domain, while U became −2 m/sec over the right
due to a strong vertical mixing as discussed previously. This yielded a
strong convergence zone between grids with different inversion height as
between points A and B in Fig. 17. Since the inversion served nearly as
a "wall" to any disturbances below, downward motion set in. This is
illustrated by Fig. 18 and Fig. 19. The positions of points A and B on
these diagrams are the same as in Fig. 17.

The order of magnitude for the downward motion is about 3×10^{-6}
sec^{-1}, which corresponds to 1 cm/sec. This agrees well with the observed
prevailing downward motion over the AMTEX region as shown in Fig. 20.
It is also noted that the maximum downward motion occurs near the base
of the inversion. Due to its adiabatic warming effect, the center of
the downward motion coincided with the "breaks" of liquid water content
in Fig. 11. A portion of cloud evaporated when subsidence warming took
place.

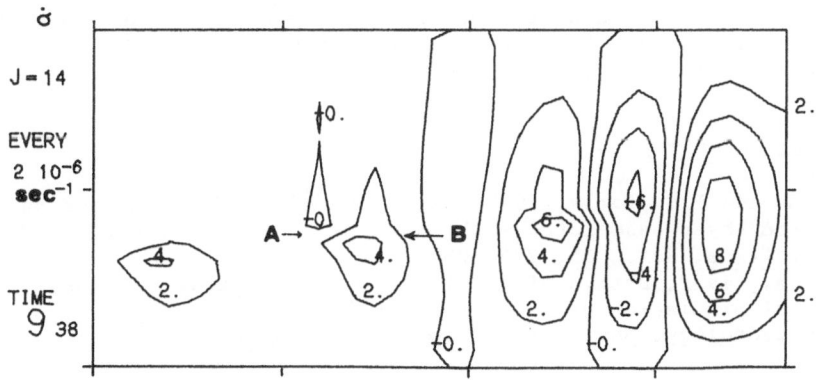

Fig. 19. As Fig. 10, but for vertical motion $\dot{\sigma}$.

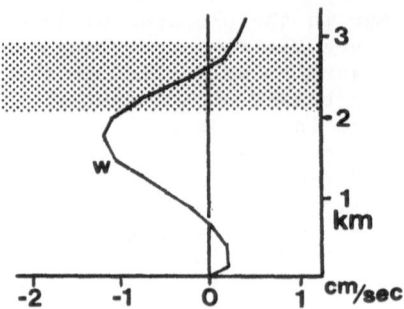

Fig. 20. As Fig. 2, but for vertical profile of
vertical velocity.

SUMMARY

A three-dimensional model has been developed to simulate the air
mass modification which occurs during a wintertime cold air outbreak
over the regions of the Kuroshio Current and Gulf Stream. The model has
a relatively high resolution in the lower troposphere to treat the
vertical turbulent transfer processes in the boundary layer. The physical processes included are: (1) surface exchange of heat, moisture, and
momentum based on the similarity theory, (2) vertical turbulent transfer
of heat, moisture, and momentum based on the modified Deardorff's (1980)
closure scheme and (3) the explicit treatment of condensation and
evaporation.

The preliminary numerical results reproduced the observed marine
type planetary boundary layer fairly well. This boundary layer consists
of a shallow superadiabatic layer just above the water surface; a well
mixed layer in which wind, temperature, and moisture are quite uniform;
and a cloud layer atop. The results also show that the thickness of the
cloud layer and mixed layer increases along downwind direction. The
vertical transport of the horizontal momentum also increases along the
downstream, therefore, it produces a prevailing downward motion in the
lower layer as observed during the Air Mass Transformation Experiment.

ACKNOWLEDGMENTS

The authors wish to thank Ms. V. Ewing for typing the manuscript.
This work was supported by the National Science Foundation under Grant
ATM-821965 by the Global Atmospheric Research Program, and ATM-8313418
by Meteorology Programs.

REFERENCES

Agee, E.M. and F.E. Lomax, 1978, Structure of the mixed layer and inver-
     sion layer associated with patterns of MCC during AMTEX 75.
     J. Atmos. Sci., 35, 2281:2301.
Arakawa, A., 1972, Design of the UCLA general circulation model. Numer-
     ical Simulation of Weather and Climate, Tech. Rept. 7, Dept. of
     Meteorology, Univ. of California, Los Angeles.

Businger, J.A., J.C. Wyngaard, Y. Izumi and E.F. Bradley, 1971, Flux-profile relationships in the atmospheric surface layer. <u>J. Atmos. Sci.</u>, <u>28</u>, 181:189.

Crowley, W.P., 1968, Numerical Advection experiments. <u>Mon. Wea. Rev.</u>, <u>96</u>, 1:11.

Deardorff, J.W., 1980, Stratocumulus-capped mixed layers derived from a three-dimensional model. <u>Boundary Layer Meteor.</u>, <u>18</u>, 495:527.

Gadd, A.J., 1978, A split explicit intergration scheme for numerical weather prediction. <u>Quart. J. Roy. Meteor. Soc.</u>, <u>104</u>, 569-582.

Ninomiya, K., 1975, Large-scale aspects of air-mass transformation over East China Sea during AMTEX '74. <u>J. Meteor. Soc. Japan</u>, <u>53</u>, 285:303.

Ninomiya, K., and T. Akiyama, 1976, Structure and heat energy budget of mixed layer capped by inversion during the period of polar outbreak over Kuroshio region, <u>J. Meteor. Soc. Japan</u>, <u>54</u>, 160:174.

Nitta, T., 1976, Large-scale heat and moisture budgets during the air mass transformation experiment. <u>J. Meteor. Soc. Japan</u>, <u>54</u>, 1:14.

Satio, N., 1975, A synoptic study of the inversion during the AMTEX '74, <u>Papers in Meteorology and Geophysics</u>, <u>26</u>, 121:147.

Sun, W.Y. and Y. Ogura, 1980, Modeling the evolution of the convective planetary boundary layer. <u>J. Atmos. Sci.</u>, <u>37</u>, 1558:1572.

Sun, W.Y., 1984, Numerical analysis for hydrostatic and nonhydrostatic equations of inertial-internal gravity waves. <u>Mon. Wea. Rev.</u>, vol. 112, no. 2, 259:268.

LARGE SCALE FLOWFIELD SIMULATION USING THE CYBER 205

Arthur Rizzi and   Michael Hodous

FFA The Aeronautical   Control Data Corporation
Research Institute   Arden Hills, Minnesota
16111 Bromma, Sweden   USA

ABSTRACT

The large-scale numerical simulation of fluid flow is described as a discipline within the field of software engineering. As an example of such work, a vortex flowfield is analyzed for its essential physical flow features, an appropriate mathematical description is presented (the Euler equations with an artificial viscosity model), a numerical algorithm to solve the mathematical equations is described, and the programming methodology is discussed in order to attain a very high degree of vectorization on the CYBER 205. Two simulated flowfields with vorticity are presented to verify the realism of the simulation model. The computed solutions show all the qualitative features that are expected in these flows.

INTRODUCTION

All aircraft industry has a practical interest in designing transonic and supersonic vehicles for both military and commercial purposes. This type of aircraft vehicle needs a very complete analysis of the design in order to establish the feasibility of that design. Today's practical airplane designer relies on wind tunnel data in order to predict flight characteristics. This paper describes a comprehensive numerical simulation technique which can be used to predict flight characteristics in conjunction with the wind tunnel measurements.

Realistic numerical simulations require three dimensional models of significant size. The recent construction for test purposes of a 16 million word memory for the CYBER 205 (or the NASA-VPS32) has allowed tests of one million grid point models on a practical basis within reasonable elapsed times. It was hoped that this type of computation could be performed with disk I/O, but experience shows that heavy time penalties will be incurred due to the nature of the algorithm as described. Therefore, good memory management techniques within a sufficient working set of real memory is important.

The three aspects of large scale flow simulation to be considered are the application analysis, the development of the simulation model and the testing and verification of that model. Well known aerodynamic designs with well established flight characteristics are to be compared with computational results.

Applications of large scale flow simulations range from marine technology for ships and submarines, automobile aerodynamics, turbomachine performance and aircraft/missile design. An example of the practical problems involved can be quickly seen by examining just one type of vehicle, the advanced tactical fighter. This vehicle is to be designed for use in subsonic, transonic and supersonic flight regimes, as well as to provide reasonably stable take-off, maneuvering and landing characteristics. These aircraft will employ a close coupled canard forward of the wing and will require well ordered engine intake flows over a wide flight envelope. A maximum amount of lift with minimum drag must be provided by the aircraft designer.

Now that the concept of the design assignment is well understood, at least in this case of the advanced tactical fighter, we can prepare for the software engineering of fluid flow problems. The physical model of the flow over the delta wing is to be examined in the aerodynamic sense. The flow of air over the wing forms a primary vortex and a secondary vortex, each of which serves to increase lift and minimize drag effects. Behavior of that physical model can be described by the equations of fluid mechanics, the continuum model in mass, momentum and energy. This continuum model is then the basis for the development of the flow field simulation.

PHYSICAL ANALYSIS OF DELTA WING VORTEX FLOW

The type of flowfield that we want to simulate is drawn schematically in Figure 1. A delta-shaped wing meets an oncoming stream of air at a high angle of attack. The flow separates from the wing in a shear layer along the entire length of the sharp leading edge, and under the influence of its own vorticity coils up to form a vortex over the upper surface of the wing. The high velocities in the vortex create a low pressure region under it which gives the wing a nonlinear lift. If the Reynolds number of the flow is not very large (say under 10^7), and the flow is laminar instead of turbulent, then the likelihood is great that a secondary vortex will form under the primary vortex because the boundary layer will separate due to the large positive pressure gradient there. The flow also separates in a shear layer from the trailing edge of the wing which forms into a wake vortex. The primary and wake vortex then interact with each other.

DEVELOPMENT OF MATHEMATICAL MODEL

Continuum Equations

The appropriate model to describe the fluid mechanics of this vortex flow is the compressible Navier Stokes equations:

$$\frac{\partial}{\partial t} \rho + \text{div } \rho \underset{\sim}{V} = 0$$

$$\frac{\partial}{\partial t} \rho \underset{\sim}{V} + \text{div } \rho \underset{\sim}{V} \underset{\sim}{V} + \text{grad } p = \Gamma \tag{1}$$

$$\frac{\partial}{\partial t} e + \text{div } (e + p) \underset{\sim}{V} = \Upsilon$$

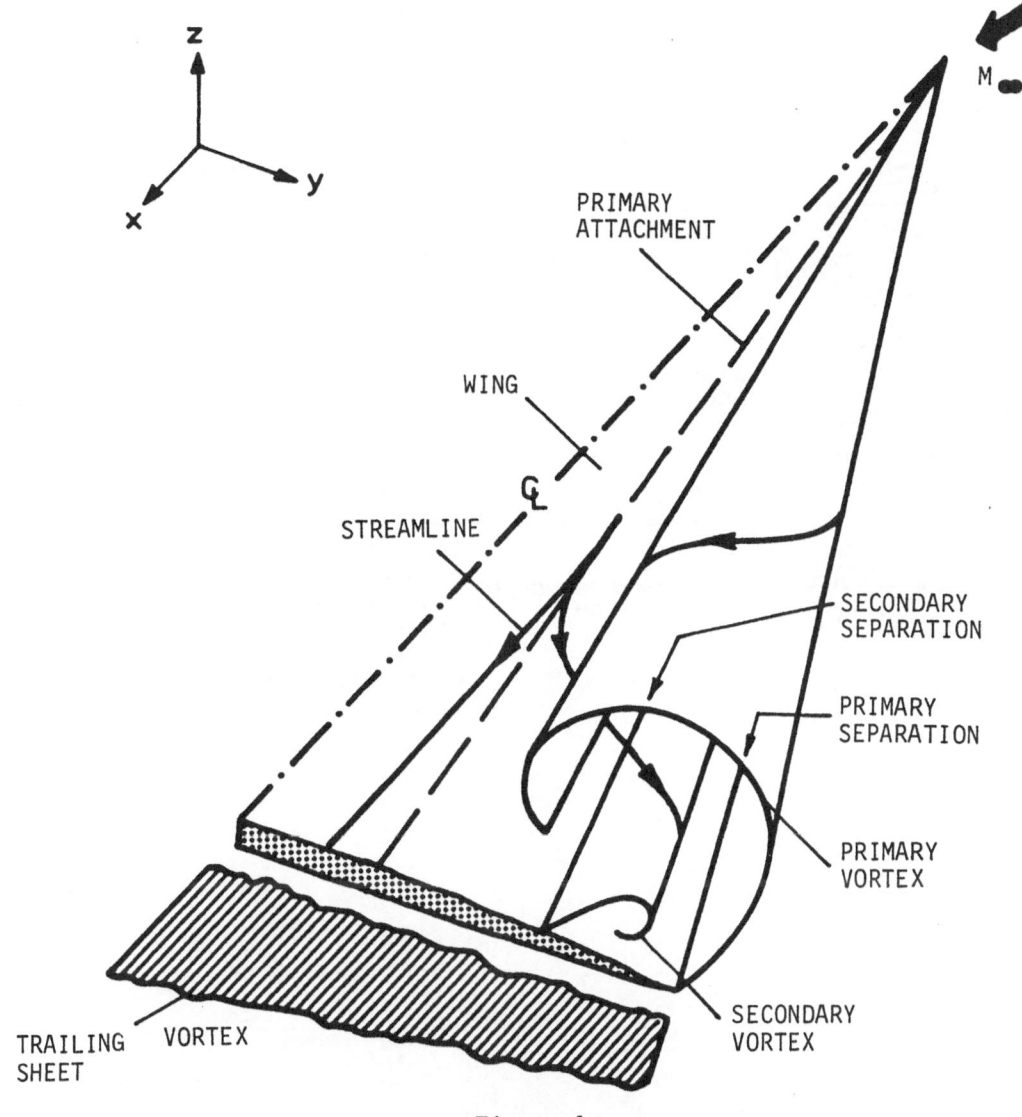

Figure 1

Schematic features of the flow example. A shear layer separates from
the leading edge and rolls up into the primary vortex above the wing.
This vortex in turn may induce a secondary vortex to develop under it.

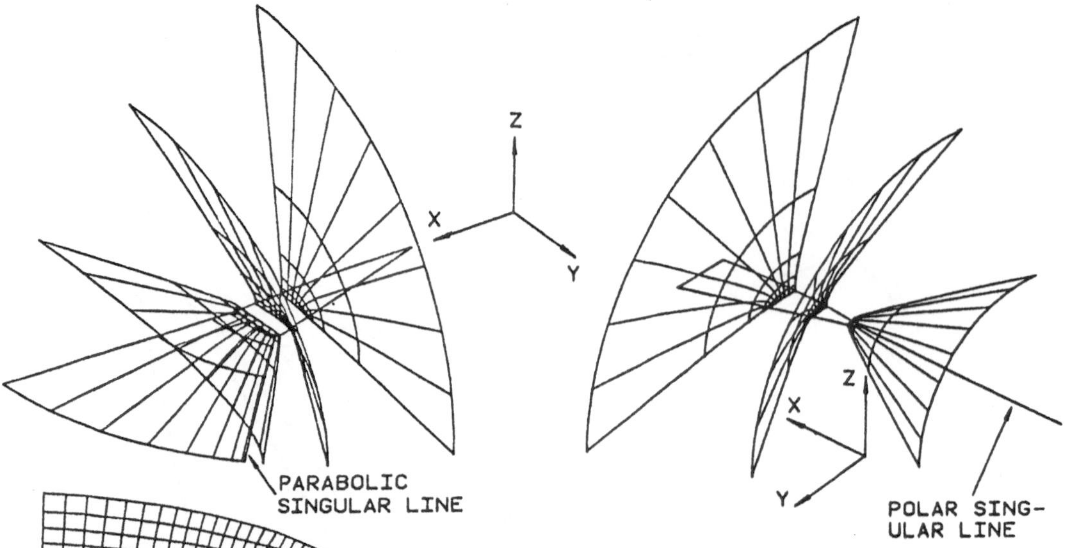

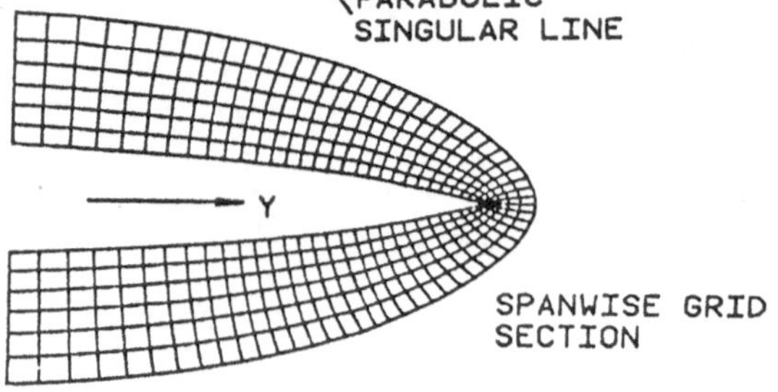

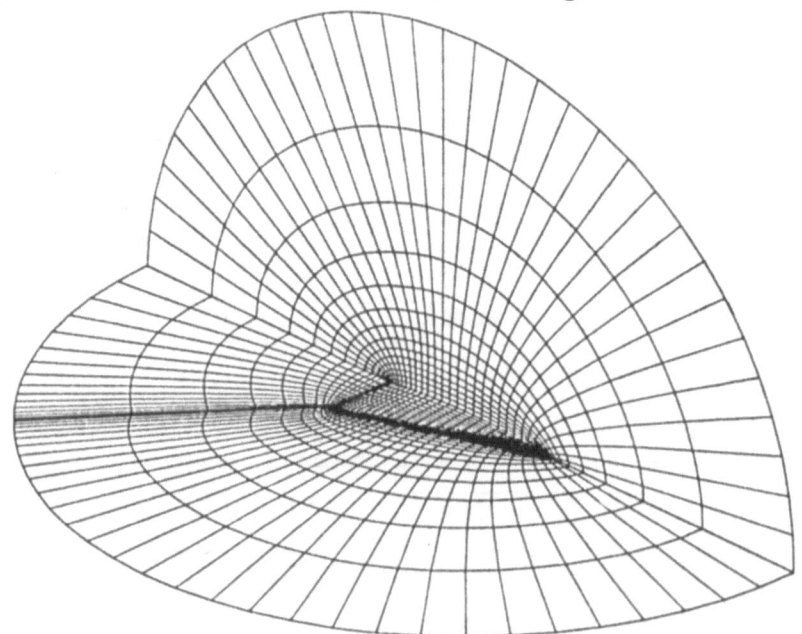

PARABOLIC
SINGULAR LINE

SPANWISE GRID
SECTION

POLAR SING-
ULAR LINE

Figure 2 (a)
Grid generated around a delta-shaped
small aspect ratio wing has an O-O
topology. The polar singular line
produces a dense and nearly conical
distribution of points at the apex
which is needed to resolve the rapidly varying flow there. This mesh is
well-suited for computing the flow
around wings of combat aircraft.

Figure 2 (b)
Three-dimensional view of the delta wing mesh.

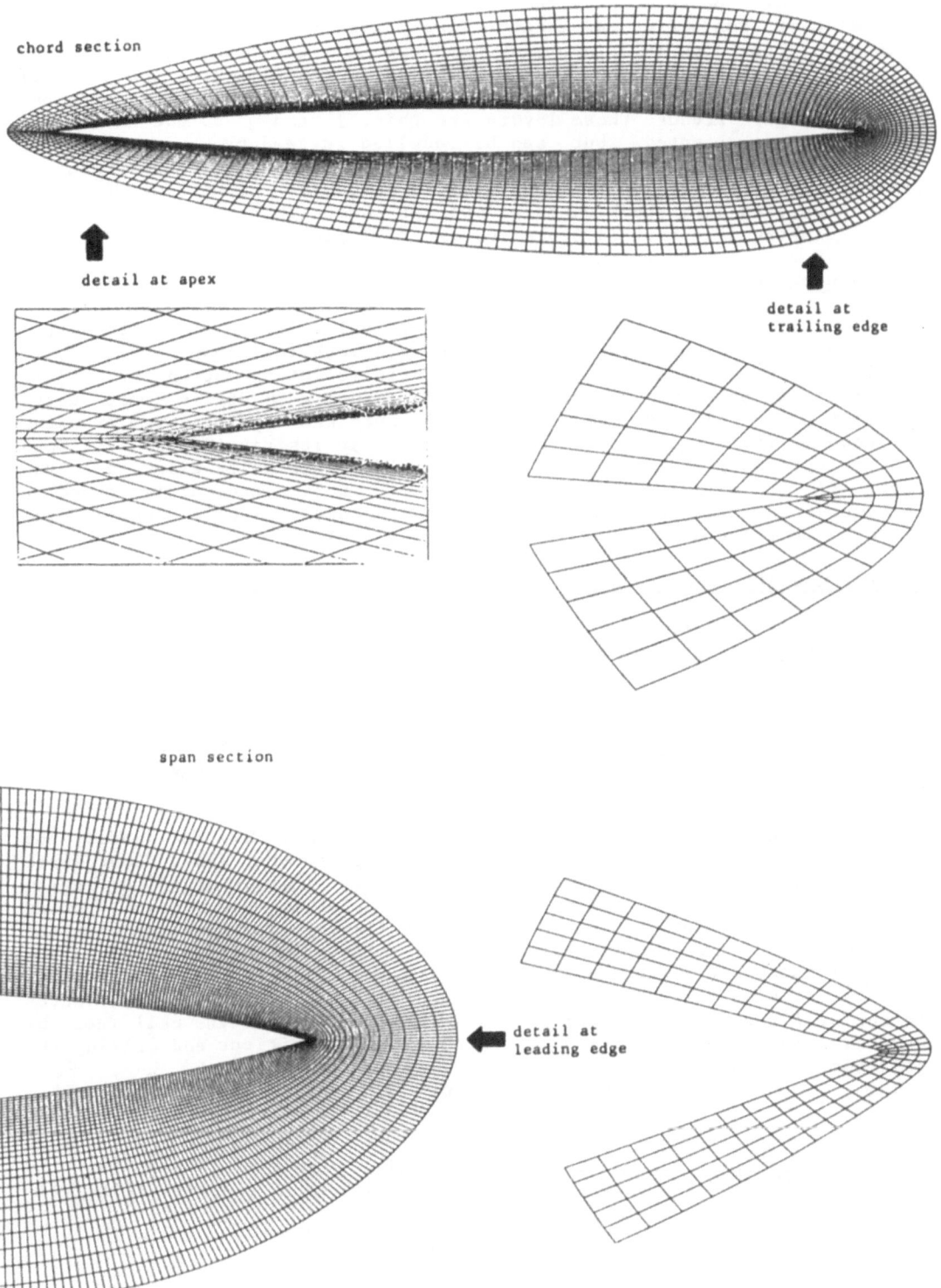

Figure 2 (c)
Partical chordwise and spanwise views of the 193x57x97 mesh.

where ρ is the density, p the pressure, e the energy, and $\underset{\sim}{V}$ the velocity of the fluid. The term Γ is the shear stresses acting on the fluid, and Υ is heat conduction and viscous dissipation terms. If the Reynolds number of the flow is large, say over one million, the terms Γ and Υ have an effect only in very thin layers of the flow near the surface of the wing and across the shear layers that separate from the leading and trailing edge. Because these layers are thin, Γ and Υ do not have to be accurately represented, but can be modelled instead by a non-physical expression. Equations (1) then are called the Euler equations with an artificial viscosity model. The flow is inviscid, except across thin discontinuities like shock waves and vortex sheets that are allowed in the solution where irreversible processes can take place.

Mesh Generation

The first step in carrying out a numerical solution of equations (1) is to discretize the flowfield by creating a mesh that spans the entire region. We use Eriksson's method of transfinite interpolation [Ref. 1] to construct a boundary-conforming O-O grid around a 70 degree swept delta wing, the so-called Dillner wing. This mesh has a polar singular line attached to its apex and a parabolic singular at its tip (Figure 2). Such an arrangement produces a natural focussing of mesh cells along all the edges of the wing. This particular type of pole placement is termed a three-pole O-O mesh (when one considers the second parabolic singular line mirrored on the other side of the symmetry plane).

Numerical Solution Procedure

An equivalent form of the Euler equations (1) can be expressed as an integral balance of the conservation laws:

$$\frac{\partial}{\partial t} \iiint q \ d \ vol + \iint \underset{\sim}{H}(q) \cdot \underset{\sim}{n} \ dS \qquad \begin{array}{l} \text{artificial} \\ = \text{viscosity} \\ \text{model} \end{array} \qquad (2)$$

where q is the vector with elements $[\rho, \rho u, \rho v, \rho w]$ for density and Cartesian components of momentum with reference to the fixed coordinate system x, y, z. The flux quantity $\underset{\sim}{H}(q) \cdot n = [q\underset{\sim}{V} + (0, e_x, e_y, e_z)p] \cdot \underset{\sim}{n}$ represents the net flux of q transported across, plus the pressure p acting on, the surface S surrounding the volume of fluid. The mesh segments the flowfield into very many small six-sided cells, in each of which the integral balance (2) must hold. The finite volume method then discretizes (2) by assuming that q is a cell-averaged quantity located in the center of the cell, and the flux term $\underset{\sim}{H}(q) \cdot \underset{\sim}{n}$ is defined only at the cell faces by averaging the values on each side. With these definitions and calling the cell surfaces in the three coordinate directions of the mesh $\underset{\sim}{S}_I$, $\underset{\sim}{S}_J$, and $\underset{\sim}{S}_k$, we obtain the finite volume form for cell ijk:

$$\frac{\partial}{\partial t} q_{ijk} + \left[\delta_I (\underset{\sim}{H} \cdot \underset{\sim}{S}_I) + \delta_J (\underset{\sim}{H} \cdot \underset{\sim}{S}_J) + \delta_k (\underset{\sim}{H} \cdot \underset{\sim}{S}_k) \right]_{ijk} / vol \qquad \begin{array}{l} \text{artificial} \\ = \text{viscosity} \\ \text{model} \end{array} \qquad (3)$$

where $\delta_I (\underset{\sim}{H} \cdot S_I) = (\underset{\sim}{H} \cdot \underset{\sim}{S}_I)_{i + \frac{1}{2}} - (\underset{\sim}{H} \cdot \underset{\sim}{S}_I)_{i - \frac{1}{2}}$ is the difference operator. To this the boundary conditions for the particular application must be specified. They occur at the six bounding faces of the computational domain and are listed in Figure 3. When included, we can write equation (3):

BOUNDARY CONDITIONS

Three Types

- nonreflecting farfield

- periodic coordinate cuts

- solid surfaces
    1. zero flux-normal mom. eq.
    2. trailing edge Kutta
       condition? not needed
       if sharp

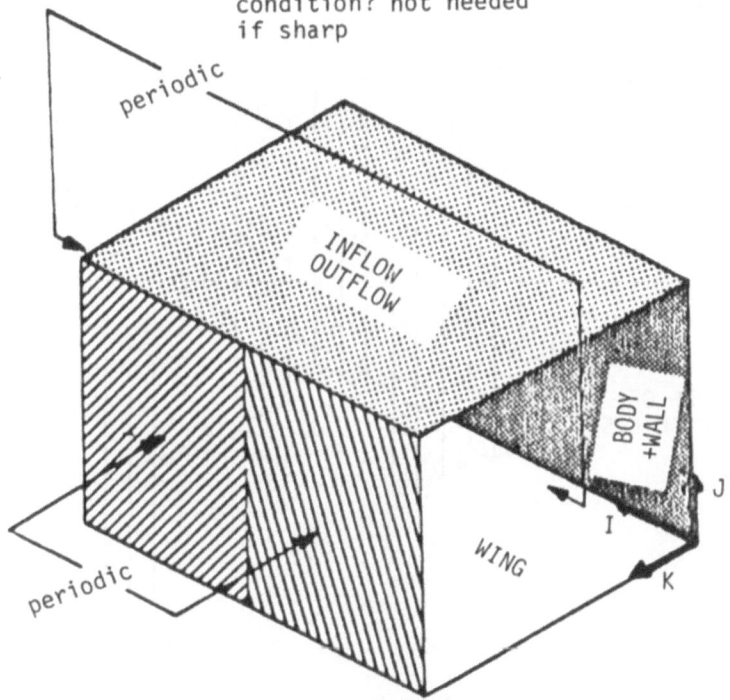

Figure 3
The boundaries of the physical domain are mapped by the mesh to the six faces of the computational cube. The appropriate flow conditions must be set on these faces.

CONTIGUOUS CELLS IN MEMORY

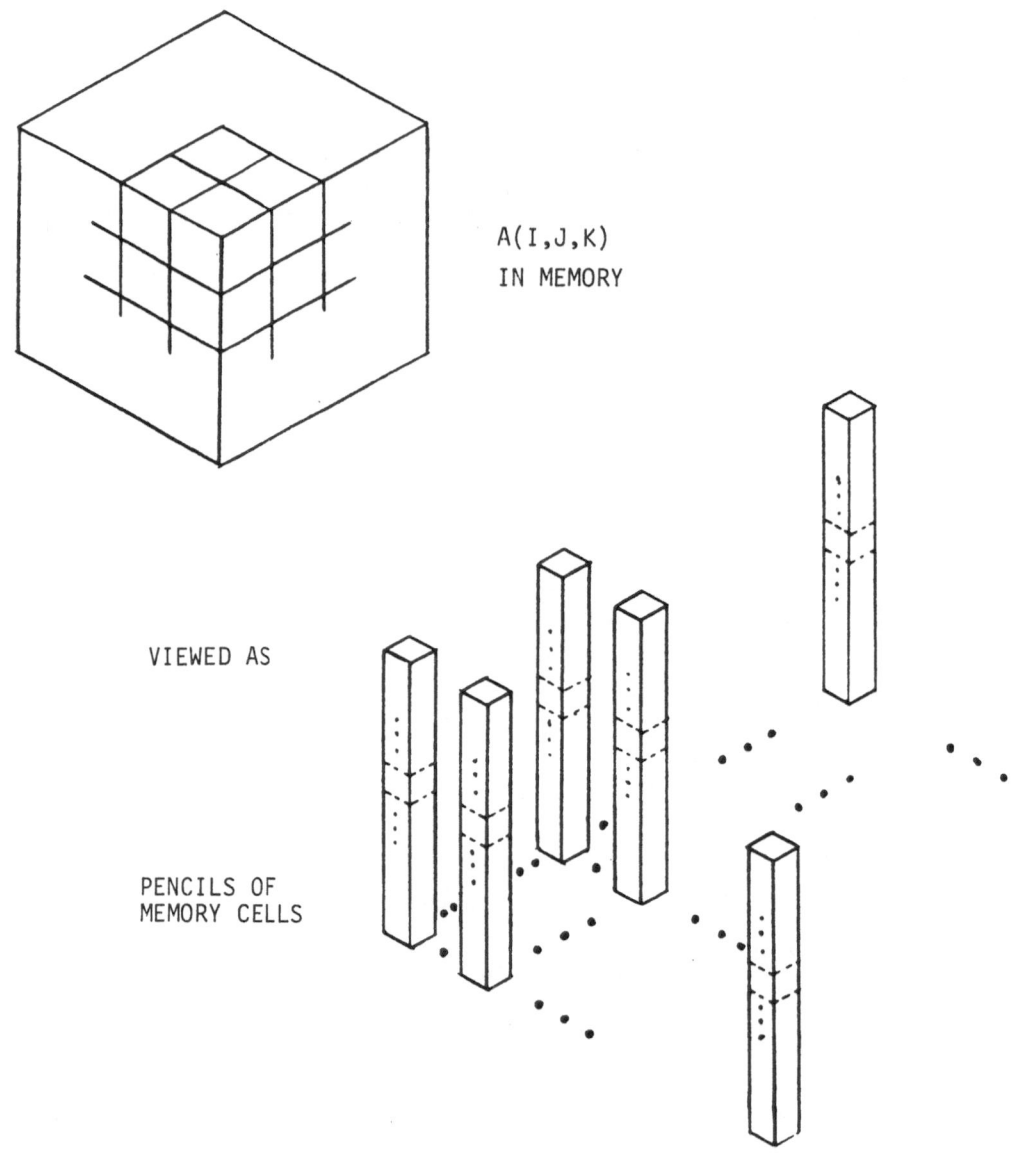

A(I,J,K)
IN MEMORY

VIEWED AS

PENCILS OF
MEMORY CELLS

Figure 4
Consideration of the way three-dimensional data is stored in contiguous memory locations.

NATURAL GROUPING FOR DIFFERENCING "SHIFT"

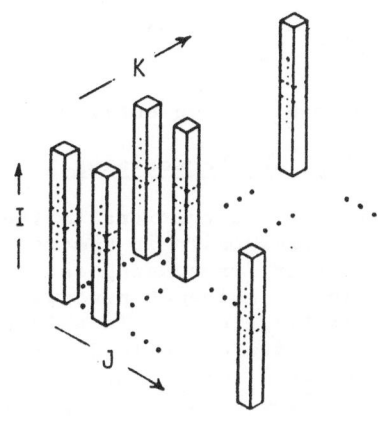

DIFFERENCES

I DIRECTION   J DIRECTION   K DIRECTION

GROUP: SINGLE   PENCILS   PLANES
         ELEMENTS

Figure 5

The natural contiguous groups for taking vector differences in each of the
three computational directions.

$$\frac{\partial}{\partial t} q_{ijk} + FD \qquad = \text{artificial viscosity model + boundary conditions} \quad (4)$$

where FD represents all of the flux differences. A more detailed description of the method is given in Ref. 2 and 3. With the spatial indices suppressed, we integrate this last equation with the two-level multistage scheme in parameter $\theta = 1/2$ or 1:

$$
\begin{aligned}
q_0 &:= q & (5)\\
q &:= q_0 + \Delta t\, FD(q_0)\\
q &:= q_0 + \Delta t[\ (1-\theta)FD(q_0)\ + \theta\, FD(q)]\\
q &:= q_0 + \Delta t[\ (1-\theta)FD(q_0)\ + \theta\, FD(q)]
\end{aligned}
$$

that steps the solution forward in time until a steady state is reached.

Data Structure and Methodology for Vectorizing

The nature of the vector instructions in the CYBER 205 emphasizes rapid operations upon contiguous cells in memory. A three dimensional structure suitable for vector processing is correctly visualized as consisting of a collection of adjacent pencils of memory cells with suitable boundary conditions (Figure 4). This leads to the natural grouping for the formation of differences in J by means of differencing adjacent pencils. The natural differencing in the K-direction thus becomes the difference of a plane of adjacent pencils. Differences in the I direction are formed by use of single elements offset one against the next within the pencil. See Figure 5 for these relationships which allow for the complete vectorization of all of the internal relationships.

External relationships are defined on the boundary. Computation of these external, i.e., boundary conditions interrupt the flow of vectorization. Our particular choices in conjunction with the data-base design lead to the following key features of the procedure: 1) separate storage arrays are assigned for the dependent variables q, flux component F, and flux differences FD; 2) one extra unit is dimensioned for each computational direction to hold the boundary conditions, and 3) flux differences are taken throughout the entire field by off-setting the starting location of the flux vector F. In this way, all of the work in updating interior points is exclusively vector operations without any data motion, accomplished at the expense of just a single scalar loop for the boundary conditions on one computational direction and a doubly-nested loop for another, while the boundary conditions for the third direction vectorize completely.

The pyramidal algorithm for total vector differences is built on inner computations computed as one long vector, computations with 2 boundary planes in K, computations in 2*J pencils each of length I, and computations in 2 cells each of quantity J*K. Four variables are updated at each boundary element. These boundary conditions are set three times for each iteration. In each boundary condition, the starting and ending boundary values are established first on the value q and second on the value of flux F. The example in K of direction differences shows that F is a function of q. The variable q is then eventually updated as a function of the differences in F (i.e., FD), and so forth. See Figure 6 for a deeper understanding.

We can see a FORTRAN example of the differences in I in Figure 7. These equations illustrate the FORTRAN methodology. The CYBER 205 hardware has a maximum vector length of 65,535 elements, which is easily exceeded by the three-dimensional vector differencing scheme described above. The technique for circumventing the possible limitation is called

EXAMPLE: K - DIRECTION DIFFERENCES

$$M = (I+1)*(J+1)$$
$$N = M*K$$

- SET RIGHT BOUNDARY VALUES IN q(1,1,K+1;M)

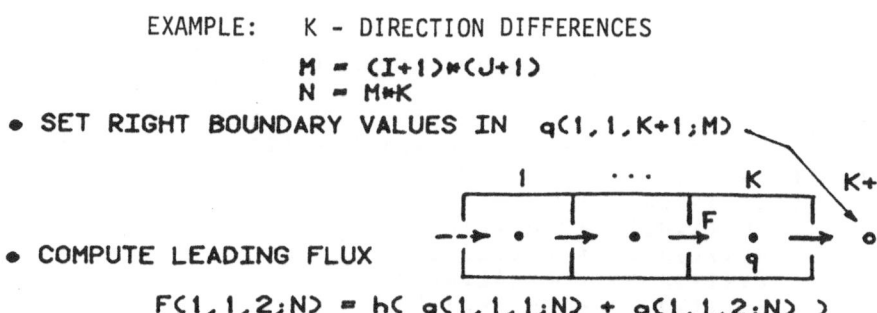

- COMPUTE LEADING FLUX

$$F(1,1,2;N) = h(q(1,1,1;N) + q(1,1,2;N))$$

- COMPUTE "LEFT" BOUNDARY FLUX F(1,1,1;M)

- DIFFERENCE FLUX

$$FD(1,1,1;N) = F(1,1,2;N) - F(1,1,1;N)$$

DIAGRAMMATICALLY

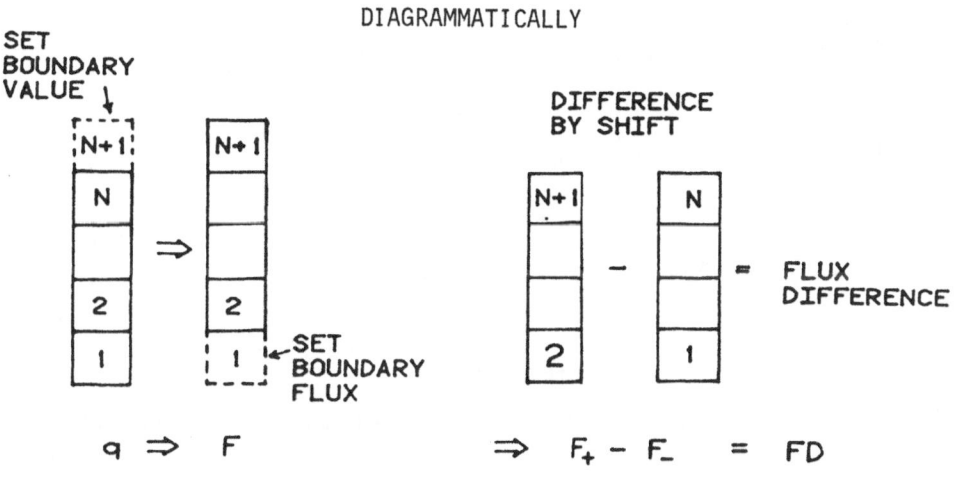

————— STEPS IN DIFFERENCING ————⟶

Figure 6

Algorithm for interleaving the boundary conditions into the vector differencing of the entire three-dimensional flux field, displayed here schematically
for the K-direction differences only.

ACTUAL I - DIFFERENCE CODE

SET RIGHT BOUNDARY CONDITION IN q

```
C      *  I - C I R E C T I C N   F L U X   D I F F E R E N C E  *
C            SYMMETRY FLUX BCLADARY CCADITIONS AT  I=IL
      DC 3C K=1,K2
         CC 3C J=1,J2
            RC(IL,J,K) = RO(I2,J,K)
            RL(IL,J,K) = RL(I2,J,K)
            RV(IL,J,K) =-RV(I2,J,K)
            RW(IL,J,K) = RW(I2,J,K)
   3C CCNTIALE
```


COMPUTE METRIC

$$F(2,1,1;N) =.5*((Y(2,2,1;N)-Y(2,1,2;N))*(X(2,2,2;N)-X(2,1,1;N))$$
$$-(X(2,2-1;N)-X(2,1,2;N))*(Y(2,2,2;N)-Y(2,1,1;N)))$$

CALCULATE FLUX F

$$F(2,1,1;N) = OS(2,1,1;N)*(RU(2,1,1;N)+RU(1,1,1;N))$$
$$+P(2,1,1;N)* F(2,1,1;N)$$

SET BOUNDARY VALUE F

```
      DC 31 K=1,K2
         CC 31 J=1,J2
   31        F(1,J,K) = C.
```


DIFFERENCE

$$FC2(1,1,1;N) = FC2(1,1,1;N) + F(2,1,1;N) -F(1,1,1;N)$$

ADVANCE IN TIME

$$FC2(1,1,1;N) = CEVMASK(FD2(1,1,1;N),0..C(1,1,1;N); FD2(1,1,1;N))$$
$$RL(1,1,1;N) = RUO(1,1,1;N) +FT*FD2(1,1,1;N)$$

Figure 7

Actual FORTRAN coding of the I-direction part of the total vector differencing of the three-dimensional flux field.

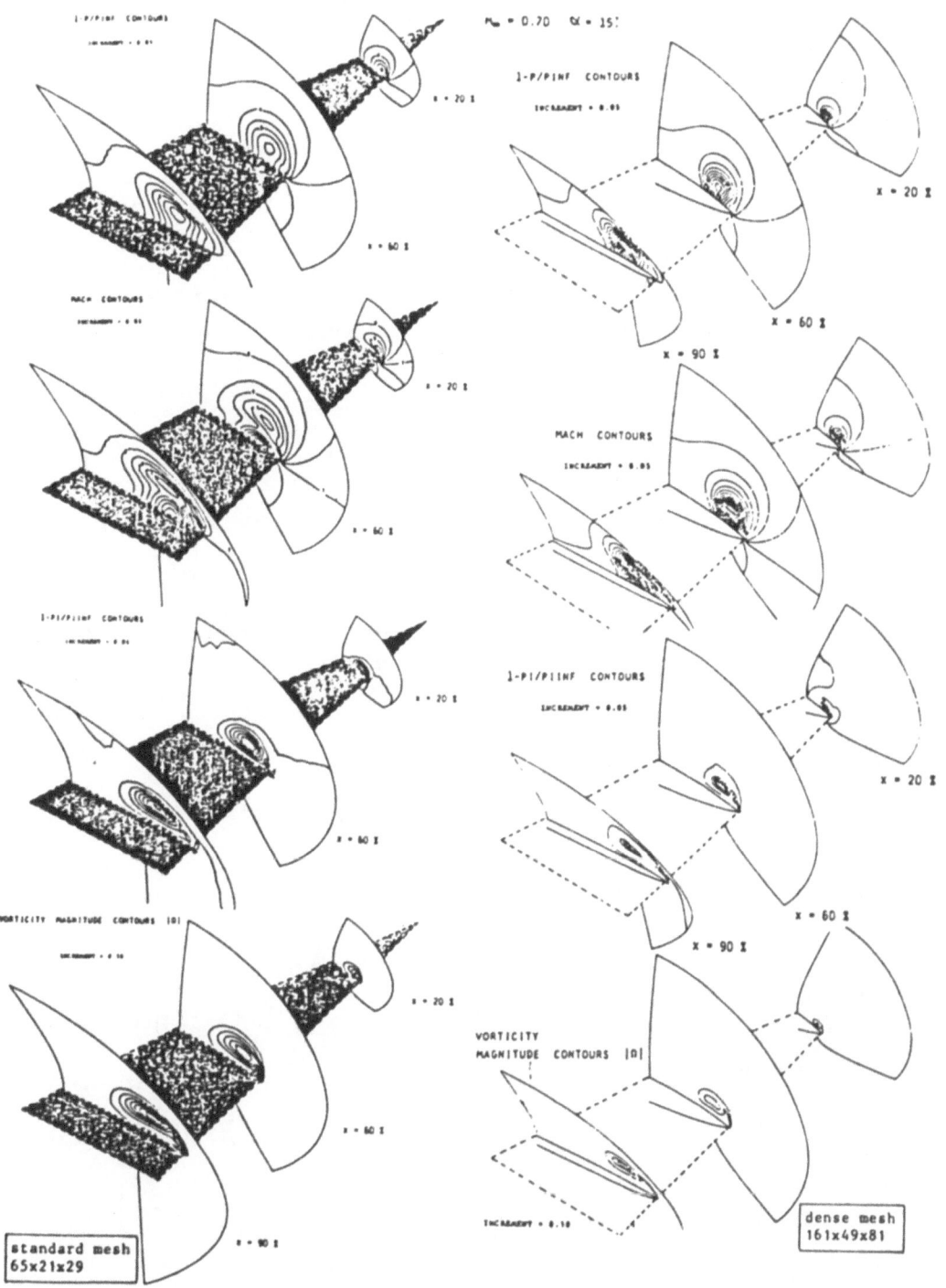

Figure 8

Comparison of standard-mesh and dense-mesh solutions in three mesh surfaces at the 20%, 60%, and 90% chord stations of the Dillner Wing. $M_\infty = 0.70$, $\alpha = 15$ deg.

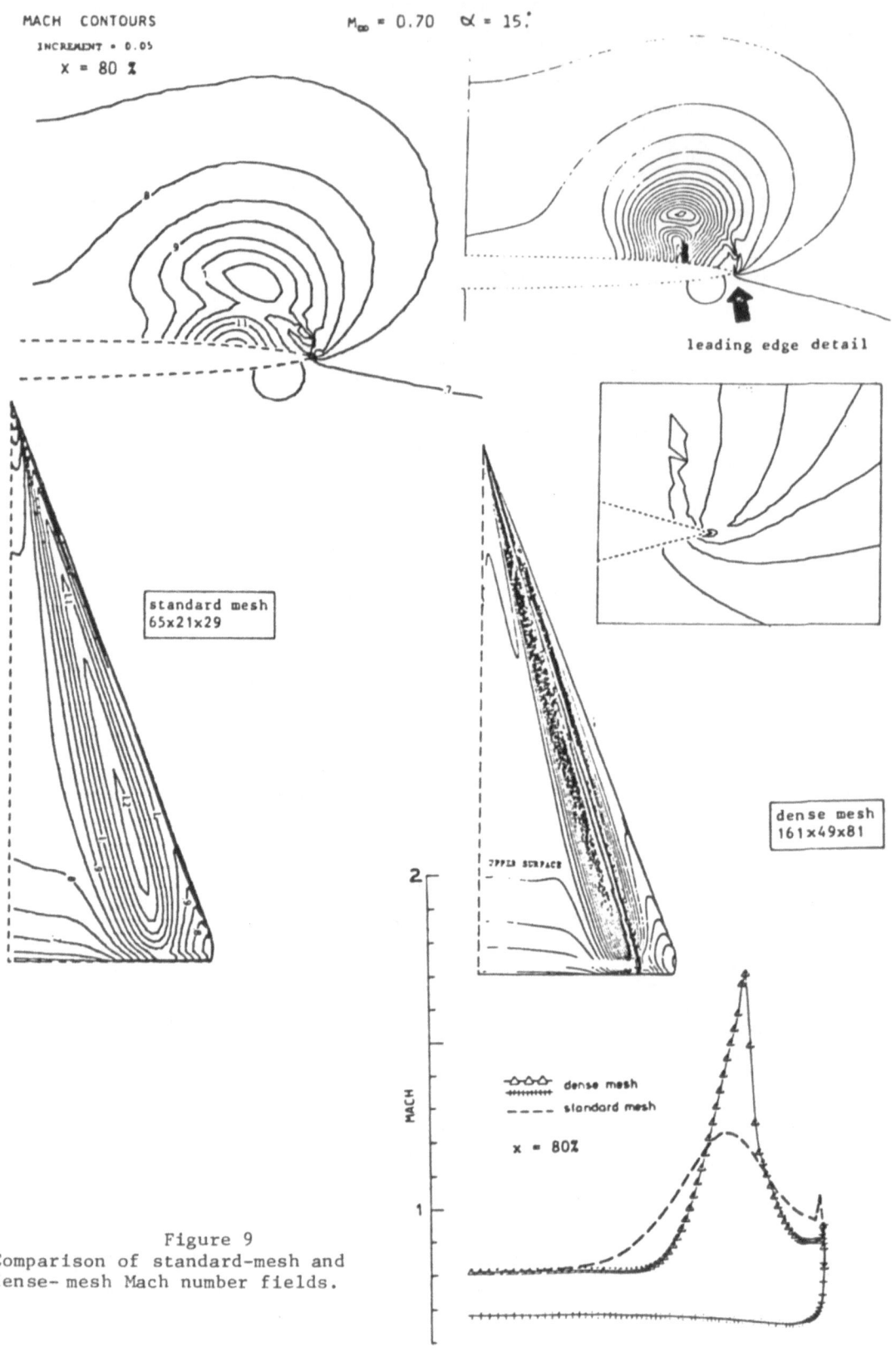

Figure 9
Comparison of standard-mesh and
dense-mesh Mach number fields.

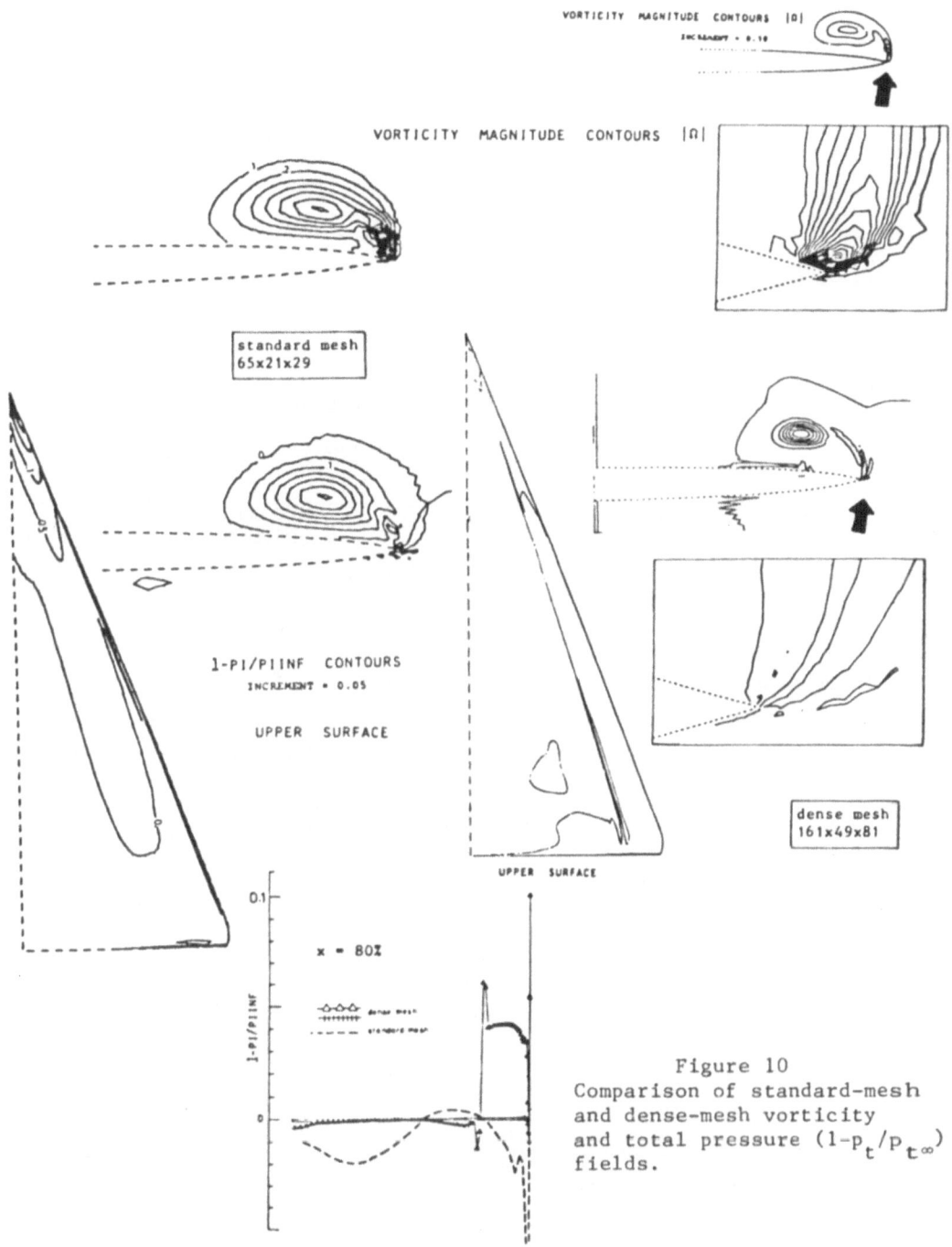

DILLNER DELTA WING

$M_\infty = 0.70 \quad \alpha = 15°$

VORTICITY MAGNITUDE CONTOURS |Ω|

VORTICITY MAGNITUDE CONTOURS |Ω|

standard mesh
65x21x29

1-PI/PIINF CONTOURS
INCREMENT = 0.05

UPPER SURFACE

dense mesh
161x49x81

UPPER SURFACE

x = 80%

Figure 10
Comparison of standard-mesh
and dense-mesh vorticity
and total pressure $(1-p_t/p_{t\infty})$
fields.

EULER FLOW CODE PERFORMANCE MODEL

(AS OF SEPT. '84)

Incompressible   8.5 microsec./
                                        gridpoint/iteration

Subsonic   12.5 microsec.

Supersonic   7.5 microsec.

Thus time for 1000 iterations at 1 million grid points;

Supersonic   2.1 hours

Figure 11

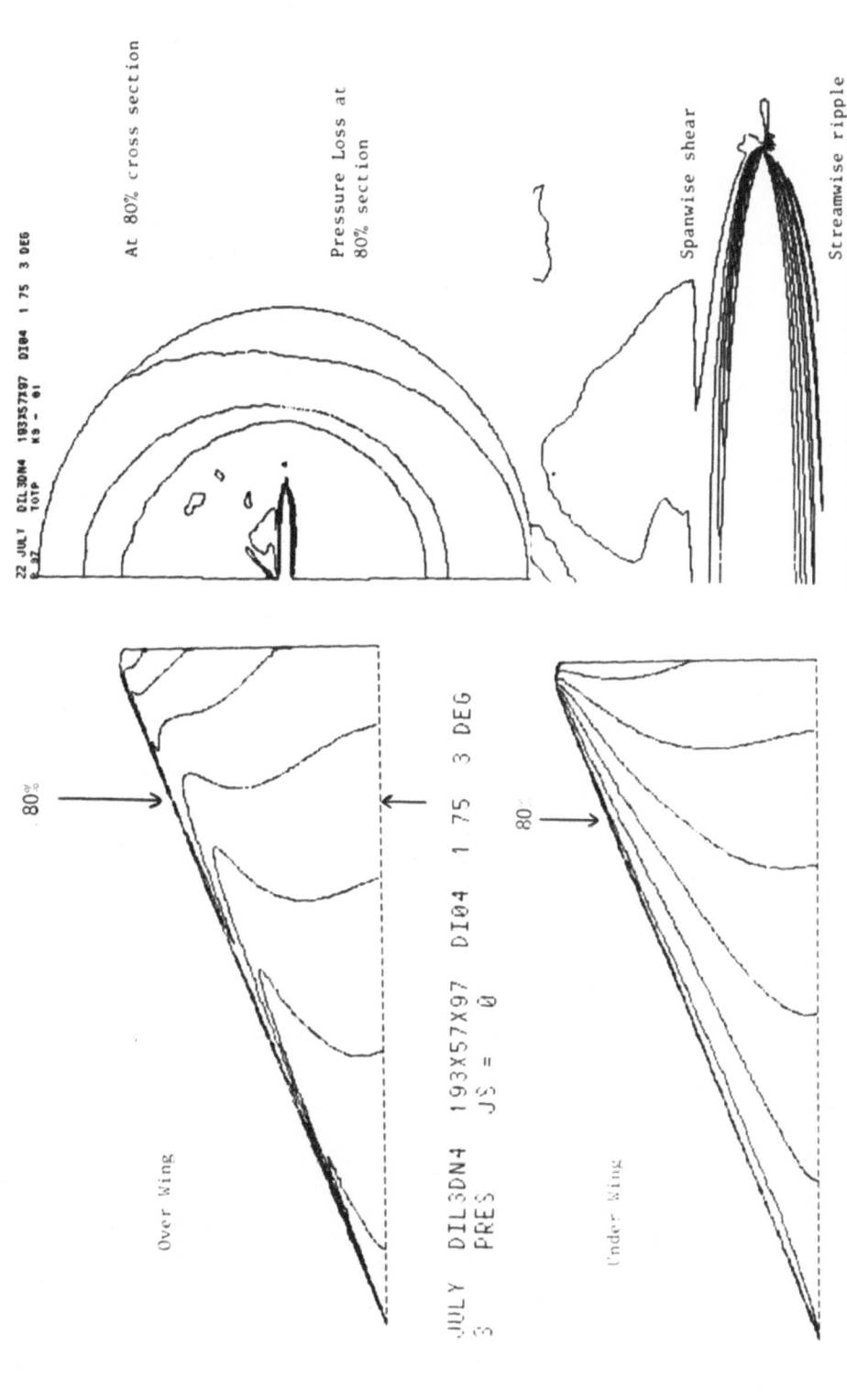

Figure 12

Pressure over/under wing versus vorticity at the 80% cross section for Dillner Wing ar M=1.75 and angle of attack of 3°.

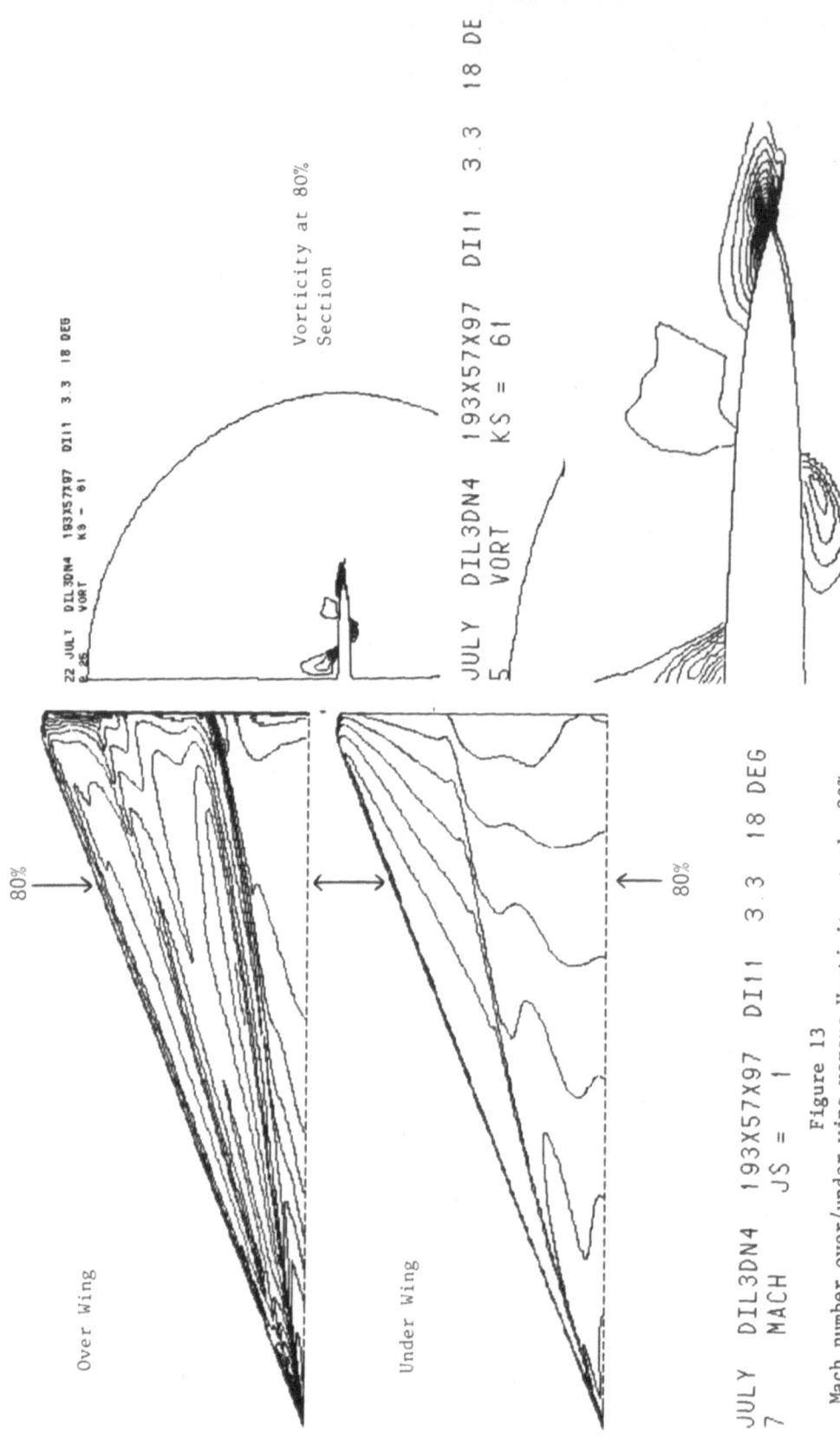

Over Wing

Under Wing

80%

80%

80%

Vorticity at 80%
Section

Shock induced vortex
especially at the tip.

JULY DIL3DN4 193X57X97 DI11 3.3 18 DE
5   VORT   KS = 61

JULY DIL3DN4 193X57X97 DI11 3.3 18 DEG
7   MACH   JS = 1

Figure 13

Mach number over/under wing versus Vorticity at the 80% cross
section for Dillner Wing at M=3.3 and angle of attack of 18°.

strip-mining for the CYBER 205. Each portion of the data set is described as a three dimensional subset (or slab) of the final data ensemble. This method creates internal boundary conditions between slabs which are accommodated within the computational process. A scalar do-loop is used to process each slab in turn at an average vector length approaching 65,000 elements. This strip-mining technique is processed sequentially (slab by slab) in the CYBER 205, but could be processed concurrently on a machine with multiple parallel processors.

The generalized mapping from data to slabs is managed via common block declarations within FORTRAN. The common blocks are grouped into large pages by the loader. The large pages are stored on the disk system across a series of independent temporary files, each of which is accessed and mapped into the executable space.

SIMULATED FLOWFIELDS

Actual experience in simulating flowfields via computed results has been obtained in water turbine cascades, fundamental 3-D vortex flow and in practical canard/delta designs. The illustrations to be examined focus on the verification of Euler flow models of theoretical interest, that is, the vortex flow on delta wings in three dimensions. The effectiveness of the method is shown in the delta wing charts of pressure loss and Mach number (Figure 8). The effect of the large real memory of the 16 million word CYBER 205 is shown in the comparisons of Mach number or vorticity between coarse and fine grids (Figures 9 and 10 compare results for a specific grid of 40,000 cells with those for the same grid with 600,000 cells). Performance as obtained in September, 1984 is shown (Figure 11).

Additional simulations have been performed in transonic regimes over various geometries. Considerable analysis remains in order to interpret these results in a meaningful engineering sense. Can separation be seen? What causes vorticity under the wing (Figures 12 and 13)?

REFERENCES

1. Ericksson, L. E.: Generation of Boundary-Conforming Grids Around Wing-Body Configurations Using Transfinite Interpolation, AIAA Journal, Vol. 20, October, 1982, pp. 1313-1320.

2. Rizzi, A. W. and Ericksson, L. E.: Computation of Flow Around Wings Based on the Euler Equations, Journal Fluid Mechanics, Vol. 142, October, 1984.

3. Rizzi, A.: Vector Coding the Finite-Volume Procedure for the CYBER 205, in Lecture Series Notes 1983-04, von Karman Institute, Brussels, 1983.

NUMERICAL SIMULATION OF THREE-DIMENSIONAL

FLOWFIELDS USING THE CYBER 205

Julie M. Swisshelm and Gary M. Johnson

Institute for Computational Studies
Fort Collins, Colorado 80522

SUMMARY

A procedure for the accelerated solution of the three-dimensional, compressible Navier-Stokes and Euler equations is described. The convergence of an explicit fine-grid scheme is enhanced through the use of a multiple-grid technique on a collection of coarser grids. The coarse-grid scheme is itself fully explicit and is independent of such details of the fine-grid problem as the formulation of viscous or damping terms and the specification of boundary conditions. Furthermore, this multiple-grid technique may be used, without modification, with a variety of fine-grid algorithms. Results are presented for the flow through a cascade of finite-span, swept blades. The Euler equations are solved for both subcritical and shocked, supercritical flows. The Navier-Stokes computations include laminar and turbulent, attached and separated flows. The procedure is vectorized for use on a CDC CYBER 205 computer and an algorithm version suitable for use on a multiple instruction-multiple data computer is mentioned.

INTRODUCTION

In order to numerically simulate complex aerodynamic flow phenomena, algorithms must be developed to rapidly solve the steady three-dimensional Euler and Navier-Stokes equations. The foundations of such techniques lie in extensive research in two dimensions. The evolution of current generation computers makes extensions to three dimensions more practicable.

A steady flow solution to the unsteady equations of motion, given steady boundary conditions and the existence of a unique steady solution, may be computed using a time-marching procedure. Explicit methods offer simplicity and low operations count but have slow convergence. However, they may be accelerated by various techniques, including, for example, local time-stepping applied to an otherwise time-accurate scheme. Implicit methods, which are unconditionally linearly stable, offer another alternative.

The method proposed here is based on the two-dimensional algorithm developed by Johnson [1,2]. It uses a collection of coarse grids to accelerate the convergence of an explicit fine-grid solution procedure. In this general sense, it is a multiple-grid method. However, the details of its implementation are a good deal simpler than is the case with the conventional multigrid approach. The coarse-grid acceleration scheme presented here is quite modular. It may be used without modification in conjunction with a number of different fine-grid solution procedures and with any set of flow equations in the hierarchy ranging from the Euler equations to the full Navier-Stokes equations. It is also independent of the formulation of fine-grid damping terms and boundary conditions. The coarse-grid scheme is, furthermore, fully explicit and the entire procedure has been vectorized.

EQUATIONS OF MOTION

The nondimensional equations of motion may be written in conservation-law form as

$$q_t = -(F_x + G_y + H_z) \tag{1}$$

where, for the full Navier-Stokes equations,

$$F = f - Re^{-1}p \qquad G = g - Re^{-1}r \qquad H = h - Re^{-1}s$$

while, for their thin-layer version,

$$F = f \qquad G = g \qquad H = h - Re^{-1}d$$

and, for the Euler equations,

$$F = f \qquad G = g \qquad H = h$$

where:

$$
q = \begin{bmatrix} \rho \\ \rho u \\ \rho v \\ \rho w \\ E \end{bmatrix}
\qquad
f = \begin{bmatrix} \rho u \\ \rho u^2 + p \\ \rho uv \\ \rho uw \\ (E+p)u \end{bmatrix}
\qquad
g = \begin{bmatrix} \rho v \\ \rho uv \\ \rho v^2 + p \\ \rho wv \\ (E+p)v \end{bmatrix}
\qquad
h = \begin{bmatrix} \rho w \\ \rho uw \\ \rho vw \\ \rho w^2 + p \\ (E+p)w \end{bmatrix}
$$

$$
p = \begin{bmatrix} 0 \\ \tau_{xx} \\ \tau_{yx} \\ \tau_{zx} \\ \beta_x \end{bmatrix}
\qquad
r = \begin{bmatrix} 0 \\ \tau_{xy} \\ \tau_{yy} \\ \tau_{zy} \\ \beta_y \end{bmatrix}
\qquad
s = \begin{bmatrix} 0 \\ \tau_{xz} \\ \tau_{yz} \\ \tau_{zz} \\ \beta_z \end{bmatrix}
\qquad
d = \begin{bmatrix} 0 \\ \mu u_z \\ \mu v_z \\ (\lambda + 2\mu)w_z \\ \gamma k Pr^{-1} e_z + (\lambda + 2\mu)ww_z \end{bmatrix}
$$

$$\tau_{xx} = \lambda(u_x + v_y + w_z) + 2\mu u_x \qquad\qquad \beta_x = \gamma k Pr^{-1} e_x + u\tau_{xx} + v\tau_{xy} + w\tau_{xz}$$

$$\tau_{yy} = \lambda(u_x + v_y + w_z) + 2\mu v_y \qquad\qquad \beta_y = \gamma k Pr^{-1} e_y + u\tau_{yx} + v\tau_{yy} + w\tau_{yz}$$

$$\tau_{zz} = \lambda(u_x + v_y + w_z) + 2\mu w_z \qquad\qquad \beta_z = \gamma k Pr^{-1} e_z + u\tau_{zx} + v\tau_{zy} + w\tau_{zz}$$

$$\tau_{xy} = \tau_{yx} = \mu(u_y + v_x)$$

$$\tau_{xz} = \tau_{zx} = \mu(u_z + w_x)$$

$$\tau_{yz} = \tau_{zy} = \mu(v_z + w_y)$$

Here ρ, u, v, w, p and e are respectively density, velocity components in the x-, y- and z-directions, pressure and total energy per unit volume. This final quantity may be expressed as

$$E = \rho(e + \tfrac{1}{2}(u^2 + v^2 + w^2))$$

where the specific internal energy, e, is related to the pressure and density by the simple law of a calorically perfect gas

$$p = (\gamma - 1)\rho e$$

with γ denoting the ratio of specific heats. The coefficient of thermal conductivity, k, and the viscosity coefficients, λ and μ, are assumed to be functions only of temperature. Furthermore, by invoking Stokes' assumption of zero bulk viscosity, λ may be expressed in terms of the dynamic viscosity μ as

$$\lambda = -\frac{2}{3}\mu$$

Re and Pr denote the Reynolds and Prandtl numbers, respectively.

Although, for simplicity, the equations of motion are presented here written in Cartesian coordinates, Viviand [3] has shown that their strong conservation law form may be maintained under an arbitrary time-dependent transformation of coordinates. Explicit detail concerning the generalized coordinate version of these equations, which is employed in the computations to be discussed subsequently, has been provided by Pulliam and Steger [4] and need not be repeated here.

We note that the thin-layer approximation, in the words of Baldwin and Lomax [5], '... evolves directly from a realistic assessment of what is really being computed in a typical high Reynolds number Navier-Stokes simulation.' A highly stretched mesh is used to resolve the large flow gradients normal to the vorticity-generating surface. Consequently, because of limitations on computer capacity, the diffusion terms involving derivatives parallel to the surface are not resolved well enough to merit their computation.

Similar viscous terms are also neglected in the classical boundary layer approximation. However, while the boundary layer approximation replaces the normal momentum equation with the assumption that the normal pressure gradient is zero across the viscous layer, all momentum equations are retained in the thin-layer approximation and no assumptions are made concerning the pressure. Consequently, the separation point is not a singularity of the thin-layer model equations nor do the problems associated with matching a boundary layer solution to an inviscid outer flow occur when they are used.

In practice, the thin-layer assumption is implemented by using a body-fitted coordinate system and neglecting the viscous terms in the coordinate direction along the body. For Cartesian coordinates, with x and y representing the body-conforming coordinates, the thin-layer version of the Navier-Stokes equations is as given above.

The effects of turbulence are simulated by means of a two-layer algebraic eddy viscosity model. In the stress terms of the Navier-Stokes equations, the coefficient of dynamic viscosity, μ, is replaced by $\mu + \mu_t$, where μ_t is the coefficient of eddy viscosity. Similarly, in the heat flux terms, the coefficient of thermal conductivity, k, is replaced by $k+c_p\mu_t/Pr_t$, where Pr_t, is the turbulent Prandtl number. The eddy viscosity is determined by the method of Baldwin and Lomax [5] which is patterned after that of Cebeci [6] with modifications to avoid the necessity of finding the boundary layer edge. Details may be found in the work of Baldwin and Lomax.

SOLUTION PROCEDURE

Given a basic fine grid on which a numerical solution of Eqn.(1) is required and an explicit, conditionally stable solution procedure, the coarse-grid acceleration concept is to construct a collection of coarser grids by means of which the fine grid solution may be rapidly advanced, while respecting the stability limits on all grids. One cycle of the procedure consists of an application of a fine-grid integration step followed by one step of the coarse-grid scheme on each coarser grid. Information flow variations for this procedure are illustrated in Fig. 1. Further details will be provided subsequently.

Fine-Grid Scheme

The fine-grid integration scheme used here is the two-step Lax-Wendroff method due to MacCormack [7]. The forward predictor – backward corrector version of this method may be written as

$$\Delta q_{i,j,k} = -\frac{\Delta t}{\Delta x}(F^n_{i+1,j,k} - F^n_{i,j,k}) - \frac{\Delta t}{\Delta y}(G^n_{i,j+1,k} - G^n_{i,j,k})$$

$$-\frac{\Delta t}{\Delta z}(H^n_{i,j,k+1} - H^n_{i,j,k})$$

$$\delta q_{i,j,k} = -\frac{\Delta t}{2\Delta x}\left[(F^n_{i+1,j,k} - F^n_{i,j,k}) + (F'_{i,j,k} - F'_{i-1,j,k})\right] \quad (2)$$

$$-\frac{\Delta t}{2\Delta y}\left[(G^n_{i,j+1,k} - G^n_{i,j,k}) + (G'_{i,j,k} - G'_{i,j-1,k})\right]$$

$$-\frac{\Delta t}{2\Delta z}\left[(H^n_{i,j,k+1} - H^n_{i,j,k}) + (H'_{i,j,k} - H'_{i,j,k-1})\right]$$

where:

$$\delta q_{i,j,k} = \left[\ q(t + \Delta t) - q(t) \ \right]_{i,j,k}$$

$$q'_{i,j,k} = q^n_{i,j,k} + \Delta q_{i,j,k}$$

$$F'_{i,j,k} = F(q'_{i,j,k}) \qquad G'_{i,j,k} = G(q'_{i,j,k}) \qquad H'_{i,j,k} = H(q'_{i,j,k})$$

First derivatives in the viscous terms are backward differenced in the predictor and forward differenced in the corrector.

This version of MacCormack's scheme is used here for convenience. Any of its many variants could also be used, as could any other one or two-step Lax-Wendroff scheme [1]. In fact the class of fine-grid schemes with which the coarse-grid scheme may be applied appears to be quite large, including schemes not of Lax-Wendroff type [8].

Coarse-Grid Scheme

Given the fine-grid corrections, δq, we wish to use successively coarser grids to propagate this information throughout the computational domain, thus accelerating convergence to the steady state while maintaining the accuracy determined by the fine-grid discretization. Given a basic fine grid with the number of points in each direction expressible as $n(2^p)+1$ for p and n integers such that $p \geq 0$ and $n \geq 2$, where p is the number of grid coarsenings and n is the number of coarsest-grid intervals, let successively coarser grids be defined by successive deletion of every other point in each coordinate direction.

The coarse-grid scheme is derived from a one-step Lax-Wendroff method written in a coarse-grid setting. For example, if we choose a coarse-grid spacing which is double that of the fine grid, the one-step scheme is written as

$$\delta q_{i,j,k} = \frac{1}{8} \left\{ \left[(I + \frac{\Delta t}{\Delta x} A + \frac{\Delta t}{\Delta y} B + \frac{\Delta t}{\Delta z} C) \Delta q \right]_{i-1,j-1,k-1} \right.$$

$$+ \left[(I + \frac{\Delta t}{\Delta x} A - \frac{\Delta t}{\Delta y} B + \frac{\Delta t}{\Delta z} C) \Delta q \right]_{i-1,j+1,k-1}$$

$$+ \left[(I + \frac{\Delta t}{\Delta x} A - \frac{\Delta t}{\Delta y} B - \frac{\Delta t}{\Delta z} C) \Delta q \right]_{i-1,j+1,k+1}$$

$$+ \left[(I + \frac{\Delta t}{\Delta x} A + \frac{\Delta t}{\Delta y} B - \frac{\Delta t}{\Delta z} C) \Delta q \right]_{i-1,j-1,k+1} \tag{3a}$$

$$+ \left[(I - \frac{\Delta t}{\Delta x} A + \frac{\Delta t}{\Delta y} B + \frac{\Delta t}{\Delta z} C) \Delta q \right]_{i+1,j-1,k-1}$$

$$+ \left[(I - \frac{\Delta t}{\Delta x} A + \frac{\Delta t}{\Delta y} B - \frac{\Delta t}{\Delta z} C) \Delta q \right]_{i+1,j-1,k+1}$$

$$+ \left[(I - \frac{\Delta t}{\Delta x} A - \frac{\Delta t}{\Delta y} B - \frac{\Delta t}{\Delta z} C) \Delta q \right]_{i+1,j+1,k+1}$$

$$\left. + \left[(I - \frac{\Delta t}{\Delta x} A - \frac{\Delta t}{\Delta y} B + \frac{\Delta t}{\Delta z} C) \Delta q \right]_{i+1,j+1,k-1} \right\}$$

where:

$$\Delta q_{i+1,j+1,k+1} = -\Delta t (F_x + G_y + H_z)_{i+1,j+1,k+1} \tag{3b}$$

$$A = \frac{\partial F}{\partial q} \qquad B = \frac{\partial G}{\partial q} \qquad C = \frac{\partial H}{\partial q}$$

Now, instead of discretizing the flux balance (3b) on the coarse grid, we approximate it with a restriction of the most recently computed fine-grid correction

$$\Delta q_{coarse} \simeq R(\delta q_{fine}) .$$

The effect of this restricted fine-grid correction is then distributed according to Eqn.(3a) to obtain a coarse-grid correction. This is, in turn, prolonged to the fine grid to become the new fine-grid correction and update the fine-grid solution (see Fig. 1a). In two dimensions, Fig. 2 contrasts the one-step scheme written on a fine grid with the coarse-grid scheme written on a grid with double the spacing. We observe that in the basic integration scheme a correction at one grid point

R — Restriction of Latest Fine–Grid δq
as Coarse–Grid Δq

P — Prolongation of Coarse–Grid δq
to Fine Grid

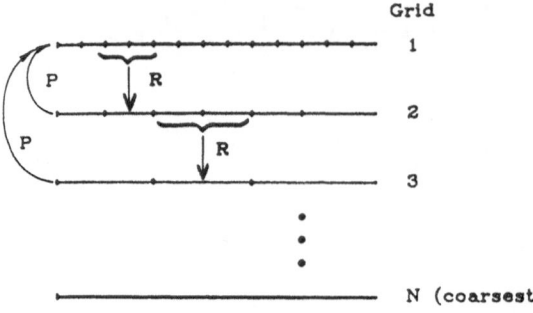

a.   Serial Algorithm

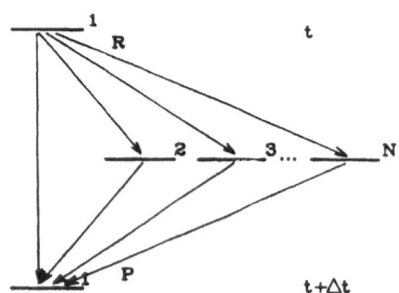

b.   Parallel Coarse–Grid
     Algorithm

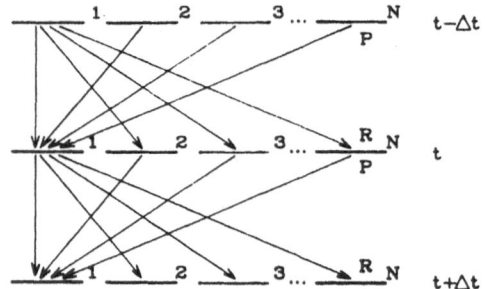

c.   Fully Parallel Algorithm

FIGURE 1. – Multiple–Grid Information Flow

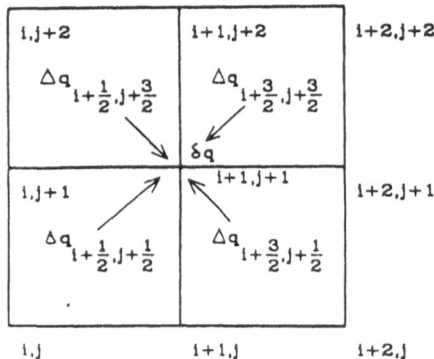

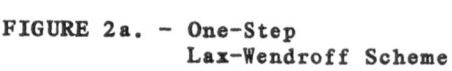

FIGURE 2a. – One–Step
           Lax–Wendroff Scheme

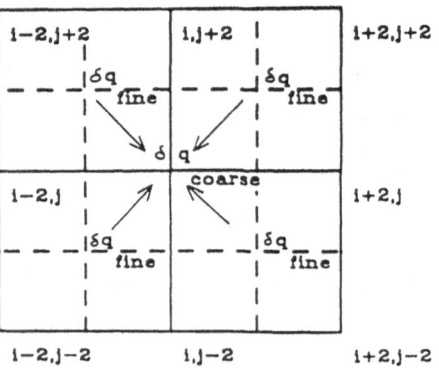

FIGURE 2b. – Coarse–Grid
           Scheme

affects only its nearest neighbors, while in a k-level multiple-grid scheme the same correction affects all points up to 2^{k-1} mesh spacings distant. Furthermore, since the coarse-grid scheme simply propagates the effects of fine-grid corrections, fine grid accuracy is maintained.

Convective Scheme. In the full coarse-grid scheme, as described above, the Jacobian matrices, A, B and C, are computed from the complete flux vectors, F, G and H, respectively. Consideration of the physical processes being modelled in a viscous flow computation lead to the formulation of an alternative coarse-grid scheme. Dissipative effects have a local character and their influence need not be taken into account in the construction of coarse-grid distribution formulae. Rather, it is the convective terms, with their global character, which are the key element in coarse-grid propagation. Hence, a coarse-grid scheme for viscous flow computations may be formulated on the basis of the inviscid equations of motion. Such a convective coarse-grid scheme is inherently more efficient than the full coarse-grid scheme because of the diminished computational effort associated with forming the Jacobian matrices of the Euler flux vectors rather than those of the viscous flux vectors. An additional benefit is that the convective coarse-grid scheme leads to a multiple-grid convergence acceleration procedure which is independent of the nature of the dissipative terms retained in the viscous model equations. That is to say: the coarse-grid scheme based on the Euler equations may be employed, without modification, to accelerate the convergence of viscous flow computations based on the Navier-Stokes equations, the thin-layer equations, or any other viscous model equations which contain the full inviscid Euler equations.

Flux-Based Scheme. Observe that a further simplification of the coarse-grid schemes is possible. We call the above coarse-grid schemes Jacobian-based schemes because of the presence of the flux vector Jacobian matrices. Their computation and storage and multiplications with them may be interpreted as causing inefficiencies in the coarse-grid schemes. The flux-based scheme described below is presented as a more efficient alternative.

A Lax-Wendroff-type coarse-grid scheme may be expressed as

$$\delta q_{coarse} = \Delta t q_t + \frac{\Delta t^2}{2} q_{tt}$$

By introducing Eqn.(1), this may be rewritten as

$$\delta q_{coarse} = -\Delta t(F_x + G_y + H_z) - \frac{\Delta t^2}{2}(F_x + G_y + H_z)_t$$

Recalling that

$$\Delta q = -\Delta t(F_x + G_y + H_z)$$

and temporally differencing the second-order term, we obtain

$$\delta q_{coarse} = \Delta q - \frac{\Delta t}{2}\left[(F_x + G_y + H_z)^{n+1} - (F_x + G_y + H_z)^n\right] \tag{4}$$

This is a flux-based coarse-grid scheme when Δq is approximated by a restriction of the latest fine-grid value of δq and

$$F^n = F(q^n) \quad , \quad G^n = G(q^n) \quad , \quad H^n = H(q^n)$$

$$F^{n+1} = F(q^n + \Delta q) \quad , \quad G^{n+1} = G(q^n + \Delta q) \quad , \quad H^{n+1} = H(q^n + \Delta q)$$

In this way the computation and storage of A, B and C and multiplications with them are eliminated. In place of these operations, we need only compute F^{n+1}, G^{n+1} and H^{n+1}, which may be immediately obtained from $q^n + \Delta q$. All other quantities are known.

Simultaneous Updating. An alternative implementation of the coarse-grid scheme has been devised to take advantage of multiple instruction-multiple data computer architectures. Instead of performing the coarse-grid computations sequentially (as illustrated in Fig. 1a), these grids can be treated independently (see Fig. 1b) on separate processors executing in parallel. Then the execution time required to do computations on the collection of coarse grids will be equivalent to the time required for the computation on only the first, or finest, coarse grid. This version of the simultaneous grid updating algorithm has been successfully simulated on the CYBER 205. A more highly parallel version consists of lagging the coarse-grid computations by one cycle so that, while the fine-grid computations are being executed on one processor, the coarse-grid updating of the previous fine-grid results is performed on other processors (see Fig. 1c). Hence, using this version, the multiple-grid acceleration of any fine-grid scheme adds essentially no cost in execution time.

Overrelaxation. Additional convergence acceleration may be obtained by weighting the coarse-grid corrections prior to their addition to the fine grid. This weighting has the effect of an overrelaxation factor.

188

Implementation

Boundary conditions are only enforced on the fine grid. This has the advantage of decoupling the coarse-grid scheme from both the physical and numerical nature of these boundary conditions. That is to say: the coarse-grid scheme always sees a Dirichlet problem. Any numerical damping terms which may be necessary are also only applied on the fine grid. This enhances the modularity of the coarse-grid scheme.

The computation of the Jacobian matrices used in the Jacobian-based coarse-grid schemes has a non-trivial influence on the coarse-grid operations count. We have found that the Jacobians need not be updated at each time cycle, but may be lagged by a substantial margin without any adverse effect on the resultant convergence history. Storage of the Jacobian matrices presents a problem due to presently available computer capacity. At every point in the computational domain, a 5x5 matrix of Jacobians is needed, and this additional storage results in serious paging problems on a 2-million word CYBER 205. This can be avoided by using the flux-based coarse-grid scheme.

In the present computations, injection is used as the restriction operator and linear interpolation is chosen as the prolongation operator. We note that these choices may not be optimal for use on highly stretched grids. We further note that, given the convective nature of the coarse-grid scheme used here, it is entirely plausible that, particularly in viscous flow computations, the coarser grids should have a more uniform structure than is obtained by successive deletion of every other grid line from a highly stretched fine grid.

As both the fine-grid solution procedure and the coarse-grid acceleration scheme used here are explicit, the resultant multiple-grid algorithm is vectorizable. Such vectorization has been performed for computation on a CDC CYBER 205. First the code was rewritten to take full advantage of the automatic vectorization which is performed by the CYBER 205 compiler. For the conservation vector q and the flux vectors F, G and H five quantities must be computed at every point in the three-dimensional domain. These vectors are stored in four-dimensional arrays. The array indices were arranged in decreasing length from left to right, and the DO loops containing these indices were likewise ordered so that the innermost loop corresponds to the longest dimension (i.e., the first index), the second innermost loop corresponds to the next longest dimension (the second index), and so on. These modifications enable access to contiguous locations in the vectors so that the loops automatically vectorize. When nested DO loops result in access to every point in the three-dimensional domain, including the boundaries, entire matrices are treated as single long vectors containing the whole flowfield (with lengths on the order of the maximum vector length of the CYBER 205). With these changes, the code compiled with the automatic vectorization option ran approximately 2 to 4 times as fast as the code using only scalar optimization.

Further vector speedup was obtained by implementing CYBER 205 explicit vector Fortran. Bit vectors were created for use in WHERE blocks to control storage for vectorized computations that involved only the interior points of the domain. Dynamic storage was introduced so that temporary vectors could be used to reduce the operations count. Vector intrinsic functions (such as Q8VCMPRS, Q8VMERG, etc.) were used to build contiguous vectors from the array elements needed on the coarse grids.

RESULTS

Extensive preparatory testing of the multiple-grid procedure has been carried out in two dimensions using scalar code [2]. The multiple-grid speedup in these test cases ranged from about 2 to 8 over a broad span of conditions, including subsonic and transonic inviscid flows and separated and attached, laminar and turbulent subsonic viscous flows with Reynolds numbers from 8.4×10^3 to 2.0×10^5.

Three-dimensional computations have been carried out for the geometry illustrated in Fig. 3, a rectilinear cascade of finite-span, swept blades mounted between endwalls. The sweep angle ranges from 0 to 26 degrees. The blade thickness to chord ratio ranges from 0.0 to 0.2. The subcritical computations are performed at an isentropic inlet Mach number of 0.5. The Mach number for the supercritical computations is 0.675. Cross-sections of typical fine grids used in this study are shown in Fig. 4.

Results are available for the same range of flow conditions previously tested in two dimensions. Sample three-dimensional inviscid results are shown in Figs. 5 and 6 for the subcritical and supercritical cases, respectively. Surface Mach number distributions are shown at the root, mid-span and tip cross-sections for the 26 degree sweep angle. Sample laminar and turbulent viscous separated flow results computed

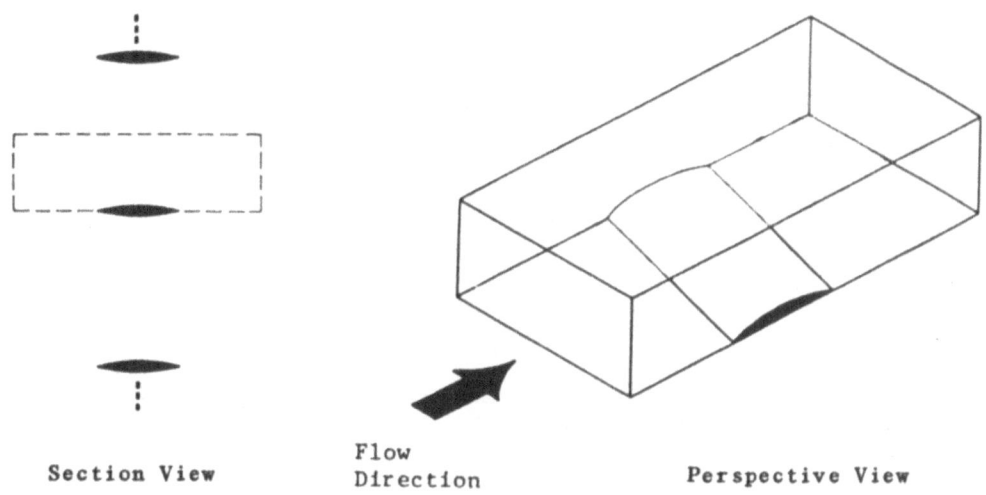

Section View   Flow
                   Direction   Perspective View

FIGURE 3. – Computational Domain for Three–Dimensional Cascade

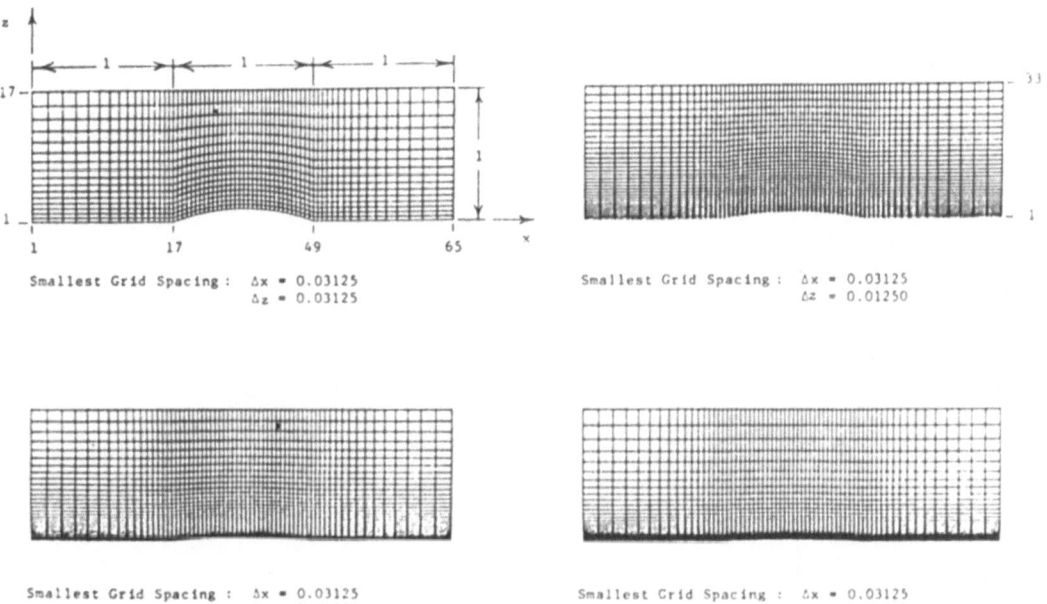

Smallest Grid Spacing : Δx = 0.03125   Smallest Grid Spacing : Δx = 0.03125
                       Δz = 0.03125   Δz = 0.01250

Smallest Grid Spacing : Δx = 0.03125   Smallest Grid Spacing : Δx = 0.03125
                       Δz = 0.00625   Δz = 0.00250

FIGURE 4. – Fine Grid Cross–Sections

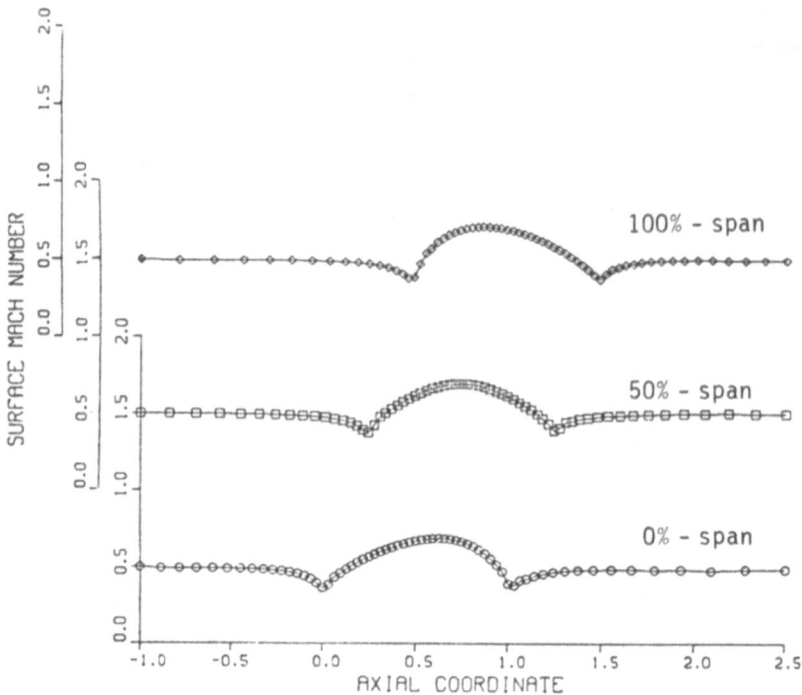

FIGURE 5. – Subcritical Inviscid Flow, 26 Deg. Sweep

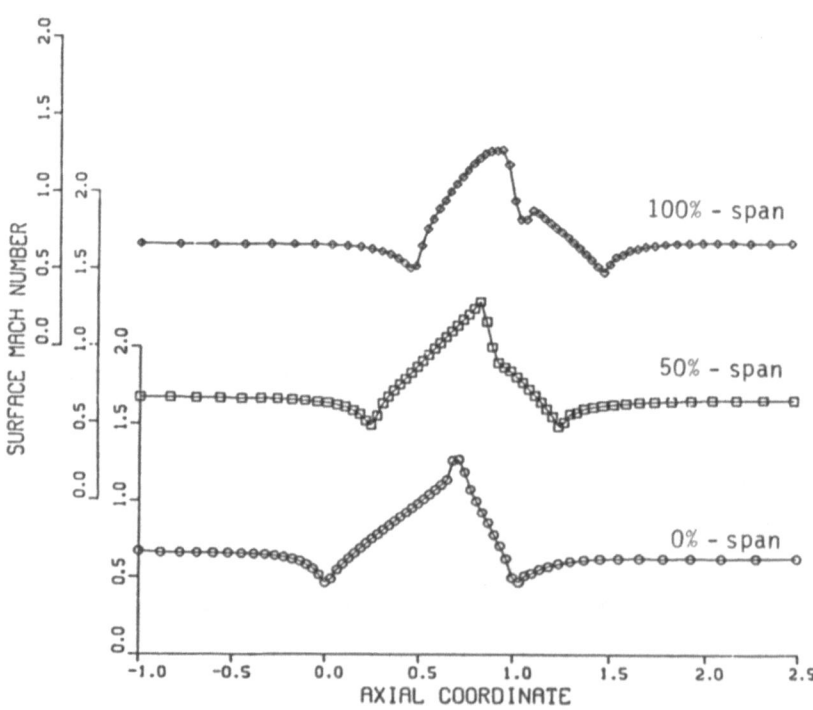

FIGURE 6. – Supercritical Inviscid Flow, 26 Deg. Sweep

using the thin-layer Navier-Stokes equations for blades with a 26 degree sweep angle are shown in Figs. 7 and 8.

Multiple-grid speedups for the three-dimensional test cases ranged from 2.5 to 4.7. Automatic vectorization, implemented on a two-pipe CYBER 205, improved performance over scalar execution by factors ranging from 2.6 to 4.2. These factors were further enhanced by explicit vectorization which yielded speedups between 3.6 and 5.7 over scalar performance (see Table I).

CONCLUSIONS

An explicit multiple-grid algorithm for the accelerated solution of the steady, three-dimensional Euler and Navier-Stokes equations has been presented. Computational results have been obtained for subsonic and transonic inviscid flows and for separated and attached, laminar and turbulent subsonic viscous flows.

The coarse-grid scheme used to accelerate convergence is compatible with a variety of fine-grid algorithms and may be used with any set of flow equations in the hierarchy ranging from the Euler equations to the full Navier-Stokes equations.

Several variants of the coarse-grid scheme have been presented. The flux-based version is more efficient and storage-conservative than its Jacobian-based analog. The coarse-grid computations may be executed on all grids in parallel, instead of sequentially, to take advantage of MIMD architectures.

Multiple-gridding has yielded speedups from 3 to 5 over a fairly broad range of flow conditions.

The entire multiple-grid procedure is explicit, and it has been vectorized for use on a CDC CYBER 205 computer.

Explicit vectorization has resulted in a version of the code which executes 4 to 6 times faster than the scalar version.

ACKNOWLEDGEMENT
The research reported here was funded by the Institute for Computational Studies. This support is gratefully acknowledged.

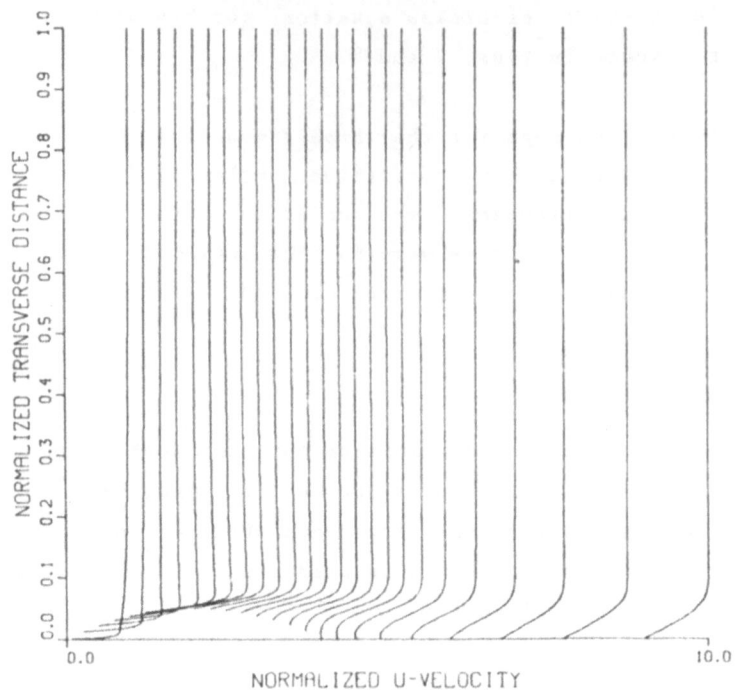

FIGURE 7. – Laminar Flow, Re = 3.4×10^4, 26 Deg. Sweep, 50% Span

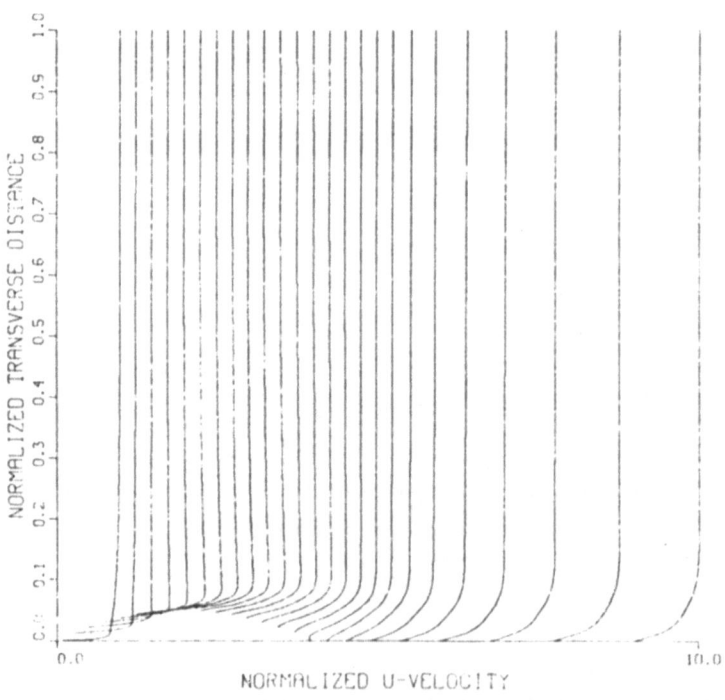

FIGURE 8. – Turbulent Flow, Re = 3.4×10^4, 26 Deg. Sweep, 50% Span

TABLE I. - SUMMARY OF THREE-DIMENSIONAL ALGORITHM PERFORMANCE

Vectorization Speedup

	Scalar	Automatic Vectorization	Explicit Vectorization
Inviscid Subcritical	1.0	4.2	5.7
Inviscid Supercritical	1.0	3.1	4.9
Turbulent Viscous	1.0	2.6	3.6

Multiple-Grid Speedup

Inviscid Subcritical	4.7
Inviscid Supercritical	2.5
Turbulent Viscous	4.4

REFERENCES
1. Johnson, G.M.- Multiple-Grid Acceleration of Lax-Wendroff Algorithms. NASA TM-82843, 1982.
2. Johnson, G.M.- Multiple-Grid Convergence Acceleration of Viscous and Inviscid Flow Computations. J. Appl. Math. Comp., Vol. 13, Nos. 3-4, pp. 375-398, Nov. 1983.
3. Viviand, H.- Formes Conservatives des Equations de la Dynamique des Gaz. Rech. Aerosp., No. 1, pp. 65-66, Jan.-Feb. 1974.
4. Pulliam, T.H. and Steger, J.L.- Implicit Finite-Difference Simulation of Three-Dimensional Compressible Flow. AIAA J., Vol. 18, No. 2, pp. 159-167, 1980.
5. Baldwin, B.S. And Lomax, H.- Thin-Layer Approximation and Algebraic Model for Separated Turbulent Flows. AIAA Paper 78-257, 1978.
6. Cebeci, T.- Calculation of Compressible Turbulent Boundary Layers with Heat and Mass Transfer. AIAA Paper 70-741, 1970.
7. MacCormack, R.W.- The Effect of Viscosity in Hypervelocity Impact Cratering. AIAA Paper 69-354, 1969.
8. Stubbs, R.M.- Multiple-Gridding of the Euler Equations with an Implicit Scheme. AIAA Paper 83-1945, 1983.

EXPLICITLY VECTORIZED FRONTAL ROUTINE FOR
HYDRODYNAMIC STABILITY AND BIFURCATION ANALYSIS BY
GALERKIN/FINITE ELEMENT METHODS

A. G. Boudouvis and L. E. Scriven

Supercomputer Institute and Department of
Chemical Engineering & Materials Science
University of Minnesota
Minneapolis, Minnesota 55455

ABSTRACT

Frontal routines are used for the solution of large, sparse and generally unsymmetric systems of linear equations arising in certain applications of the Galerkin/finite element approach to the iterative solution of
nonlinear boundary value problems. Based on Gauss elimination, these methods have advantages over banded matrix methods in that core requirements
and computation time can be considerably reduced. We present an explicitly
vectorized frontal routine for in-core solution of the equation set with
the CYBER 205 vector processor; the routine exploits the large central
memory, high execution speed and pipelined functional units of this supercomputer. Nodal arrays of the original algorithm of Hood are replaced by
long vectors of node labels to get the benefit of contiguity in memory; the
CYBER 200 FORTRAN vector syntax is used to reduce the number of non-vectorizable loops; the elimination procedure is modified to take advantage of
longer vectors and the associated large iteration counts. The algorithm is
enhanced by efficient threshold pivoting and determinant evaluation for
stability and bifurcation analysis. The assembly procedure is designed
with the flexibility to accommodate extra equations and unknowns for continuation in parameters and eigenanalysis. A listing of the code is provided.

INTRODUCTION

The method of subdomains with finite element basis functions combined
with the Galerkin method of weighted residuals and Newton iteration has
become well established for solving distributed problems such as the Navier-
Stokes equation together with appropriate boundary and interface equations
(e.g. the Young-Laplace equation) and other field equations (e.g. temperature, concentration, electric, magnetic). Moreover, the structure of the
solution families in parameter space, in particular the stability and multiplicity of solutions, can be analyzed by joining these methods with modern
computer-aided functional analysis (Keller, 1977; Brown et al., 1980).

In cases of time-independent problems the Galerkin/finite element
approach reduces the governing equations to a set of algebraic equations
for the coefficients of the basis functions; the coefficients are the values
of the unknown variables at the nodes of the finite element subdomains. The

technique of choice for solving the nonlinear equation set is Newton iteration (or a more cost-effective modification of the full Newton process), and the approximation to the solution at each iteration is found by Gauss elimination in the linearized equation set.

Frontal algorithms for Gauss elimination were originally developed for computers with limited core memory (Hood, 1976; Irons, 1970). The contributions from each equation to the entries of the Jacobian matrix of Newton iteration are calculated for the nodes in each element of the domain; those contributions are stored in a submatrix called the element stiffness matrix. The element stiffness matrices are added ("assembled") into the Jacobian matrix in an order dictated by element numbering. The assembly procedure is continued until a pre-set limited core area is filled. Within this assembled part of the Jacobian, pivots are selected from among those rows and columns that have been completed ("fully summed"), i.e. those to which no further contributions will arise in subsequent assembly of element stiffness matrices. The pivot row is used for Gauss elimination and then it is stored out of the core. The frontal algorithm alternates between assembly and elimination until all the element stiffness matrices are assembled and elimination is finished; then the solution is found by successively substituting the newly found unknowns in the previously eliminated rows in reverse order (back substitution).

Numerous modifications of Hood's original frontal algorithm can be found in the literature of sparse matrix technology. Duff (1984a,b) and Duff and Reid (1984) present frontal codes for nonsymmetric matrices and emphasize pivoting strategies, design of the user interface, storage requirements, isolation of machine dependencies and efficient manipulation of different data structures. Thompson and Shimazaki (1980) analyze a frontal procedure which allows for compact skyline-like storage and reduces the transfers to and from auxiliary storage; Walters (1980) and Coyle (1984) advance means of varying the front size to reduce both computation time and storage requirements. Ida and Lord (1980) develop a frontal solver best suited for solution of medium- to large-sized problems on minicomputers.

Although the chief advantage of frontal algorithms is that the main storage requirements are low, they can also reduce computation time appreciably if they function entirely in the main memory. It appears that this advantage is particularly useful on the CYBER 205 for direct solution of large equation sets, for example those originating in two-dimensional and three-dimensional boundary value problems.

We present here a modified frontal solver for in-core solution of large equation sets with the CYBER 205 vector processor. The solver is explicitly vectorized: manually introducing vector syntax into the code is advantageous because Hood's algorithm contains a large number of DO loops which, although they possess vectorizable structure, do not qualify for automatic vectorization or, in some cases, have relatively small iteration counts. Using the highly optimized SYSLIB functions and the vector syntax of CYBER 200 FORTRAN (1981) lowers the number of non-vectorizable loops and takes fewer scalar operations.

In what follows we describe recently emerged methods for computer-aided nonlinear analysis, which motivated development of the solver, and then describe in detail the explicitly vectorized frontal algorithm. The efficiency of our solver is illustrated with Galerkin/finite element analyses of stability and bifurcation of equilibrium shapes of fluid interfaces in external electric and magnetic fields. Unless an iterative solution technique is indicated, frontal methods appear to be well suited for bifurcation and stability analyses, because both boil down to the solution of large, sparse systems $\underline{\underline{A}}\underline{x}=\underline{b}$ and to exploitation of the by-products of the solution procedure.

198

A complete listing of our code and a glossary of terms are presented in the Appendix.

COMPUTER-AIDED ANALYSIS

Nonlinear partial differential equations that govern physical systems can be advantageously discretized and solved in terms of finite element basis functions (Strang and Fix, 1973), which are constructed to represent solutions subdomain-by-subdomain and to facilitate computerized determination of the coefficients of all of the local basis functions.

The physical domain is divided into subdomains and within each the solution, call it $f(x,y)$, is approximated by

$$f(x,y) = \sum_{i=1}^{N} a_i \phi^i(x,y)$$

where a_i are coefficients to be found and the ϕ^i are known finite element basis functions, both N in number.

Galerkin's method of weighted residuals consists of weighting the governing equations by each of the basis functions in turn, integrating over the entire domain, and setting each of the resulting algebraic expressions to zero. Galerkin's method produces a finite set of nonlinear algebraic equations for the coefficients a_i, denoted by

$$R_i(\underline{a};\underline{p}) = 0 \quad i = 1, 2, \ldots N \tag{1}$$

where $\underline{a}$ is the vector of the unknown coefficients a_i and $\underline{p}$ is the vector of parameters that appear in the problem.

The equation set (1) is most advantageously solved for the unknown coefficients by Newton's iteration process (or a variant) starting from an initial estimate that falls in the domain of convergence of the iteration. The approximate solution $\underline{a}^{(k+1)}$ at the (k+1)-st iteration is related to that at the k-th by

$$[\underline{\underline{J}}(\underline{a}^{(k)})] \, [\underline{a}^{(k+1)} - \underline{a}^{(k)}] = -\underline{R}(\underline{a}^{(k)}) \tag{2}$$

where $\underline{\underline{J}}$ is the Jacobian of equation set (1), i.e. the matrix of derivatives $J_{ij} = \partial R_i/\partial a_j$. Newton iteration is superior because its rate of convergence is asymptotically quadratic and because the Jacobian has valuable by-products: the information in the Jacobian can be used for analysis of the sensitivity of the solution to changes in parameter values, to analyze the stability and bifurcation of solutions, and to generate improved initial estimates for continuation in parameter values in order to compute whole families of solutions.

At each iteration equation (2) is linear and is to be solved by direct factorization of $\underline{\underline{J}}$. Because $\underline{\underline{J}}$ is, in many finite element applications, large, sparse and structured the matrix equation solvers of choice are frontal algorithms.

The solution at a new value, $p+\Delta p$, of a parameter is estimated from the converged solution at p by a continuation method. At regular points $(\det \underline{\underline{J}} \neq 0)$, first-order continuation gives good initial estimates:

$$\underline{a}^{(o)}(p+\Delta p) = \underline{a}(p) + \underline{a}_p \Delta p$$

Here the tangent vector $\underline{a}_p$ is calculated from the set of linear equations for the directional derivative of (1) along the solution family at p:

$$\underline{\underline{J}}(p) \ \underline{a}_p = - \ \partial\underline{R}/\partial p \tag{3}$$

This is solved simultaneously with the last Newton iteration at p. Because (1) is nonlinear it may have more than one solution — or even no solution (as when a solution family turns back with respect to the parameter). At bifurcation points two or more solution families intersect and the Jacobian is singular (det$\underline{\underline{J}}$=0); beyond limit (turning) points in a solution family no solutions of that family exist and at those points too the Jacobian is singular (see, for example Brown and Scriven, 1980). One of the two important classes of bifurcation points can be detected by change in sign of det$\underline{\underline{J}}$ in continuation along a solution family. The bifurcating family branches off in the direction of the null vector $\underline{z}$ of the Jacobian. A nearby solution $\underline{a}*$ in the bifurcating family can be found from the initial approximation (Rheinboldt, 1978)

$$\underline{a}*(p) = \underline{a}(p) + \varepsilon\underline{z}$$

where $\underline{a}(p)$ is the solution closest to the bifurcation point in the original family and ε is a small quantity found by trial. Limit points are circumvented by employing a suitable parameter, analogous to arc-length, that increases monotonically along the solution family (Keller, 1977). The technique is to choose adaptively as the continuation parameter $p*$ the most rapidly changing, between two successive iterations, of the unknowns $\underline{a}$ as well as the original parameter p (Abbott, 1978); another residual is defined which specifies the adaptive choice of $p*$:

$$R_{N+1} \equiv p* - p_o* - \Delta p* = 0 \tag{4}$$

Thus $p*$ is the value of the continuation parameter at a known solution $\underline{a}*\equiv(\underline{a},p)$ on the family of solutions and the parameter step size $\Delta p*$ is the chosen increment to the point on the same family where the new solution of the augmented system

$$R_i(\underline{a}*) = 0 \ , \ i = 1, 2, \ldots, N+1$$

is to be found.

The stability of each solution (state) with respect to solutions (states) that are adjacent, whether in the same solution family or in bifurcating families, is evaluated by monitoring sign changes of det$\underline{\underline{J}}$ and taking into account the connectivity of the solution families (Iooss and Joseph, 1980). To evaluate the stability to all admissible disturbances, including time periodic ones (i.e. Hopf bifurcation, the second important class) the procedure is to solve the generalized eigenproblem $\underline{\underline{J}}(\underline{a})\underline{x}_i = \lambda_i\underline{\underline{M}}\underline{x}_i$ for the eigenpairs $(\underline{x}_i,\lambda_i)$, which are normal modes and their exponential time factors. Here $\underline{\underline{J}}(\underline{a})$ is the Jacobian at a solution and $\underline{\underline{M}}$ is the basis function overlap matrix ("mass matrix"), i.e. its entries are the inner products of basis functions.

The robustness of the results to general refinement of the discretization is established by reducing the size of the subdomains. This is usually the most expensive part of computer-aided analysis, and the part where the cost-effectiveness of state-of-the-art machines is most evident.

DESCRIPTION OF THE ALGORITHM

Basic Principles

The essentials of the frontal method were laid out in the original paper of Hood (1976). We review the basic concepts to the extent necessary for appreciation of the modifications we have made in the explicitly vectorized frontal algorithm for in-core solution of the matrix equations that arise in Galerkin/finite element analysis.

The frontal method is to solve the set of linear equations $\underline{A}x = \underline{b}$ by Gauss elimination, i.e. LU-decomposition of $\underline{A}$, where $\underline{L}$ and $\underline{U}$ are lower and upper triangular matrices respectively; thus $\underline{A} = \underline{LU}$. The numerical stability of LU-decomposition is maintained with a suitable pivoting procedure, which might introduce interchanges of rows and of columns of the matrix $\underline{A}$ (Dahlquist and Björck, 1974).

The matrix $\underline{A}$ is formed from contributions from all elements, i.e. $\underline{A}$ is the sum

$$\underline{A} = \sum_{\ell} \underline{B}^{(\ell)}$$

where $\underline{B}^{(\ell)}$ is the stiffness matrix of element ℓ; each coefficient of the matrix $\underline{A}$ is thus the sum

$$a_{ij} = \sum_{\ell} \partial R_i^{(\ell)} / \partial a_j \tag{5}$$

where $R_i^{(\ell)}$ is the contribution from element ℓ to the residual associated with node i and a_j is the unknown associated with node j. A coefficient of the matrix $\underline{A}$ is called "fully summed" if the summation in (5) for this coefficient is completed, i.e. there are no more contributions to come from subsequent assembly of element stiffness matrices. A row (or column) of $\underline{A}$ is fully summed if all the coefficients of this row (or column) are fully summed. If at some stage in the assembly procedure the m-th row and n-th column are fully summed then the coefficient a_{mn} can be used as pivot to eliminate the m-th row and n-th column of $\underline{A}$ in the Gauss elimination scheme:

$$a_{ij} \leftarrow a_{ij} - a_{mj} a_{in} / a_{mn} \tag{6}$$

This can be done even if the coefficient a_{ij} is not yet fully summed. Thus rows and columns can be eliminated at early stages in the assembly procedure and a frontal algorithm takes advantage of this by never storing (or indeed generating) the whole $\underline{A}$ at one time. Instead, the contributions from individual elements to the global matrix $\underline{A}$ are assembled in succession in core until an allocated core matrix — the matrix EQ — reaches its pre-set dimension. Within this assembled part of the global matrix, pivots are selected among those rows and columns which are fully summed and the fully summed equations are sequentially eliminated and stored. The only equations which must remain in core at any stage are those in each assembled element which are not yet fully summed; these equations define a "front" that can be seen to propagate through the mesh of finite element nodes in the spatial domain as assembly/elimination progresses. All equations associated with nodes behind the front have either been eliminated or are awaiting selection as pivot equation in order to be eliminated. When the front has moved through the mesh of nodes and elimination is finished the solution is obtained by solving the two triangular systems $\underline{Ly} = \underline{b}$ and $\underline{Ux} = \underline{y}$.

Frontal algorithms for solving sparse and structured matrix equations are superior to traditional banded matrix solvers because they need less core memory for a given problem. The reason is the way they marshall the assembly/elimination process. In banded matrix algorithms an assembled equation associated with a node in an element must remain in core and cannot be eliminated, even if it is fully summed, until all the equations associated with all the nodes in the same element are fully summed. Hence the ordering of equations and thus the numbering of nodes are crucial to keep the bandwidth as small as possible in the assembly/elimination procedure. In contrast, in frontal algorithms a nodal equation can be eliminated as soon as it is fully summed; that is, it is the element numbering which must be chosen to minimize the frontwidth at each stage in the assembly/elimination procedure and not the nodal numbering. Because the frontwidth depends solely on element numbering, frontal algorithms are more efficient than band solvers for mesh refinement, a particularly valuable aspect because of the need to establish solutions that are insensitive to refining the discretization into subdomains. The refining does not require altering the nodal numbering in the original mesh, whereas a band algorithm may demand extensive re-numbering to preserve a small bandwidth (Irons, 1970).

Program Details

Hood's (1976) routine is here called FRONTAL and our new routine, VFRONTAL. In this section we describe the latter in parallel with FRONTAL and bring out the basic differences between the two.

Prefront process. The functions of prefront are (1) to set the order in which the equations in each element are assembled in the global matrix and (2) to establish the correspondence between the rows and the columns of the global matrix and those of the element stiffness matrices, i.e. to indicate the position in the global matrix at which each element stiffness matrix has to be assembled. These things are done in the subroutine VPREFR, which is called in the main subroutine VFRONT.

Each equation in an element stiffness matrix, like the basis function that gives rise to it in Galerkin's method, is associated with a node in the finite element mesh; there are as many equations associated with a node (nodal equations) as unknowns at this node (nodal unknowns). There may be more than one nodal unknown or "nodal degree of freedom" (ndf) associated with a node (for example the unknown nodal pressure and velocity components that arise in the discretization of the Navier-Stokes equation); in this case there are "multiple nodal degrees of freedom" at each node and the dimension of the element stiffness matrix is equal to the total number of ndf in the element. The position of each nodal equation (i.e. its row number) in an element stiffness matrix is indicated by the number (label) of the node in the element (local nodal numbering), whereas the position at which the same equation is assembled in the global matrix is dictated by the number of the node in the finite element mesh (global nodal numbering). When more than one ndf is associated with a node, the nodal degrees of freedom at this node are put in a standard order (e.g. pressure, x-component of velocity, y-component of velocity, etc.) and the corresponding nodal equations are assembled in both the element stiffness matrix and the global matrix in the same order. In Figure 1 a finite element mesh is shown that consists of four elements (in the order of the circled numbers) and nine nodes with one ndf per node. The global nodal numbering is depicted in Figure 1 while the local nodal numbering for the first element is shown in Figure 2. The nodal numberings at the global and local levels for the sample mesh in Figure 1 indicate the positions of each equation (row) in the global matrix and in the element stiffness matrix, respectively. For example, the contribution from the first element to the equation associated with node 4 (global position) is assembled in the third row of the stiffness matrix of element ① , whereas in the global matrix it is assembled in the fourth row.

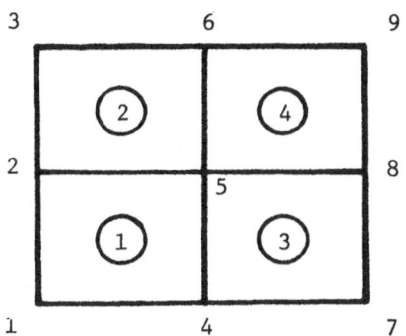

Figure 1. A sample finite element mesh showing element ordering and global nodal numbering

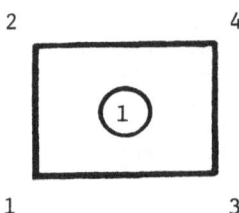

Figure 2. Local nodal numbering in the first element

The nodal numbering at both local and global levels can be arbitrary, but both must be defined by the programmer, as in Figures 1 and 2. The ways the nodal numbering is supplied by the programmer differ in FRONTAL and VFRONTAL. In FRONTAL, the global position of each node was stored in a two-dimensional array named NOP. The first argument in NOP (I,J) is the element number and the second is the local number of each node; thus for the example in Figure 1 NOP (2,4) = 6. The multiplicity of the degrees of freedom at each node was recorded in a separate vector named MDF; if the multiplicity of ndf was K at the node I (global position) then MDF(I) = K, and in the subsequent assembly procedure K rows were provided in the element stiffness matrix and the global matrix too for the assembly of the corresponding nodal equations. In contrast, in VFRONTAL the global positions of the nodes are stored in a vector named NT. For each element in turn the global numbers of the nodes are recorded in the order of the local nodal numbering; for the example in Figure 1 the NT array is

NT: 1,2,4,5;   2,3,5,6;   4,5,7,8;   5,6,8,9;

    element #1   element #2   element #3   element #4

Whatever the number of ndf at each node, they are put in a standard order, labelled, and then all are listed in sequence in the vector NT. This way of recording the nodal labels and the ndf multiplicities is superior to the one used in FRONTAL because the assembly order of each nodal equation in the global matrix is read immediately in the vector NT as the element stiffness matrices are assembled in succession in core. The NT vector is redundant; the global label of each ndf appears as many times as there are elements containing the corresponding node.

In the sequence of elements, the last one that contains a given ndf is noted in the prefront process, so that during the subsequent assembly/elimination process the completion of a row in the global matrix can be detected. This was done in FRONTAL by making negative the corresponding entry in the NOP array. This procedure complicates coding, during which the programmer has to refer to (positive) labels of nodal variables. In VFRONTAL, we perform a pre-pass on the NT vector which generates a single, integer, one-dimensional array named LEAP that records for each ndf the call at which it appears for the last time. An associated BIT-type vector (CYBER 200 FORTRAN, 1981) called B1 signals when a ndf appears for the first time in the NT vector.

The heading vectors KHED and LHED indicate the rows and columns of the global matrix in which each nodal equation is to be assembled. These vectors are set up in the prefront process by "compressing" (Q8VCMPRS vector command) the vector NT to eliminate the redundancies in ndf positions.

The construction of the element destination vectors — to point the rows and columns in matrix EQ where each element stiffness matrix is to be assembled — begins in prefront: the position in EQ for each row and column of the element stiffness matrices is shown by the entries in the vectors LKR and LKC; each of these vectors is constructed by picking up entries selectively from the NT vector (an operation controlled by the B1 BIT vector), arranging them appropriately by the Q8VEQI vector function and then merging (Q8VMERG) the intermediate position vectors in a single, long, destination vector.

The prefront procedure in VFRONTAL is explicitly vectorized. The nodal labels are recorded in the NT vector in the order they will appear in the subsequent assembly of the element stiffness matrices so that during the assembly they will be accessed in contiguous positions in memory, which is essential in vector processing. The vectors constructed during the prefront procedure (NT, LEAP, etc.) are redundant; their length, however, enhances the vectorability of the code because a large number of DO loops with small iteration counts that appear in FRONTAL are replaced in VFRONTAL by vector operations between long vectors. Moreover, this tactic avoids the DO loops of the FRONTAL code which incorporated numerical IF statements and so did not qualify for automatic vectorization. In VFRONTAL those DO loops are replaced by vector functions which exploit the BIT patterns (B1 vector) created in prefront.

Assembly. The element stiffness matrices are assembled by the main subroutine VFRONT after the prefront procedure is finished. The entries in each element stiffness matrix are calculated in the subroutine ABFIND; this subroutine has to be supplied by the programmer, because the composition of each element stiffness matrix depends, through the discretized governing equations, on the problem to be solved.

Each element stiffness matrix (ESTIFM) is assembled into a core area — the EQ matrix — in the position indicated by the corresponding element destination vectors. The size of the EQ matrix must be chosen to exceed the largest frontwidth by at least the number of element equations to be assembled next. The assembly order depends on the geometry and connectivity of the underlying problem and should be chosen to keep the frontwidth (and hence the size of EQ) small so as to reduce the arithmetic operations and storage requirements. In the present code the ordering is supplied by the programmer but it can be automated by means of procedures from graph theory and network analysis to arrive at optimal, or at least suboptimal ordering of the elements in the mesh (a step not yet undertaken).

The heading and destination vectors of the equations to be assembled were formed in FRONTAL in non-vectorizable nested DO loops. The same vectors are created in VFRONTAL in a sequence of explicitly vectorized operations: the number of rows and of columns to be added to the EQ matrix (because of the assembly of a new element stiffness matrix) are calculated; then the destination vectors are assigned to the equations from the LKR and LKC vectors that have been constructed in the prefront process; and the heading vectors for the equations are read from the vectors KHED and LHED formed in subroutine VPREFR.

This assembly procedure is designed to have the flexibility to accommodate a few relatively full rows, e.g. those arising from integral constraints like the fixed volume constraint from the incompressibility of a given liquid mass. Equations arising from integral constraints become fully summed at a late stage in the assembly procedure and cannot be eliminated until the contributions from all the elements in the domain of integration are assembled. Such equations we treat separately from the other discretized equations by assigning "artificial" nodes at the elements that contribute to the integral constraint; these extra nodes we number in such a way that the corresponding, relatively full rows are assembled last in the Jacobian and eliminated after finishing the assembly and elimination of all the other nodal equations.

The extra equation that describes Abbott's adaptive continuation (equation (4)) and the extra unknown — the continuation parameter p — are accommodated in the assembly procedure; the number of ndf is increased by one in the elements which contain equations that depend on the parameter p and the NT vector is augmented appropriately. The extra (N+1)-st row in the augmented Jacobian has zeros everywhere except at the entry that multiplies the adaptive continuation parameter p^* in the set $\underline{\underline{A}}\underline{x} = \underline{b}$; the value of that entry we set to unity (cf. Benner, 1983).

The sensitivities of the residuals R_i (i = 1,2,...N) to the parameter p, i.e. $\partial R_i / \partial p$, are calculated in ABFIND and they are needed for two purposes: in adaptive continuation, because they are the entries in the last column of the augmented Jacobian, and in getting initial estimates with first-order continuation (right-hand side of equation (3)).

The assembly procedure is suspended temporarily when the size of the allocated core matrix exceeds a pre-set critical value called NCRIT; then elimination begins.

Elimination and Back Substitution. Gauss elimination follows, in subroutine VFRONT, the search for assembled equations which have been fully summed. In FRONTAL this was done by checking, in DO loops over all rows and columns in EQ, which entries in the heading vectors were coded with negative sign; this procedure was non-vectorizable because the check was done with numerical IF statements. In VFRONTAL we find the fully summed rows and columns of EQ by using the information stored in the vectors KHED, LHED and LEAP; by-products of this procedure, which is explicitly vectorized, are the number and the position of the rows and columns of EQ from which pivots can be chosen.

In both codes the equations which are fully summed and correspond to assigned boundary values are modified. All entries of the corresponding rows are deleted, the values of the diagonals are set to unity, and the boundary condition values are inserted in the vector $\underline{b}$.

Pivots are selected among the fully summed rows and columns of the matrix EQ. Depending on the degree of assymetry of the Jacobian, different pivoting strategies can be used. In FRONTAL full pivoting was adopted. In VFRONTAL a threshold pivoting strategy is followed: the value of the first fully summed entry is checked and if it is greater than a certain pre-set value (e.g. 10^{-4}) it is accepted as a pivot; otherwise the pivot is sought among all the other fully summed rows and columns of EQ. This tactic proves to have no observable effect on the accuracy of solutions of a wide range of fluid mechanics problems and it significantly reduces computation time. The pivot row is normalized by division with the pivot; then elimination follows:

$$a_{i,j} \leftarrow a_{i,j} - \frac{a_{m,j}}{a_{m,n}} a_{i,n}$$

$$b_i \leftarrow b_i - \frac{b_m}{a_{m,n}} a_{i,n}$$

Here the indices m and n indicate the pivot row and column, respectively.

A by-product of the Gauss elimination procedure is the value of the determinant of the Jacobian, det$\underline{J}$, the absolute value of which is the product of the absolute values of the pivots (Dahlquist and Björck, 1974). In order to find the *sign* of det $\underline{J}$, any permutations of rows and of columns that have been introduced by the search for pivots are taken into account. When we iterate with an augmented Jacobian, as in arc-length continuation, the determinant of the original Jacobian can be found in the following way: after a converged solution is gotten, all the entries in the last row of the augmented Jacobian are set to zero except the diagonal one, which is set to unity; the determinant of the so-modified augmented Jacobian is found by direct factorization by the frontal method and it is equal to the determinant of the original Jacobian.

The elimination procedure is followed by the deletion of the pivot row and column from EQ. This is done by repositioning appropriately the rows and columns that remain in EQ. The heading vectors are then adjusted to account for the compression of the matrix; this is done in the main subroutine VFRONT.

The deleted pivot row and column have to be stored for use in back substitution. In FRONTAL, which was designed for computers with limited main memory, they were stored in disc, a low speed-access storage space. In VFRONTAL the normalized pivot row and the heading vectors for the pivot column are written in the vector ECV and stored *in the core*, a rapid-access space; this storage mode is cost-effective on CYBER 205 and other machines with large core.

The elimination continues until the number of the assembled rows falls below NCRIT. To decide whether to proceed with further assembly or to begin back substitution, the number of the last assembled element is checked. If there are no more elements remaining to be assembled, elimination is completed, and subroutine BACSUB is called. BACSUB starts by inserting the values of the assigned boundary conditions into the vector SK. Each pivot row is transferred in turn from the vector ECV and the unknown in that row is computed and stored in the solution vector SK.

EXAMPLE OF APPLICATION

The efficiency of the new frontal routine is illustrated with Galerkin/ finite element analysis of two cases drawn from electrohydrostatics and magnetohydrostatics. In the first case the evolution of the equilibrium shape of the interface between conducting fluid and less dense dielectric fluid is examined as the strength of a vertically applied electric field is increased. In the second case the evolution of the equilibrium shape of the surface of a pool of ferromagnetic liquid is examined as the strength of a vertically applied magnetic field is increased. The governing equations are those of electrostatics, magnetostatics and capillary hydrostatics. Both situations constitute *free boundary problems* because the interface position is unknown and has to be found by solving the equations for interface shape and electric or magnetic field distributions simultaneously. The contributions of the surface tension effects and of the electrostatic or magnetostatic pressure to the stress balance at the interface (the augmented Young-Laplace equation) make the problems nonlinear. The presence of the free boundary leads, through the dependence of field equations on its position, to unsymmetric Jacobian matrices in the Newton iteration.

Solutions, i.e. interface shapes and electric or magnetic field distributions, are computed and their stability and bifurcation behavior are examined by the Galerkin/finite element method. The physical domain is discretized, or tessellated, into quadrilateral elements and the solution is represented in terms of biquadratic basis functions. For the examples reported here relatively coarse discretizations proved adequate; the most refined consisted of 12 x 12 = 144 elements. The codes were written in standard FORTRAN, with explicit vector syntax, and executed on the CDC CYBER 205 supercomputer with the CYBER 200 FORTRAN compiler. The front-end was a CYBER 175 under the NOS operating system. With 652 unknowns each iteration consumed around 1.5 sec of CPU time, and the Newton process converged in three or four iterations. For purposes of comparison, the calculations were repeated on the CDC CYBER 170 scalar machine with the MNF compiler at the University of Minnesota. In this case auxiliary storage (disc) was used for the eliminated equations. The CYBER 170 CPU-time was ten times more than that required by the CYBER 205.

The execution time on CYBER 205 appears to depend not only on the size of the problems but also on the mode of pivoting. The CPU time is almost halved by the threshold pivoting described above (in comparison with full pivoting). The influence of problem size on the performance of the explicitly vectorized frontal routine will not be entirely clear until more problems of larger sizes have been tested. We anticipate that beyond some critical problem size, the vectorized frontal code will be still more cost-effective than for the present examples. This prediction is based on past experience with the performance of vectorized versus scalar codes (cf. Stanat and Nolen, 1982) and rests on the fact that execution speed increases with vector lengths (and hence with the number of unknowns) because of the amortization of the startup time for a vector operation (CYBER 205, 1982) over a larger number of calculations.

CONCLUDING REMARKS

An explicitly vectorized frontal routine is presented for the solution of large, sparse and generally unsymmetric systems of linear equations that arise in a broad class of Galerkin/finite element analyses of nonlinear boundary value problems. It is more cost-effective than Hood's original, scalar frontal routine because it exploits the pipelined functional units and high execution speed of the CYBER 205 supercomputer and can take advantage of the large storage capacity of central memory. Moreover, it is well suited, in combination with Newton iteration or more cost-effective modifications thereof, to the analysis of the solution space of the nonlinear problems, namely the sensitivities of solutions to changes in parameters, the stability of solutions to small disturbances and the bifurcations of families of solutions. These aspects are often more important in applications than are the individual solutions themselves.

It appears that skyline algorithms (Hasbani and Engelman, 1979) and other frontal-like routines can be similarly vectorized. Such algorithms reportedly can compete favorably with frontal ones, particularly when they are implemented with certain modifications of Newton iteration (Engelman et al., 1981). Frontal methods can also be tailored to the matrix structure of systems of equations that arise in finite difference discretizations of nonlinear differential equations (Duff, 1984a). Thus frontal methods seem sure to take an important place in the library of direct matrix solvers for cost-effective use of vector processors in large-scale scientific computation.

ACKNOWLEDGEMENTS

This work is supported by the University of Minnesota Computer Center, the Control Data Corporation through the Institute for Mathematics and Its Applications at the University of Minnesota, and a grant-in-aid for research in Fluid Physics from the Exxon Foundation.

REFERENCES

Abbott, J. P., 1978, An efficient algorithm for the determination of certain bifurcation points, J. Comput. Appl. Math., 4:19.

Benner, R. E., 1983, Equilibria, Stability and Bifurcations in the Physics of Fluid Interfaces, Ph.D. Thesis, Univ. of Minnesota, Minneapolis.

Brown, R. A., and Scriven, L. E., 1980, The shape and stability of captive rotating drops, Phil. Trans. Roy. Soc. Lond., A297:51.

Brown, R. A., Scriven, L. E., and Silliman, W. J., 1980, Computer-aided analysis of nonlinear problems in transport phenomena, in: "New Approaches to Nonlinear Problems in Dynamics," P. J. Holmes, ed., SIAM, Philadelphia, PA.

Coyle, D. J., 1984, The Fluid Mechanics of Roll Coating: Steady Flows, Stability, and Rheology, Ph.D. Thesis, University of Minnesota, Mpls.

CYBER 200 FORTRAN, 1981, Reference Manual, Version 2, Control Data Corporation, Sunnyvale, CA.

CYBER 205, 1982, Users Guide, Control Data Corporation, Minneapolis, MN.

Dahlquist, G., and Björck, A., 1974, "Numerical Methods," Prentice-Hall, Englewood Cliffs, NJ.

Duff, I. S., 1984a, Design features of a frontal code for solving sparse unsymmetric linear systems out-of-core, SIAM J. Sci. Stat. Comput., 5:270.

Duff, I. S., 1984b, Direct methods for solving sparse systems of linear equations, SIAM J. Sci. Stat. Comput., 5:605.

Duff, I. S., and Reid, J. K., 1984, The multifrontal solution of unsymmetric sets of linear equations, SIAM J. Sci. Stat. Comput., 5:633.

Engelman, M. S., Strang, G., and Bathe, K. J., 1981, The application of quasi-Newton methods in fluid mechanics, Int. J. Num. Meth. Engng., 17:707.

Hasbani, Y., and Engelman, M. S., 1979, Out-of-core solution of linear equations with non-symmetric coefficient matrix, Comput. Fluids, 7:13.

Hood, P., 1976, Frontal solution program for unsymmetric matrices, Int. J. Num. Meth. Engng., 10:379.

Ida, N., and Lord, W., 1984, Solution of linear equations for small computer systems, Int. J. Num. Meth. Engng., 20:625.

Iooss, G., and Joseph, D. D., 1980, "Elementary Stability and Bifurcation Theory," Springer-Verlag, New York.

Irons, B. M., 1970, A frontal solution program for finite element analysis, Int. J. Num. Meth. Engng., 2:5.

Keller, H. B., 1977, Numerical solution of bifurcation and nonlinear eigenvalue problems, in: "Applications of Bifurcation Theory," P. Rabinowitz, ed., Academic Press, New York.

Rheinboldt, W. C., 1978, Numerical methods for a class of finite dimensional bifurcation problems, SIAM J. Num. Anal., 15:1.

Stanat, P. L., and Nolen, J. S., 1982, Performance Comparisons for Reservoir Simulation Problems in Three Supercomputers, Proceedings of the Symposium on CYBER 205 Applications, August 12-13, Institute for Computational Studies, Colorado State University, Fort Collins, Colorado.

Strang, G., and Fix, G. J., 1973, "An Analysis of the Finite Element Method," Prentice-Hall, Englewood Cliffs, NJ.

Thompson, E., and Shimazaki, Y., 1980, A frontal procedure using skyline storage, Int. J. Num. Meth. Engng., 15:889.

Walters, A., 1980, The frontal method in hydrodynamics simulations, Comput. Fluids, 8:265.

```
          SUBROUTINE VFRONT
C*****  SUBROUTINE FOR FRONTAL ELIMINATION *****
          COMMON/STABLE/DET
          ROWWISE EQ(90,90)
          BIT B1(1400),BL(90)
          COMMON/VM1/NP,NE,NCN(100),NELL,NTRA,ESTIFM(14,14)
          COMMON/VM2/NCOD(464),BC(464),R1(464),SK(8100)
          COMMON/V14/ECV(100000),KHED1(464),LHED1(464),QQ(90),NT(1400),NH2
          COMMON/V12/KHED(464),LEAP(1400),LHED(464),LKC(1400),LKR(1400),B1
          DIMENSION IQ(464),IS(464),NRS(464),NCS(464),LKC1(1400),LKR1(1400)
          DIMENSION LDEST(14),KDEST(14),JMOD(90),PVKOL(90),KPIV(90),LPIV(90)
          DIMENSION S1HED(90),SHED(90),PIVO(90),LX(90),ICH(14),ICCR(14)
          EQUIVALENCE(SK(1),EQ(1,1))
          NMAX=90
          NCRIT=NMAX-50
          ICE=1
          NELL=0
          NRS(1;NP)=0
          NCS(1;NP)=0
          IF(NTRA.EQ.0) GO TO 14
          CALL VPREFR
   14     LHED1(1;NP)=LHED(1;NP)
          KHED1(1;NP)=KHED(1;NP)
          LKC1(1;NH2)=LKC(1;NH2)
          LKR1(1;NH2)=LKR(1;NH2)
          NTRA=0
          IQ(1;NP)=Q8VCMPRS(LEAP(1;NH2),B1(1;NH2);IQ(1;NP))
          IS(1;NP)=IQ(1;NP)
C*****  ASSEMBLY *****
          KEL=1
          LCOL1=0
          KROW1=0
          EQ(1,1;NMAX*NMAX)=0.0
   18     CONTINUE
          IF(NELL.NE.0)  KEL=KEL+NCN(NELL)
          NELL=NELL+1
          CALL ABFIND
          N=NELL
          KC=0
C*****  SET UP HEADING VECTORS *****
          LEND=NCN(N)
          KETL=KEL+LEND-1
          LCOL=Q8SMAX(LKC1(1;KETL))
          KROW=Q8SMAX(LKR1(1;KETL))
          LDEST(1;LEND)=LKC1(KEL;LEND)
          KDEST(1;LEND)=LKR1(KEL;LEND)
          I67=LCOL-LCOL1
          ICH(1;I67)=Q8VCMPRS(NT(KEL;LEND),B1(KEL;LEND);ICH(1;I67))
          ICCR(1;I67)=Q8VCMPRS(LEAP(KEL;LEND),B1(KEL;LEND);ICCR(1;I67))
          LHED1(LCOL1+1;I67)=ICH(1;I67)
          KHED1(KROW1+1;I67)=ICH(1;I67)
          IQ(LCOL1+1;I67)=ICCR(1;I67)
          IS(KROW1+1;I67)=ICCR(1;I67)
          WHERE(IQ(1;LCOL).EQ.NELL)
          LHED1(1;LCOL)=(-1)*LHED1(1;LCOL)
          END WHERE
          WHERE(IS(1;KROW).EQ.NELL)
          KHED1(1;KROW)=(-1)*KHED1(1;KROW)
          END WHERE
          IF(KROW.LE.NMAX.AND.LCOL.LE.NMAX) GO TO 54
          NERROR=2
          WRITE(6,417) NERROR
          STOP
   54     LEND=NCN(N)
          DO 56 L=1,LEND
          LL=LDEST(L)
          KEND=NCN(N)
          DO 56 K=1,KEND
          KK=KDEST(K)
   56     EQ(KK,LL)=EQ(KK,LL)+ESTIFM(K,L)
          LCOL1=LCOL
          KROW1=KROW
          IF(KROW.LT.NCRIT.AND.NELL.LT.NE) GO TO 18
C*****  FIND OUT WHICH MATRIX ELEMENTS ARE FULLY SUMMED *****
   60     SHED(1;KROW)=VREAL(KHED1(1;KROW);SHED(1;KROW))
          S1HED(1;KROW)=VABS(SHED(1;KROW);S1HED(1;KROW))
          S1HED(1;KROW)=(-1)*S1HED(1;KROW)
          LX(1;KROW)=Q8VEQI(SHED(1;KROW),S1HED(1;KROW);LX(1;KROW))
          BL(1;KROW)=LX(1;KROW).NE.KROW
          LX(1;KROW)=LX(1;KROW)+1
          KR=Q8SCNT(BL(1;KROW))
          KPIV(1;KR)=Q8VCMPRS(LX(1;KROW),BL(1;KROW);KPIV(1;KR))
          IR=0
          SHED(1;LCOL)=VREAL(LHED1(1;LCOL);SHED(1;LCOL))
          S1HED(1;LCOL)=VABS(SHED(1;LCOL);S1HED(1;LCOL))
          S1HED(1;LCOL)=(-1)*S1HED(1;LCOL)
```

```
            LX(1;LCOL)=Q8VEQI(SHED(1;LCOL),S1HED(1;LCOL);LX(1;LCOL))
            BL(1;LCOL)=LX(1;LCOL).NE.LCOL
            LX(1;LCOL)=LX(1;LCOL)+1
            LC=Q8SCNT(BL(1;LCOL))
            LPIV(1;LC)=Q8VCMPRS(LX(1;LCOL),BL(1;LCOL);LPIV(1;LC))
            DO 68 K=1,KROW
            KT=KHED1(K)
            IF(KT.GE.0) GO TO 68
            KRO=IABS(KT)
            IF(NCOD(KRO).NE.1) GO TO 68
            IR=IR+1
            JMOD(IR)=K
            NCOD(KRO)=2
            R1(KRO)=BC(KRO)
      68    CONTINUE
C***** MODIFY EQUATIONS WITH APPLIED BOUNDARY CONDITIONS *****
            IF(IR.EQ.0) GO TO 71
            DO 70 IRR=1,IR
            K=JMOD(IRR)
            KH=IABS(KHED1(K))
            DO 70 L=1,LCOL
            EQ(K,L)=0
            LH=IABS(LHED1(L))
      70    IF(LH.EQ.KH) EQ(K,L)=1.
      71    CONTINUE
            IF(KR.GT.0.AND.LC.GT.0)  GO TO 72
            NERROR=3
            WRITE(6,418) NERROR,NELL
      72    CONTINUE
C***** THRESHOLD PIVOTING *****
            KPIVRO=KPIV(1)
            LPIVCO=LPIV(1)
            PIVOT=EQ(KPIVRO,LPIVCO)
            IF(ABS(PIVOT).GT.1.E-4) GO TO 211
            PIVOT=0.0
            DO 74 L=1,LC
            LPIVC=LPIV(L)
            DO 74 K=1,KR
            KPIVR=KPIV(K)
            PIVA=EQ(KPIVR,LPIVC)
            IF(ABS(PIVA).LT.ABS(PIVOT)) GO TO 74
            PIVOT=PIVA
            LPIVCO=LPIVC
            KPIVRO=KPIVR
      74    CONTINUE
C***** NORMALIZE PIVOT ROW *****
     211    KPIC=LPIVCO
            KPIR=KPIVRO
            KRO=IABS(KHED1(KPIVRO))
            LCO=IABS(LHED1(LPIVCO))
            NHLP=KRO+LCO+NRS(KRO)+NCS(LCO)
            DET=DET*PIVOT*(-1)**NHLP
            NRS(KRO+1;NP-KRO)=NRS(KRO+1;NP-KRO)-1
            NCS(LCO+1;NP-LCO)=NCS(LCO+1;NP-LCO)-1
            IF(ABS(PIVOT).LT.1.E-11) WRITE(6,476) NELL,KRO,LCO,PIVOT
            IF(ABS(PIVOT).LT.1.E-11) STOP
            QQ(1;LCOL)=EQ(KPIVRO,1;LCOL)/PIVOT
            RHS=R1(KRO)/PIVOT
            R1(KRO)=RHS
            PVKOL(KPIVRO)=PIVOT
C***** ELIMINATE AND DELETE THE PIVOT ROW & COLUMN *****
            IF(KPIVRO.EQ.1) GO TO 104
            KPIVR=KPIVRO-1
            DO 100 K=1,KPIVR
            KRW=IABS(KHED1(K))
            FAC=EQ(K,LPIVCO)
            PVKOL(K)=FAC
            IF(LPIVCO.EQ.1.OR.FAC.EQ.0) GO TO 88
            LPIVC=LPIVCO-1
            EQ(K,1;LPIVC)=EQ(K,1;LPIVC)-FAC*QQ(1;LPIVC)
      88    IF(LPIVCO.EQ.LCOL) GO TO 96
            LPIVC=LPIVCO+1
            LPI2=LPIVC-1
            EQ(K,LPI2;LCOL-LPI2)=EQ(K,LPIVC;LCOL-LPI2)-FAC*QQ(LPIVC;LCOL-LPI2)
      96    R1(KRW)=R1(KRW)-FAC*RHS
     100    CONTINUE
     104    IF(KPIVRO.EQ.KROW) GO TO 128
            KPIVR=KPIVRO+1
            DO 120 K=KPIVR,KROW
            KRW=IABS(KHED1(K))
            FAC=EQ(K,LPIVCO)
            PVKOL(K)=FAC
            IF(LPIVCO.EQ.1) GO TO 112
            LPIVC=LPIVCO-1
            EQ(K-1,1;LPIVC)=EQ(K,1;LPIVC)-FAC*QQ(1;LPIVC)
     112    IF(LPIVCO.EQ.LCOL) GO TO 120
            LPIVC=LPIVCO+1
            LPI3=LCOL-LPIVC+1
            EQ(K-1,LPIVC-1;LPI3)=EQ(K,LPIVC;LPI3)-FAC*QQ(LPIVC;LPI3)
     120    R1(KRW)=R1(KRW)-FAC*RHS
     128    CONTINUE
C***** WRITE PIVOT EQUATION ON ECV *****
            ECV(ICE;LCOL)=QQ(1;LCOL)
```

210

```
          ICE=ICE+LCOL
          ECV(ICE;LCOL)=LHED1(1;LCOL)
          ICE=ICE+LCOL
          ECV(ICE)=KRO
          ECV(ICE+1)=LCOL
          ECV(ICE+2)=LPIVCO
          ICE=ICE+3
          DO 129 K=1,KROW
  129     EQ(K,LCOL)=0.
          EQ(KROW,1;LCOL)=0.
C*****  REARRANGE THE HEADING VECTORS *****
          LCOL=LCOL-1
          KROW=KROW-1
          IF(LPIVCO.EQ.LCOL+1) GO TO 136
          LHED1(LPIVCO;LCOL-LPIVCO+1)=LHED1(LPIVCO+1;LCOL-LPIVCO+1)
          IQ(LPIVCO;LCOL-LPIVCO+1)=IQ(LPIVCO+1;LCOL-LPIVCO+1)
  136     CONTINUE
          IF(KPIVRO.EQ.KROW+1) GO TO 144
          KHED1(KPIVRO;KROW-KPIVRO+1)=KHED1(KPIVRO+1;KROW-KPIVRO+1)
          IS(KPIVRO;KROW-KPIVRO+1)=IS(KPIVRO+1;KROW-KPIVRO+1)
  144     CONTINUE
          WHERE(LKC1(1;NH2).GT.KPIC)
          LKC1(1;NH2)=LKC1(1;NH2)-1
          END WHERE
          WHERE(LKR1(1;NH2).GT.KPIR)
          LKR1(1;NH2)=LKR1(1;NH2)-1
          END WHERE
          LCOL1=LCOL
          KROW1=KROW
C*****  DETERMINE WHETHER TO ASSEMBLE, ELIMINATE OR BACKSUBSTITUTE *****
          IF(KROW.GT.NCRIT) GO TO 60
          IF(NELL.LT.NE) GO TO 18
          IF(KROW.GT.1) GO TO 60
          LCO=IABS(LHED1(1))
          KPIVRO=1
          PIVOT=EQ(1,1)
          KRO=IABS(KHED1(1))
          LPIVCO=1
          NHLP=KRO+LCO+NRS(KRO)+NCS(LCO)
          DET=DET*PIVOT*(-1)**NHLP
          QQ(1)=1.0
          IF(ABS(PIVOT).LT.1.E-11) WRITE(6,476) NELL,KRO,LCO,PIVOT
          R1(KRO)=R1(KRO)/PIVOT
          ECV(ICE)=QQ(1)
          ICE=ICE+1
          ECV(ICE)=LHED1(1)
          ICE=ICE+1
          ECV(ICE)=KRO
          ECV(ICE+1)=LCOL
          ECV(ICE+2)=LPIVCO
          ICE=ICE+3
          CALL BACSUB(ICE)
          RETURN
  417     FORMAT(/8H NERROR=,I5/2X,*NMAX-NCRIT NOT LARGE ENOUGH FOR *
         $*ASSEMBLY OF THE NEXT ELEMENT---INCREASE NMAX OR LOWER NCRIT*/)
  418     FORMAT(/8H NERROR=,I5,6H NELL=,I5/2X,*NO MORE ROWS FULLY SUMMED *
         $*DUE TO (1) INCORRECT CODING OF NT ARRAY; OR (2) INCORRECT
         $ VALUE OF NCRIT.*)
  476     FORMAT(44H WARNING-MATRIX SINGULAR OR ILL CONDITIONED ,
         $ *NELL=*,I4,2X,*KRO=*,I5,2X,*LCO=*,I5,2X,*PIVOT=*,E10.2)
          END


          SUBROUTINE VPREFR
          BIT B1(1400),B0(1400)
          COMMON/VM1/NP,NE,NCN(100),NELL,NTRA,ESTIFM(14,14)
          COMMON/V12/KHED(464),LEAP(1400),LHED(464),LKC(1400),LKR(1400),B1
          COMMON/V14/ECV(100000),KHED1(464),LHED1(464),QQ(90),NT(1400),NH2
          DIMENSION LB0(1400),LS0(1400),LCD(464),SB0(1400),
         $SB1(1400),LS1(1400),IC(1400),RNT(1400),TNR(1400),IN2(1400)
          NH2=Q8SSUM(NCN(1;NE))
          CALL NOPNT        *
          RNT(1;NH2)=VREAL(NT(1;NH2);RNT(1;NH2))
          TNR(1;NH2)=Q8VREV(RNT(1;NH2);TNR(1;NH2))
          LKC(1;NH2)=Q8VEQI(RNT(1;NH2),TNR(1;NH2);LKC(1;NH2))
          IN2(1;NH2)=(-1)*LKC(1;NH2)+NH2
          K=1
          DO 151 I=1,NE
          K1=NCN(I)
          LKC(K;K1)=I
  151     K=K+K1
          LEAP(1;NH2)=Q8VGATHR(LKC(1;NH2),IN2(1;NH2);LEAP(1;NH2))
          LCD(1;NP)=Q8VINTL(1,1;LCD(1;NP))
          SB1(1;NP)=VREAL(LCD(1;NP);SB1(1;NP))
          LKR(1;NP)=Q8VEQI(SB1(1;NP),RNT(1;NH2);LKR(1;NP))
          LKR(1;NP)=LKR(1;NP)+1
          NP1=NP+1
          N22=NH2-NP
          LKC(1;NP)=LKR(1;NP)
          LKC(NP1;N22)=LKR(1)
          IC(1;NH2)=1
          IN2(1;NH2)=0
          IN2(1;NH2)=Q8VSCATR(IC(1;NH2),LKC(1;NH2);IN2(1;NH2))
          B1(1;NH2)=IN2(1;NH2).EQ.1
```

```
           BØ(1;NH2)=IN2(1;NH2).NE.1
           LKC(1;NH2)=Q8VXPND(LCD(1;NP),B1(1;NH2);LKC(1;NH2))
           LS1(1;NP)=Q8VCMPRS(LKC(1;NH2),B1(1;NH2);LS1(1;NP))
           LHED(1;NP)=Q8VCMPRS(NT(1;NH2),B1(1;NH2);LHED(1;NP))
           KHED(1;NP)=LHED(1;NP)
           LBØ(1;N22)=Q8VCMPRS(NT(1;NH2),BØ(1;NH2);LBØ(1;N22))
           SBØ(1;N22)=VREAL(LBØ(1;N22);SBØ(1;N22))
           SB1(1;NP)=VREAL(LHED(1;NP);SB1(1;NP))
           LSØ(1;N22)=Q8VEQI(SBØ(1;N22),SB1(1;NP);LSØ(1;N22))
           LSØ(1;N22)=LSØ(1;N22)+1
           LKC(1;NH2)=Q8VMERG(LS1(1;NP),LSØ(1;N222),B1(1;NH2);LKC(1;NH2))
           LKR(1;NH2)=LKC(1;NH2)
           RETURN
           END

           SUBROUTINE BACSUB(ICE)
           ROWWISE EQ(90,90)
           COMMON/VM1/NP,NE,NCN(100),NELL,NTRA,ESTIFM(14,14)
           COMMON/VM2/NCOD(464),BC(464),R1(464),SK(8100)
           COMMON/V14/ECV(100000),KHED1(464),LHED1(464),QQ(90),NT(1400),NH2
           EQUIVALENCE(SK(1),EQ(1,1))
           DO 4 I=1,NP
     4     SK(I)=BC(I)
           DO 32 IV=1,NP
           ICE=ICE-3
           KRO=ECV(ICE)
           LCOL=ECV(ICE+1)
           LPIVCO=ECV(ICE+2)
           IF(IV.EQ.1) GO TO 300
           ICE=ICE-LCOL
           DO 201 III=1,LCOL
     201   LHED1(III)=ECV(ICE-1+III)
           ICE=ICE-LCOL
           DO 202 III=1,LCOL
     202   QQ(III)=ECV(ICE-1+III)
           GO TO 301
     300   ICE=ICE-1
           LHED1(1)=ECV(ICE)
           ICE=ICE-1
           QQ(1)=ECV(ICE)
     301   CONTINUE
           LCO=IABS(LHED1(LPIVCO))
           IF(NCOD(LCO).GT.0) GO TO 24
           GASH=0.0
           QQ(LPIVCO)=0.0
           DO 16 L=1,LCOL
     16    GASH=GASH-QQ(L)*SK(IABS(LHED1(L)))
           SK(LCO)=R1(KRO)+GASH
           GO TO 32
     24    NCOD(LCO)=1
     32    CONTINUE
           RETURN
           END

           SUBROUTINE NOPNT
           ROWWISE NOP(100,14)
           COMMON/VMØ/NOP
           COMMON/VM1/NP,NE,NCN(100),NELL,NTRA,ESTIFM(14,14)
           COMMON/V14/ECV(100000),KHED1(464),LHED1(464),QQ(90),NT(1400),NH2
           K=1
           DO 890 I=1,NE
           DO 891 J=1,NCN(I)
     891   NT(K+J-1)=NOP(I,J)
     890   K=K+NCN(I)
           RETURN
           END
```

Glossary of program variables

Throughout this glossary an asterisk denotes that the variable must be given values by the programmer. The dimension for each array is given in parentheses.

BC*(NP)   For each ndf this is assigned zero value, unless a boundary condition is applied, in which case it is given the value of the boundary condition.

B1(NH2)   A BIT vector; unity is assigned when an entry in NT appears for the first time.

DET   The value of the determinant of the Jacobian. Set to unity before enter in VFRONT.

ECV   The vector where data needed for back substitution (pivot columns, associated heading vectors, pivot row number, number of occupied columns in EQ and the pivot column number) are stored. The dimension of ECV must be provided by the programmer and must be set large enough to contain all the values stored in ECV.

EQ(NMAX,NMAX)   Submatrix of $\underline{A}$ (in $\underline{A}\,\underline{x}=\underline{b}$) into which all element stiffness matrices are assembled.

ESTIFM*(NCN,NCN)   The element stiffness matrix formed in the call to subroutine ABFIND.

KDEST(NCN)   The destination vector, to indicate the rows in EQ where the entries of each element stiffness matrix are to be assembled.

KHED(NP)   The heading vector, to indicate the position (row) of each nodal equation in $\underline{A}$.

LDEST(NCN)   The destination vector, to indicate the columns in EQ where the entries of each element stiffness matrix are to be assembled.

LEAP(NH2)   A vector which indicates the element in which each entry in NT appears for the last time.

LHED(NP)   The heading vector, to indicate the position of each nodal unknown in the vector of unknowns.

LKC(NH2) (LKR(NH2))   A vector which counts how many columns (rows) in $\underline{A}$ are occupied during the assembly procedure.

NCN *(NE)   The total number of ndf in each element.

NCOD*(NP)   The entries of this array are coded zero for a ndf with no applied boundary condition, and unity for an applied boundary condition.

NCRIT*   The critical number of rows of EQ. If this value is exceeded assembly is suspended. It must be chosen to be smaller than NMAX by the number of ndf in an element. The minimum value is the number of ndf in the largest frontwidth.

NE*   The number of elements in the mesh.

NELL   Element number of the stiffness matrix currently being assembled.

NH2 $NH2 = \sum_{I=1}^{NE} NCN(I)$

NMAX*   This must be chosen to be greater than the largest frontwidth + the number of the next element equations to be assembled.

NOP*(NE,NCN)   The nodal numbering array. It is not needed in VFRONTAL, but it is used in subroutine NOPNT to illustrate how the NT vector can be constructed from the NOP array.

NP*   The total number of ndf, i.e. the total number of equations to be solved.

NT*(NH2)   The nodal numbering vector; each ndf is recorded in the order of the local nodal numbering.

NTRA*   This is coded unity before entry into VFRONT and zero on subsequent entry, in order to avoid the prefront process after the first cycle.

R1*(NP)   The right-hand side vector $\underline{b}$ (in $\underline{A}\,\underline{x}=\underline{b}$) assembled for each element in the call to ABFIND.

SK(NMAX*NMAX)   The solution is output in this array.

LARGE-SCALE QUANTUM MECHANICAL SCATTERING
CALCULATIONS ON VECTOR COMPUTERS

David W. Schwenke and Donald G. Truhlar

Department of Chemistry and Supercomputer Institute
University of Minnesota
Minneapolis, MN 55455

I. INTRODUCTION

Energy transfer in molecular collisions is a very fundamental problem
in chemical physics, both experimentally and theoretically. Quantitative
state-to-state cross sections for energy transfer processes are important
for understanding and modelling many kinds of systems, including lasers,
shocked gases, planetary atmospheres, and systems containing excited reaction
products of any kind. In theory these cross sections can be calculated from
first principles using quantum mechanics but in practice this involves severe
computational difficulties. The first step is the calculation of molecular
interaction potentials, which are a consequence of the electronic structure
of the interacting molecules for various geometries. There has been enormous
progress in this area in the last few years,[1-3] but we shall not consider it
further in this chapter. Instead we shall concentrate on the second step,
the dynamical problem that yields the desired inelastic cross sections from
the intermolecular potential. Only when both steps have been solved ade-
quately will the ab initio method have reached fruition, but the techniques
involved in the two steps are very specialized and to a large extent progress
on these steps may occur separately and in parallel. There are two basic
approaches to the dynamics problem. First, one can try to develop reliable
methods based on simplifying approximations, such as using semiclassical or
classical methods or low-order perturbation theory. Second, one can attempt
a direct solution of the Schroedinger equation. Since the Schroedinger equation for the systems of interest is not solvable analytically, we are forced
to use numerical methods. In this context an "exact" solution of the
Schroedinger equation is actually a "converged" solution, converged to
within some acceptable but small margin of error (e.g., a few per cent in

the cross section of interest) with respect to the relevant convergence parameters. We shall discuss a general and well studied method for obtaining
such a direct solution. This method is variously called the close coupling
or coupled-channels (CC) method.[4-6] It is based on a steady-state description
of the scattering process in terms of the time-independent Schroedinger partial differential equation, and it converts this equation into a set of
coupled ordinary differential equations by expansion of the D-dimensional
solution in (D-1)-dimensional basis functions. The resulting CC equations
may be cast in many forms. One may directly solve for the wave function, or
one may use invariant imbedding.[7,8] The particular invariant imbedding
algorithm we use is called R matrix propagation,[5,6,9-12] where the R matrix
is Wigner's derivative matrix[13] (not to be confused with another R matrix,
the reactance matrix,[4] that also plays a role in the problem). This method
involves the calculation of large numbers of multi-dimensional integrals and
the diagonalization, inversion, and multiplication of large matrices, as
well as linear-equation solving; thus it is well suited to attack on vector
computers. Other methods of solving the CC equations involve somewhat different mixes of the various matrix operations, but do not differ in any
fundamental way. We shall not consider other methods of solving the
Schroedinger equation for dynamics problems in detail. Some of these, like
full multi-dimensional finite-difference[14] or finite-element[15] solutions,
have been applied successfully to problems of artificially reduced dimensionality and are also well suited to modern vector computers, but at present
they seem less efficient for solving real problems in the three-dimensional
world.

We concentrate in this chapter on collisions in which both partners
start and remain in their ground electronic state and no rearrangements
occur. The techniques discussed here are also useful for electronically
inelastic[16,17] and reactive collisions,[9,18] as well as for electron scattering.[12,19]

The "state of the art" for converged calculations is the calculation
of vibrational-rotational transition probabilities in atom-diatom collisions involving a light diatom[4,20] (heavy molecules have more closely spaced
states and hence require larger expansion bases and bigger matrices to obtain
converged results for all the states of interest) or the calculation of probabilities for pure rotational quantum number changes in collisions of two
light diatoms assumed to be rigid.[21] Already at the level of complexity of
rigid diatom-diatom collisions, the number of arithmetic operations scales
as the twelfth power of the maximum rotational quantum number involved, and
the computational expense rapidly becomes prohibitive.[22] Here we report our

attempts to converge vibrational-rotational state-to-state transitions in diatom-diatom collisions, specifically the collision of two hydrogen fluoride (HF) molecules with enough total energy for both molecules to be vibrationally excited. This allows, even with indistinguishable molecules, one to observe vibration-to-vibration (V-V) energy transfer in which a quantum of vibrational energy originally on one molecule is transferred to the other, e.g., a process in which initially each molecule has one quantum of vibrational excitation and finally one molecule has two and the other has none. Such V-V energy transfer processes are of great interest because they provide the dominant pathway to redistribute vibrational energy in many nonequilibrium chemical mixtures. Prior to the availability of supercomputers, though, accurate quantum mechanical calculations on V-V energy transfer were prohibitively time consuming. Even with supercomputers these calculations are very expensive; thus, for the work we discuss here, we limit ourselves to the case where the total angular momentum is restricted to be zero. Although one cannot calculate observable cross sections or rate constants from this subcase, except at extremely low translational energies, it does serve as a prototype, and exact solutions of the quantal dynamics for this prototype can serve as practical benchmarks for testing approximate, but less computationally expensive, methods for attacking this kind of problem.

Section II gives an introduction to the theory, section III describes our numerical algorithm, section IV discusses the details of the inputs of the calculation, section V discusses vectorizing our program on the Cyber 205 and Cray-1 computers, and section VI compares execution times for scalar and vector versions of the code on one minicomputer and two supercomputers. Section VII gives preliminary results, and section VIII gives our conclusions. Our equations will be given in a form that is valid in any consistent set of units; however we find it convenient to quote numerical parameters in Hartree atomic units. In this system of units, the unit of length is the bohr, abbreviated a_0 (1 a_0 = 0.5291771 × 10^{-10} m), the unit of mass is the electron mass, abbreviated m_e (1 m_e = 9.109534 × 10^{-31} kg), the unit of energy is the hartree, abbreviated E_h (1 E_h = 27.21161 eV = 4.359814 × 10^{-18} J), and $\hbar$ (Planck's constant divided by 2π) has the value unity.

II. THEORY

All of the information concerning a collision of molecule A with molecule B is contained in the scattering wave function $\psi_{n_0}(x,\vec{r},E)$, which solves the Schroedinger equation

$$H\psi_{n_0} = E\psi_{n_0}.$$
(1)

The symbol n_0 stands for all of the quantum numbers necessary to specify the initial state of the system, $\underset{\sim}{x}$ is the collection of all internal coordinates, with the exception of $\vec{r}$, which is the vector connecting the center of mass of A to that of B, E is the total energy, and H is the system Hamiltonian:

$$H = - \frac{\hbar^2}{2\mu} \nabla_{\vec{r}}^2 + H_{int}(\underset{\sim}{x}) + V(\underset{\sim}{x},\vec{r}) , \qquad (2)$$

where μ is the reduced mass for relative translational motion of A with respect to B, $\nabla_{\vec{r}}^2$ is the Laplacian with respect to $\vec{r}$, H_{int} is the "internal Hamiltonian", defined as the sum of the Hamiltonians of the isolated A and B subsystems, and V is the interaction potential, which means that V vanishes in the large-r limit. For the next step it is useful to separate $\nabla_{\vec{r}}^2$ into its radial and angular parts and to partition the internal Hamiltonian into a conveniently diagonalized part, $\tilde{H}_{int}$ (which may in some cases simply be H_{int}), and the rest ΔH_{int}. Then we treat the angular part of $-\hbar^2\nabla_{\vec{r}}^2/2\mu$ together with H_{int}; their sum is called the primitive Hamiltonian H°. This yields

$$H = - \frac{\hbar^2}{2\mu}[\frac{1}{r^2} \frac{\partial}{\partial r}(r^2 \frac{\partial}{\partial r})] + H^\circ(\underset{\sim}{x},\vec{r}) + V(\underset{\sim}{x},\vec{r}) , \qquad (3)$$

where

$$H^\circ = \frac{\ell_{\hat{r}}^2}{2\mu r^2} + \tilde{H}_{int}(\underset{\sim}{x}) , \qquad (4)$$

$$V(\underset{\sim}{x},\vec{r}) = V(\underset{\sim}{x},r) + \Delta H_{int}(\underset{\sim}{x}) , \qquad (5)$$

and $\ell_{\hat{r}}^2$ is the quantum mechanical operator for the square of the orbital angular momentum of the relative translational motion of A with respect to B.

To determine ψ_{n_0}, we expand it in terms of simultaneous matrix eigenvectors of $\tilde{H}_{int}$ and $\ell_{\hat{r}}^2$ defined by

$$\int d\hat{r}\int d\underset{\sim}{x} \; X_m^*(\underset{\sim}{x},\hat{r})\tilde{H}_{int}(\underset{\sim}{x})X_n(\underset{\sim}{x},\hat{r}) = \delta_{mn}\tilde{\varepsilon}_n , \qquad (6)$$

$$\ell_{\hat{r}}^2 X_n(\underset{\sim}{x},\hat{r}) = \hbar^2\ell_n(\ell_n + 1)X_n(\underset{\sim}{x},\hat{r}) , \qquad (7)$$

and

$$\int d\hat{r}\int d\underset{\sim}{x} \; X_m^*(\underset{\sim}{x},\hat{r})X_n(\underset{\sim}{x},\hat{r}) = \delta_{mn} , \qquad (8)$$

where δ_{mn} is the Kronecker delta, and eq. (8) is an orthonormality condition. The expansion is

$$\psi_{n_0}(\underset{\sim}{x}, \vec{r}, E) = \frac{1}{r} \sum_{n=1}^{N} X_n(\underset{\sim}{x}, \hat{r}) f_{nn_0}(r, E) \quad , \tag{9}$$

and the terms in this expansion are called channels. In these equations, $\hat{r}$ is the unit vector which has the direction as $\vec{r}$. Substituting eq. (9) into eq. (1), multiplying by $rX_m^*(\underset{\sim}{x}, \hat{r})$, and integrating over $\underset{\sim}{x}$ and $\hat{r}$ yields

$$(-\frac{\hbar^2}{2\mu} \frac{d^2}{dr^2} + \frac{\ell_m(\ell_m + 1)\hbar^2}{2\mu r^2} + \tilde{\epsilon}_m) f_{mn_0}(r, E) + \sum_n V_{mn}(r) f_{nn_0}(r, E)$$

$$= E f_{mn_0}(r, E) \quad , \quad m = 1, 2, \ldots, N, \tag{10}$$

where

$$V_{mn}(r) = \int d\underset{\sim}{x} \int d\hat{r} \; X_m^*(\underset{\sim}{x}, \hat{r}) V(\underset{\sim}{x}, \hat{r}) X_n(\underset{\sim}{x}, \hat{r}) \quad . \tag{11}$$

Equations (10) are the close-coupling equations, and they can be written in the form

$$\frac{d^2}{dr^2} \underset{\sim}{f}(r, E) = \underset{\sim}{D}(r, E) \underset{\sim}{f}(r, E) \quad , \tag{12}$$

where $\sim$ under a symbol denotes a matrix (except for $\underset{\sim}{x}$, for which it denotes a collection of coordinates). The components of $\underset{\sim}{f}$ are just the f_{mn} of eq. (9), and the elements of $\underset{\sim}{D}$ are given by

$$D_{mn}(r, E) = \frac{2\mu}{\hbar^2} V_{mn}(r) + \delta_{mn}[\ell_n(\ell_n + 1)/r^2 - \tilde{k}_n^2] \quad , \tag{13}$$

where

$$\tilde{k}_n^2 = 2\mu(E - \tilde{\epsilon}_n)/\hbar^2 \quad . \tag{14}$$

$\tilde{k}_n$ is called the primitive wave number. $\underset{\sim}{D}$ is real and symmetric. Notice that the rows of $\underset{\sim}{f}$ correspond to different channels, which are coupled, and the columns to different initial conditions. The channels are ordered so that $\tilde{k}_n^2 \geq \tilde{k}_m^2$ if and only if $n \geq m$.

In order to discuss the large-r boundary conditions on eq. (12) we must consider a transformation that diagonalizes $\underset{\sim}{D}$ at large r. Furthermore, in the R matrix propagation algorithm considered in the next section, we will use transformations to basis functions that diagonalize $\underset{\sim}{D}$ at finite r. At any r we may define functions

$$Z_m(\underset{\sim}{x}, \vec{r}) = \sum_{n=1}^{N} U_{nm}(r) X_n(\underset{\sim}{x}, \hat{r}) \tag{15}$$

such that

$$\sum_{k, \ell=1}^{N} U_{kn}(r) D_{k\ell}(r, E) U_{\ell m}(r) = \delta_{nm}[\lambda_{mm}(r, E)]^2 \quad . \tag{16}$$

Notice that, because of the way that E appears in eq. (13), i.e., only on the diagonal, the eigenvalues λ_{mm}^2 depend on E, but the eigenvectors, which are the columns of $\underset{\sim}{U}$, do not. If we use the Z_m as new basis functions, called adiabatic basis functions, the expansion of the wave function becomes

$$\psi_{n_0}(x,\vec{r},E) = \frac{1}{r} \sum_{m=1}^{N} Z_m(x,\vec{r}) g_{mn_0}(r,E) \tag{17}$$

where the g_{mn_0} are easily related to the f_{mn_0}. The terms in (17), like those in (9), are called channels. The boundary conditions on the solutions to the close coupling equations may be given in many equivalent forms, but the one that is most convenient for our algorithm is

$$g_{mn} \underset{r \to 0}{\sim} 0 , \qquad\qquad 1 \leq n, m \leq N , \tag{18}$$

$$g_{mn} \underset{r \to \infty}{\sim} \begin{cases} \delta_{mn} \sin[k_m(E)r - \ell_m \pi/2] + a_{mn}(E) \cos[k_m(E)r - \ell_m \pi/2] , \\ \qquad\qquad\qquad\qquad 1 \leq n, m \leq P^o , \\ \delta_{mn} b \exp[k_m(E)r] + a_{mn}(E) \exp[-k_m(E)r] , \\ \qquad\qquad\qquad\qquad P^o + 1 \leq n, m \leq N , \end{cases} \tag{19}$$

where b is arbitrary and k_n is an asymptotic wave number defined by

$$k_n = \lim_{r \to \infty} |\lambda_{nn}(r,E)| . \tag{20}$$

"Open" channels are defined as those with positive k_n^2, and P^o is the number of open channels in the basis. Channels that are not open are called "closed". If H_{int} is the same as $\tilde{H}_{int}$, then k_n equals $|\tilde{k}_n|$ and the boundary conditions (18)–(19) apply to the f_{mn} as well as the g_{mn}.

The final result of the calculation is the $P^o \times P^o$ unitary scattering matrix $\underset{\sim}{S}$ defined by

$$\underset{\sim}{S} = [\underset{\sim}{1} + i\underset{\sim}{R}(E)][\underset{\sim}{1} - i\underset{\sim}{R}(E)]^{-1} , \tag{21}$$

where $\underset{\sim}{1}$ is the unit matrix, $i^2 = -1$, and $\underset{\sim}{R}$ is the real symmetric $P^o \times P^o$ reactance matrix with elements

$$R_{nm} = k_n^{\frac{1}{2}} a_{nm}(E) k_m^{-\frac{1}{2}} , \qquad\qquad 1 \leq n, m \leq P^o . \tag{22}$$

All physical observables may be calculated from the scattering matrix by standard formulas.

The size of N, the number of terms in eq. (9), is determined by including all of the states of interest plus a sufficient number of other states in order for eq. (9) to accurately represent ψ_{n_0}. For the present study, the quantum numbers specified by n or n_0 are[23] v_1 and v_2, the vibrational quantum numbers of the two molecules, j_1 and j_2, the rotational quantum numbers of the two molecules, j_{12}, the angular momentum quantum number associated with the vector sum of the rotational angular momenta of the two molecules, ℓ (called ℓ_n above when necessary to indicate that ℓ may be different for different n), the orbital angular momentum for relative motion, J and M, the total angular momentum and its projection on a laboratory-fixed Z axis, and η, the symmetry under interchange of the two molecules. The indistinguishability of identical molecules in quantum mechanics makes it impossible to distinguish which molecule has which set of v and j quantum numbers. The parity ζ is $(-1)^{j_1+j_2+\ell}$ and it too is specified by these quantum numbers.

Since J, M, η, and ζ are good quantum numbers, the matrix D is block diagonal in them, and using simultaneous eigenfunctions of their associated operators as our basis functions uncouples the solutions into noninteracting components and reduces the computational work. In the numerical applications considered below we shall only consider the block specified by $J = M = 0$ and $\eta = \zeta = +1$.

III. NUMERICAL METHODS

We solve eq. (12) using R matrix propagation. In this method one first subdivides the coordinate r into N_s sectors with sector midpoints $r_C^{(i)}$ and widths $h^{(i)}$ such that

$$r_C^{(i+1)} = r_C^{(i)} + [h^{(i+1)} + h^{(i)}]/2 \ . \tag{23}$$

In sector i, it is convenient to expand the wavefunction ψ_n in terms of sector-dependent functions defined by

$$z_m^{(i)}(x,\hat{r}) = \sum_{n=1}^{N} T_{nm}^{(i)} X_n(x,\hat{r}) \ , \qquad 1 \le m \le P^{(i)} \ . \tag{24}$$

The $N \times P^{(i)}$ rectangular matrix $\underset{\sim}{T}^{(i)}$ is made up of the first $P^{(i)}$ columns of the $N \times N$ matrix $\underset{\sim}{U}^{(i)}$, where $\underset{\sim}{U}^{(i)}$ diagonalizes $\underset{\sim}{D}$:

$$\sum_{k,k'=1}^{N} U_{kn}^{(i)} D_{kk'}(r_C^{(i)},E) U_{k'm}^{(i)} = \delta_{nm} [\lambda_{nn}^{(i)}(E)]^2 \ . \tag{25}$$

In terms of the new functions $z_m^{(i)}$, the wave function is

$$\psi_n(x,\vec{r},E) = \frac{1}{r} \sum_{m=1}^{P^{(i)}} z_m^{(i)}(x,\hat{r}) g_{mn}^{(i)}(r,E) \ . \tag{26}$$

The new radial functions $\underset{\sim}{g}^{(i)}$ are related to the functions $\underset{\sim}{f}$ by

$$g_{mn}^{(i)}(r,E) = \sum_{k=1}^{N} T_{km}^{(i)} f_{kn}(r,E) \, , \qquad\qquad 1 \le m, n \le P^{(i)} \, . \qquad (27)$$

The $g^{(i)}$ solve the equation

$$\frac{d^2}{dr^2} \, \underset{\sim}{g}^{(i)}(r,E) = \underset{\sim}{L}^{(i)}(r,E) \underset{\sim}{g}^{(i)}(r,E) \, , \qquad\qquad (28)$$

where

$$L_{nm}^{(i)}(r,E) = \sum_{k,\ell=1}^{N} T_{kn}^{(i)} D_{k\ell}(r,E) T_{\ell m}^{(i)} \, , \qquad 1 \le n, m \le P^{(i)} \, . \qquad (29)$$

According to eq. (28), the channels of eq. (26) are uncoupled at $r = r_C^{(i)}$.

In order to enforce eq. (18), it would be convenient if $r_C^{(1)} - h^{(1)}/2$ were equal to zero. However, in atom–molecule and molecule–molecule collisions one finds

$$V_{nn}(r) \gg E \, , \qquad\qquad \text{all } n, \; r \ll \sigma \, , \qquad (30)$$

where σ is the distance at which the two subsystems begin to repel strongly. This implies that, as a function of r in the decreasing r direction, all $f_{mn}(r)$ and hence all $g_{mn}(r)$ decrease rapidly and are totally negligible for r less than some finite nonzero value. We chose such a nonzero value of r as $r_C^{(1)}$ and thereby avoid the work of propagating the solution to the equations over the region where it is known to be essentially zero(in the present calculations we use $r_C^{(1)}$ equal to 3.0 a_0).

We begin with $P^{(1)} = N$. At large r, because of eq. (19), f_{mn} for $m > P^o$ rapidly decays to zero. In this region we use a criterion described below to allow $P^{(i)}$ to decrease. We constrain $P^{(i+1)}$ to equal $P^{(i)}$ or $P^{(i)} - 1$ in all cases to simplify the algorithm. We also enforce the constraint $P^{(i)} \ge P^o$ for all (i). The $2P^{(i)} \times 2P^{(i)}$ sector propagator $\underset{\sim}{P}^{(i)}$ is defined by

$$\underset{\sim}{G}_L^{(i)}(E) = \underset{\sim}{P}^{(i)}(E) \underset{\sim}{G}_R^{(i)}(E) \, , \qquad\qquad (31)$$

where the $2P^{(i)} \times 2P^{(i)}$ matrix $\underset{\sim}{G}^{(i)}$ is given by

$$\underset{\sim}{G}^{(i)}(E) = \begin{bmatrix} \underset{\sim}{g}^{(i)}(r,E) \\ \underset{\sim}{g}'^{(i)}(r,E) \end{bmatrix} \, , \qquad\qquad (32)$$

where $\underset{\sim}{g}'^{(i)}(r,E)$ denotes $d\underset{\sim}{g}^{(i)}/dr$ and where L and R denote the left $[r_L^{(i)} = r_C^{(i)} - h^{(i)}/2]$ and right $[r_R^{(i)} = r_C^{(i)} + h^{(i)}/2]$ sides of a sector. We partition $\underset{\sim}{P}^{(i)}(E)$ so that

$$P^{(i)}(E) = \begin{bmatrix} P_1^{(i)}(E) & P_2^{(i)}(E) \\ \\ P_3^{(i)}(E) & P_4^{(i)}(E) \end{bmatrix} , \tag{33}$$

and the matrices $P_j^{(i)}(E)$ are all square.

We use the first-order Magnus method[24] for $P^{(i)}$:

$$[P_1^{(i)}(E)]_{nm} = [P_4^{(i)}(E)]_{nm} = \begin{cases} \delta_{nm} \cosh[-h^{(i)}|\lambda_{nn}^{(i)}(E)|] , & \lambda_{nn}^{(i)^2}(E) > 0 , \\ \\ \delta_{nm} \cos[-h^{(i)}|\lambda_{nn}^{(i)}(E)|] , & \lambda_{nn}^{(i)^2}(E) < 0 , \end{cases} \tag{34}$$

$$[P_2^{(i)}(E)]_{nm} = \begin{cases} \delta_{nm}|\lambda_{nn}^{(i)}(E)|^{-1} \sinh[-h^{(i)}|\lambda_{nn}^{(i)}(E)|] , & \lambda_{nn}^{(i)^2}(E) > 0 , \\ \\ \delta_{nm}|\lambda_{nn}^{(i)}(E)|^{-1} \sin[-h^{(i)}|\lambda_{nn}^{(i)}(E)|] , & \lambda_{nn}^{(i)^2}(E) < 0 , \end{cases} \tag{35}$$

$$[P_3^{(i)}(E)]_{nm} = \begin{cases} \lambda_{nn}^{(i)^2}[P_2^{(i)}(E)]_{nm} , & \lambda_{nm}^{(i)^2}(E) > 0 , \\ \\ -\lambda_{nn}^{(i)^2}[P_2^{(i)}(E)]_{nm} , & \lambda_{nm}^{(i)^2}(E) < 0 . \end{cases} \tag{36}$$

The error of this propagator is proportional to $h^{(i)^3}dL^{(i)}/dr$.

We can choose the stepsize so that the error term is small. We estimate the error in sector $(i+1)$ by

$$\text{error} \propto h^{(i+1)^3}[\frac{1}{N} \sum_{j=1} (\frac{dD_{jj}^{(i)}}{dr})^2]^{\frac{1}{2}} . \tag{37}$$

This translates into the algorithm

$$h^{(i+1)} = \min \begin{cases} EPS[\frac{1}{N} \sum_{j=1}^{N} (\frac{D_{jj}^{(i)} - D_{jj}^{(i-1)}}{r_C^{(i)} - r_C^{(i-1)}})^2]^{-1/6} \\ \\ h_{max} \end{cases} \tag{38}$$

where EPS and h_{max} are input parameters, and $\min\{^a_b$ means the minimum of a and b.

Continuity of the functions $g^{(i)}$ across sector boundaries can be expressed by

$$g_R^{(i-1)}(E) = T(i-1,i)g_L^{(i)}(E) \tag{39}$$

$$g'_R^{(i-1)}(E) = T(i-1,i)g'_L^{(i)}(E) \tag{40}$$

where the overlap matrix $T(i-1,i)$ is defined by

$$T_{nm}(i-1,i) = \sum_{k=1}^{N} T_{kn}^{(i-1)} T_{km}^{(i)} , \qquad\qquad 1 \leq n, m \leq P^{(i)}. \qquad (41)$$

Thus when $P^{(i-1)} \neq P^{(i)}$ only the upper left $P^{(i)} \times P^{(i)}$ part of $g_R^{(i-1)}$ is used to calculate $g_L^{(i)}$. We now define the sector R matrix $r^{(i)}$ by

$$\begin{bmatrix} g_R^{(i-1)}(E) \\[2ex] g_R^{(i)}(E) \end{bmatrix} = \begin{bmatrix} r_1^{(i)}(E) & r_2^{(i)}(E) \\[2ex] r_3^{(i)}(E) & r_4^{(i)}(E) \end{bmatrix} \begin{bmatrix} g'_R^{(i-1)}(E) \\[2ex] -g'_R^{(i)}(E) \end{bmatrix} \qquad (42)$$

$$r^{(i)}(E) = \begin{bmatrix} r_1^{(i)}(E) & r_2^{(i)}(E) \\[2ex] r_3^{(i)}(E) & r_4^{(i)}(E) \end{bmatrix} , \qquad (43)$$

where the $r_j^{(i)}$ matrices are $P^{(i)} \times P^{(i)}$. It is easy to show that

$$r_1^{(i)}(E) = T(i-1,i) P_1^{(i)}(E) [P_3^{(i)}(E)]^{-1} [T(i-1,i)]^{-1} , \qquad (44)$$

$$r_2^{(i)}(E) = T(i-1,i) [P_3^{(i)}(E)]^{-1} , \qquad (45)$$

$$r_3^{(i)}(E) = [P_3^{(i)}(E)]^{-1} [T(i-1,i)]^{-1} , \qquad (46)$$

$$r_4^{(i)}(E) = [P_3^{(i)}(E)]^{-1} P_4^{(i)}(E) . \qquad (47)$$

Note that the matrix P_3 is diagonal so its inversion is not time consuming.

The global R matrix which spans from the left-hand side of the first sector to the right-hand side of sector (i) is defined by

$$\begin{bmatrix} g_L^{(1)}(E) \\[2ex] g_R^{(i)}(E) \end{bmatrix} = \begin{bmatrix} R_1^{(i)}(E) & R_2^{(i)}(E) \\[2ex] R_3^{(i)}(E) & R_4^{(i)}(E) \end{bmatrix} \begin{bmatrix} g'_L^{(1)}(E) \\[2ex] -g'_R^{(i)}(E) \end{bmatrix} , \qquad (48)$$

where the $R_j^{(i)}$ matrices are all square. It can be shown that if $[\lambda_{nn}^{(1)}]^2 \gg 2\mu E/\hbar^2$ for all n, then $R_2^{(i)}$ and $R_3^{(i)}$ are approximately zero and that all scattering information, i.e., the a_{mn}, $1 \leq m, n \leq P^o$, can be determined only from $R_4^{(N_S)}$. In the present case, it is true that $[\lambda_{nn}^{(1)}(E)]^2 \gg 2\mu E/\hbar^2$ for all n, and we only propagate $R_4^{(i)}$. This matrix depends only on $R_4^{(i-1)}$ and the $r_j^{(i)}$:

$$R_4^{(i)}(E) = r_4^{(i)}(E) - r_3^{(i)}(E) [R_4^{(i-1)}(E) + r_1^{(i)}(E)]^{-1} r_2^{(i)}(E) \qquad (49)$$

and

$$R_{\sim4}^{(1)}(E) = r_{\sim4}^{(1)}(E) \ . \tag{50}$$

When $P^{(i-1)} \neq P^{(i)}$, only the upper-left $P^{(i)} \times P^{(i)}$ part of $R_{\sim4}^{(i-1)}(E)$ is used in the calculation of $R_{\sim4}^{(i)}(E)$. The determination of the number of channels propagated in a given sector proceeds as follows. If $r_c^{(i)}$ is less than a pre-determined value ($10 \ a_0$ for the calculations presented here) or if $P^{(i)}$ has already been reduced to P^0, we set $P^{(i+1)} = P^{(i)}$. Otherwise we check whether $|(r_{\sim2}^{(i)})_{nP^{(i)}}| \leq$ EPSRED and $|(r_{\sim2}^{(i)})_{P^{(i)}n}| \leq$ EPSRED, $1 \leq n \leq P^{(i)}$, where EPSRED is a pre-set parameter (set equal to 10^{-3} for the calculations reported here). If both inequalities are satisfied, then $P^{(i+1)} = P^{(i)} - 1$; otherwise $P^{(i+1)} = P^{(i)}$.

Before applying the boundary conditions of eq. (19) to $g_{\sim}^{(i)}$ it is sometimes convenient to reorder the channels in $g_{\sim}^{(i)}$ and sometimes it is necessary to make linear combinations of the channels in $g_{\sim}^{(i)}$. It is convenient to reorder the channels if $V(\underset{\sim}{x},\vec{r})$ falls off faster than r^{-2}. This is because for large r it will be approximately true that

$$D_{nm}(r,E) = \delta_{nm}[\ell_n(\ell_n+1)/r^2 - \tilde{k}_n^2(E)] \ , \tag{51}$$

and, depending on r and r', it may be possible that $D_{nn}(r,E) > D_{mm}(r,E)$ and $D_{nn}(r',E) < D_{mm}(r',E)$. Since the subprogram that calculates $T^{(i)}$ and $[\lambda^{(i)}]^2$ orders the eigenvalues from lowest to highest, the relative position of channel n and m may change in g when going from r to r'.

It is necessary to make new linear combinations of the channels in $g^{(i)}$ if there exist degenerate channels, i.e., those with $[\lambda_{nn}^{(i)}]^2 = [\lambda_{mm}^{(i)}]^2$ and $n \neq m$. This is because our matrix diagonalization routine will mix these states. Degenerate channels will occur in the current calculations at very large r where the term $\ell(\ell+1)/2\mu r^2$ is negligible since there are channels with the same $\tilde{k}_n$ but different values of ℓ_n. (Additional accidental degeneracies would occur if we used the harmonic oscillator and rigid-rotor approximations to calculate asymptotic energies, but we do not make these approximations.) In order to sort out these effects, we make the transformation to new radial functions $h_{\sim}^{(i)}$ defined by

$$g_{mn}^{(i)}(r,E) = \sum_{\ell=1}^{N} \sum_{k=1}^{P^{(i)}} T_{\ell m}^{(i)} U_{\ell k}^o h_{kn}^{(i)}(r,E) \ , \quad 1 \leq m \leq P^{(i)}, \ 1 \leq n \leq P^{(i)} \ , \tag{52}$$

where $\underset{\sim}{U}^o$ diagonalizes H_{int} in the X_n basis, and has the channels in some fixed order that does not mix degenerate channels. This equation can be written in matrix notation as

$$g_{\sim}^{(i)}(r,E) = [\underset{\sim}{T}^{(i)}]^T \underset{\sim}{U}^o \underset{\sim}{h}^{(i)}(r,E) \ . \tag{53}$$

We then define a new global R matrix satisfying

$$
\begin{bmatrix} \underset{\sim}{h}_L^{(1)} \\[1em] \underset{\sim}{h}_R^{(i)} \end{bmatrix} = \begin{bmatrix} \underset{\sim}{\tilde{R}}_1^{(i)} & \underset{\sim}{\tilde{R}}_2^{(i)} \\[1em] \underset{\sim}{\tilde{R}}_3^{(i)} & \underset{\sim}{\tilde{R}}_4^{(i)} \end{bmatrix} \begin{bmatrix} \underset{\sim}{h'}_L^{(1)} \\[1em] -\underset{\sim}{h'}_R^{(i)} \end{bmatrix} \tag{54}
$$

where

$$
\underset{\sim}{\tilde{R}}_1^{(i)} = \{[\underset{\sim}{T}^{(1)}]^T \underset{\sim}{U}^o\}^{-1} \underset{\sim}{R}_1^{(i)} [\underset{\sim}{T}^{(1)}]^T \underset{\sim}{U}^o \tag{55}
$$

$$
\underset{\sim}{\tilde{R}}_2^{(i)} = \{[\underset{\sim}{T}^{(1)}]^T \underset{\sim}{U}^o\}^{-1} \underset{\sim}{R}_2^{(i)} [\underset{\sim}{T}^{(i)}]^T \underset{\sim}{U}^o \tag{56}
$$

$$
\underset{\sim}{\tilde{R}}_3^{(i)} = \{[\underset{\sim}{T}^{(i)}]^T \underset{\sim}{U}^o\}^{-1} \underset{\sim}{R}_3^{(i)} [\underset{\sim}{T}^{(1)}]^T \underset{\sim}{U}^o \tag{57}
$$

$$
\underset{\sim}{\tilde{R}}_4^{(i)} = \{[\underset{\sim}{T}^{(i)}]^T \underset{\sim}{U}^o\}^{-1} \underset{\sim}{R}_4^{(i)} [\underset{\sim}{T}^{(i)}]^T \underset{\sim}{U}^o \tag{58}
$$

The matrix $\underset{\sim}{a}$ of eq. (19) is determined by

$$
\underset{\sim}{a}(E) = \lim_{i \to \infty} \underset{\sim}{a}^{(i)}(E) \tag{59}
$$

where

$$
\underset{\sim}{a}^{(i)}(E) = [-\underset{\sim}{F}^{(i)}(E) + \underset{\sim}{\tilde{R}}_4^{(i)}(E) \underset{\sim}{H}^{(i)}(E)]^{-1} [\underset{\sim}{B}^{(i)}(E) + \underset{\sim}{\tilde{R}}_4^{(i)} \underset{\sim}{G}^{(i)}(E)] \Delta \tag{60}
$$

and

$$
\dot{F}_{nm}^{(i)}(E) = \delta_{nm} \begin{cases} \cos[k_m(E) r_R^{(i)} - \ell_m \pi/2], & 1 \le m \le P^o, \\[1em] \exp[-k_m(E) r_R^{(i)}], & P^o < m \le P^{(i)}, \end{cases} \tag{61}
$$

$$
B_{nm}^{(i)}(E) = \delta_{nm} \begin{cases} \sin[k_m(E) r_R^{(i)} - \ell_m \pi/2], & 1 \le m \le P^o, \\[1em] \exp[k_m(E) r_R^{(i)}], & P^o < m \le P^{(i)}, \end{cases} \tag{62}
$$

$$
H_{nm}^{(i)}(E) = k_m(E) B_{nm}^{(i)}(E), \qquad 1 \le m \le P^{(i)}, \tag{63}
$$

$$
G_{nm}^{(i)}(E) = k_m(E) F_{nm}^{(i)}(E), \qquad 1 \le m \le P^{(i)}, \tag{64}
$$

$$
\Delta_{nm} = \delta_{nm} \begin{cases} 1, & 1 \le m \le P^o, \\[1em] b, & P^o < m \le P^{(i)}. \end{cases} \tag{65}
$$

If it is true that for a given $m > P^o$, $(R_4^{(i)})_{nm} = (R_4^{(i)})_{nm} = 0$ for all $n \le P^o$, channel m is not required in the calculation of $a_{nm}^{(i)}$, $1 \le n, m \le P^o$. Our program determines the smallest $m \ge P^o$, called $P_a^{(i)}$, such that $|(\underset{\sim}{R}_4^{(i)})_{nm}| \le$ EPSDR and $|(\underset{\sim}{R}_4^{(i)})_{mn}| \le$ EPSDR for all $n \le P^o$, where EPSDR is some small number (EPSDR equals 10^{-3} in the present calculations), and then uses the upper left $P_a^{(i)} \times P_a^{(i)}$ subblock of $R_{\sim 4}^{(i)}$ to calculate $a_{\sim}^{(i)}$.

It should be noted that the total energy E appears in eq. (13) only as a multiple of the unit matrix, so that the matrices $T_{\sim}^{(i)}$ are independent of the total energy, and the eigenvalues $[\lambda_{nn}^{(i)}]^2$ at a new energy can easily be determined by

$$[\lambda_{nn}^{(i)}(E)]^2 = [\lambda_{nn}^{(i)}(E')]^2 + 2\mu(E' - E)/\hbar^2 \quad . \tag{66}$$

We use this fact to save computer time on multiple-energy runs by reusing the $\lambda_{\sim}^{(i)}$ and $T_{\sim}(i-1,i)$. This makes the calculation of $V_{nm}(r)$, $T_{\sim}^{(i)}$, and $T_{\sim}(i-1,i)$ unnecessary for second energies and results in a great savings in computational time for such energies. The possibility of reusing this information is one of the reasons for preferring the present algorithm. Notice, however, that although it decreases the computation time, it greatly increases the storage requirements. There are two ways we have implemented the second-energy calculations. In the first method, the calculations for a given energy are done completely before the calculation for the next energy begins, and to do this as efficiently as possible requires the storage of the $P^{(i)} \times P^{(i)}$ matrices $T_{\sim}(i-1,i)$ and $[T_{\sim}(i-1,i)]^{-1}$. Since in the current calculation we will require on the order of 300 sectors, this option requires a great deal of storage space when $P^{(i)}$ is large. The second method we use is to propagate all energies together, that is, the global R matrix for sector (i) is calculated for all of the energies before the global R matrix for sector (i + 1) is calculated for any of the energies. In this case it is only necessary to store the $P^{(i)} \times P^{(i)}$ matrix $R_{\sim 4}^{(i)}$ for each energy. If there are fewer energies than sectors, which is ordinarily the case, this decreases the storage requirements. For the large-scale calculations described here, we performed calculations for 2–7 energies in a given run, and we used the second method for performing second-energy runs.

IV. THE HF-HF SYSTEM

Our calculations are for the collision of two identical hydrogen fluoride molecules. We take the hydrogen mass to be 1837.15 m_e and the fluorine mass to be 34631.94 m_e. We partition the Hamiltonian into the true diatomic and three-body parts, so that $\tilde{H}_{int}$ equals H_{int}, which is given by

$$H_{int} = H_{int}^{(1)} + H_{int}^{(2)} \quad , \tag{67}$$

$$H_{int}^{(i)} = -\frac{\hbar^2}{2\mu_{HF}} \frac{1}{R_i^2} \frac{\partial}{\partial R_i}(R_i^2 \frac{\partial}{\partial R_i})] + j_{R_i}^2 / 2\mu R_i^2 + V_{vib}(R_i) \quad , \tag{68}$$

where μ_{HF_2} is the reduced mass for an HF molecule, R_i is the bond length of molecule i, $j_{\hat{R}_i}^2$ is the quantum mechanical operator for the square of the rotational angular momentum of molecule i, and V_{vib} is the vibrational potential. For V_{vib} we use the function proposed by Murrell and Sorbie,[25] which is a fit to an RKR potential curve determined from experiment. The primitive basis functions $X_n(\underset{\sim}{x},\hat{r})$ are given by[23]

$$X_n(\underset{\sim}{x},\hat{r}) = [2(1 + \delta_{v_1 v_2}\delta_{j_1 j_2})]^{-\frac{1}{2}}[\Phi_\alpha(\underset{\sim}{x},\hat{r}) + n(-1)^{j_1+j_2+j_{12}+\ell}\Phi_{\bar{\alpha}}(\underset{\sim}{x},\hat{r})] ,$$

$$\tag{69}$$

$$\Phi_\alpha(\underset{\sim}{x},\hat{r}) = (R_1 R_2)^{-1} x_{v_1 j_1}(R_1) x_{v_2 j_2}(R_2)\Theta^{JM}_{j_1 j_2 j_{12}\ell}(\hat{R}_1,\hat{R}_2,\hat{r}) , \tag{70}$$

$$\Theta^{JM}_{j_1 j_2 j_{12}\ell}(\hat{R}_1,\hat{R}_2,\hat{r}) = \sum_{\substack{m_1 m_2 \\ m_{12} m_\ell}} (j_1 m_1 j_2 m_2 | j_1 j_2 j_{12} m_{12})(j_{12} m_{12} \ell m_\ell | j_{12} \ell JM)$$

$$\times Y_{j_1 m_1}(\hat{R}_1) Y_{j_2 m_2}(\hat{R}_2) Y_{\ell m_\ell}(\hat{r}) , \tag{71}$$

where x_{vj} is a vibrational wave function that solves

$$[- \frac{\hbar^2}{2\mu_{HF}} \frac{d^2}{dR^2} + j(j+1)\hbar^2/2\mu_{HF}R^2 + V_{vib}(R)]x_{vj}(R) = \varepsilon_{vj}x_{vj}(R) , \tag{72}$$

$(\ldots \ldots \ldots \ldots | \ldots \ldots \ldots \ldots)$ is a Clebsch–Gordan coefficient,[26] and $Y_{\ell m}$ is a spherical harmonic. The unit vector $\hat{R}_i$ describes the orientation of molecule i in the laboratory-fixed frame of reference. As discussed at the end of section II, n stands for the quantum numbers v_1, j_1, v_2, j_2, j_{12}, ℓ, J, M, and η, where it is understood that it is not possible to distinguish which molecule has which set of v and j quantum numbers; in contrast, α stands for the quantum numbers v_1, j_1, v_2, j_2, j_{12}, ℓ, J, and M, where formally the two molecules are distinguished so that molecule i is known to have vibrational and rotational quantum numbers v_i and j_i, and $\bar{\alpha}$ means that the quantum numbers for the two molecules are exchanged. For our calculations we obtained the x_{vj} by solving eq. (72) by the linear variational method in a basis of harmonic oscillator functions. The eigenenergies, ε_{vj}, were obtained, however, from the experimental spectroscopic parameters of Ref. 27. [These are in good agreement with the variational ones from eq. (72).] Notice that

$$\varepsilon_n = \tilde{\varepsilon}_n = \varepsilon_{v_1 j_1} + \varepsilon_{v_2 j_2} . \tag{73}$$

In the final calculations the potential $V(x,\vec{r})$ is the recent fit of Redmon to the extensive _ab initio_ calculations of Binkley.[28] (Some of our early calculations, including the 101-channel timing runs discussed in section VI, used the function of Poulsen et al.[29] These early runs also involve somewhat different values of several of the numerical parameters of the calculations. Except when stated otherwise, the numerical parameters given in the rest of the chapter are the values used in the production runs with the Redmon-Binkley potential.) To evaluate the V matrix elements we first express $V_{nn'}$ in terms of matrix elements over the Φ_α:

$$
\begin{aligned}
V_{nn'}(r) = {}& [4(1 + \delta_{v_1 v_2}\delta_{j_1 j_2})(1 + \delta_{v_1' v_2'}\delta_{j_1' j_2'})]^{-\frac{1}{2}}[V_{\alpha\alpha'}(r) \\
& + n(-1)^{j_1+j_2+j_{12}+\ell}V_{\bar{\alpha}\alpha'}(r) + n'(-1)^{j_1'+j_2'+j_{12}'+\ell'}V_{\alpha\bar{\alpha}'}(r) \\
& + nn'(-1)^{j_{12}+j_{12}'}V_{\bar{\alpha}\bar{\alpha}'}(r)] ,
\end{aligned}
\tag{74}
$$

where

$$
V_{\alpha\alpha'}(r) = \int dx \int d\hat{r}\ \Phi_\alpha^*(x,\hat{r})V(x,r)\Phi_{\alpha'}(x,\hat{r}) ,
\tag{75}
$$

and we have used the fact that $V_{nn'}$ is diagonal in $(-1)^{j_1+j_2+\ell}$. Since $V_{nn'}$ is diagonal in n, eq. (74) implies that

$$
\begin{aligned}
V_{\alpha\alpha'}(r) + n(-1)^{j_1+j_2+j_{12}+\ell}V_{\bar{\alpha}\alpha'}(r) = {}& n(-1)^{j_1'+j_2'+j_{12}'+\ell'}V_{\alpha\bar{\alpha}'}(r) \\
& + (-1)^{j_{12}+j_{12}'}V_{\bar{\alpha}\bar{\alpha}'}(r) .
\end{aligned}
\tag{76}
$$

Thus

$$
\begin{aligned}
V_{nn'}(r) = {}& \delta_{nn'}[(1 + \delta_{v_1 v_2}\delta_{j_1 j_2})(1 + \delta_{v_1' v_2'}\delta_{j_1' j_2'})]^{-\frac{1}{2}}[V_{\alpha\alpha'}(r) \\
& + n(-1)^{j_1+j_2+j_{12}+\ell}V_{\bar{\alpha}\alpha'}(r)] .
\end{aligned}
\tag{77}
$$

One implication of eq. (77) is that only states with $n = +1$ will couple with the states with $j_1 + j_2 = 0$

The determination of $V_{\alpha\alpha'}$ requires the evaluation of eq. (75), which is an 8-dimensional integral. We proceed by making the expansion[30]

$$
V(x,r) = \sum_{q_1 q_2 \mu} v_{q_1 q_2 \mu}(R_1,R_2,r) y_{q_1 q_2 \mu}(\hat{r}_1,\hat{r}_2) ,
\tag{78}
$$

$$
y_{q_1 q_2 \mu}(\hat{r}_1,\hat{r}_2) = \frac{4\pi}{[2(1 + \delta_{\mu 0})]^{\frac{1}{2}}}[Y_{q_1 \mu}(\hat{r}_1)Y_{q_2 -\mu}(\hat{r}_2) + Y_{q_1 -\mu}(\hat{r}_1)Y_{q_2 \mu}(\hat{r}_2)] ,
\tag{79}
$$

where $\hat{r}_i$ is in the same direction as $\hat{R}_i$ but is expressed in the body-fixed frame of reference where the z axis is in the direction of $\hat{r}$. Equation (75) then becomes

$$V_{\alpha\alpha'}(r) = \sum_{q_1 q_2 \mu} B_{\beta\ell\beta'\ell'}^{q_1 q_2 \mu} C_{\gamma\gamma'}^{q_1 q_2 \mu}(r) \quad , \tag{80}$$

where

$$B_{\beta\ell\beta'\ell'}^{q_1 q_2 \mu} = \int d\hat{R}_1 \, d\hat{R}_2 \int d\hat{r} \; \theta_{j_1 j_2 j_{12} \ell}^{JM}{}^*(\hat{R}_1,\hat{R}_2,\hat{r}) Y_{q_1 q_2 \mu}(\hat{r}_1,\hat{r}_2)\theta_{j_1' j_2' j_{12}' \ell'}^{J'M'}(\hat{R}_1,\hat{R}_2,\hat{r}), \tag{81}$$

$$C_{\gamma\gamma'}^{q_1 q_2 \mu}(r) = \int dR_1 \int dR_2 \; \chi_{v_1 j_1}^{*}(R_1)\chi_{v_2 j_2}^{*}(R_2) \nu_{q_1 q_2 \mu}(R_1,R_2,r)\chi_{v_1' j_1'}(R_1)\chi_{v_2' j_2'}(R_2), \tag{82}$$

β stands for the quantum numbers j_1, j_2, j_{12}, J, and M, and γ stands for the quantum numbers v_1, j_1, v_2, and j_2. Equation (81) is a 6-dimensional integral which is independent of r and can be evaluated analytically. Equation (82) is a two-dimensional integral which must be performed at every sector for every vibrational-rotational quantum number pair $\gamma\gamma'$ and every set of $q_1 q_2 \mu$ for which $B_{\beta\ell\beta'\ell'}^{q_1 q_2 \mu}$ is nonzero. However, before eq. (82) can be evaluated, it is necessary to determine the $\nu_{q_1 q_2 \mu}$ which are given by

$$\nu_{q_1 q_2 \mu}(R_1,R_2,r) = \frac{1}{4\pi} \int_{-1}^{1} [1 - \cos^2(\phi_1 - \phi_2)]^{-\frac{1}{2}} \, d[\cos(\phi_1 - \phi_2)] \int_{-1}^{1} d(\cos \theta_1)$$

$$\times \int_{-1}^{1} d(\cos \theta_2) Y_{q_1 q_2 \mu}^{*}(\hat{r}_1,\hat{r}_2)V(\underset{\sim}{x},r) \quad , \tag{83}$$

where θ_i and ϕ_i are the inclination and azimuthal angles of $\hat{r}_i$. The three-dimensional integral in eq. (83) is independent of the channel indices β, γ, and ℓ.

The $B_{\beta\ell\beta'\ell'}^{q_1 q_2 \mu}$ are independent of sector and are evaluated as follows. Let $\tilde{\theta}_{j_1 j_2 j_{12} \Omega}^{JM}$ be the simultaneous eigenfunctions of the operators with eigenvalues $j_1(j_1 + 1)$, $j_2(j_2 + 1)$, $j_{12}(j_{12} + 1)$, $J(J + 1)$, M, and Ω, where Ω is the projection of J on the body-fixed z axis: $\tilde{\theta}_{j_1 j_2 j_{12} \Omega}^{JM}$ is given by[31]

$$\tilde{\theta}_{j_1 j_2 j_{12} \Omega}^{JM}(\hat{r}_1,\hat{r}_2,\hat{r}) = \sum_{m_1 m_2} (j_1 m_1 j_2 m_2 | j_1 j_2 j_{12} \Omega) Y_{j_1 m_1}(\hat{r}_1)$$

$$\times Y_{j_2 m_2}(\hat{r}_2)[(2J + 1)/4\pi]^{\frac{1}{2}} \mathcal{D}_{M\Omega}^{(J)}(\phi,\theta,0) \quad , \tag{84}$$

where $\mathcal{D}_{M\Omega}^{(J)}$ is a rotation matrix[26] and ϕ and θ are the azimuthal and inclination angles of $\hat{r}$. The integral of an expansion function of the potential between these body-frame basis functions is

$$\tilde{B}^{q_1 q_2 \mu}_{\beta \Omega \beta' \Omega'} = \int d\hat{r}_1 \int d\hat{r}_2 \; \tilde{\theta}^{JM\,*}_{j_1 j_2 j_{12} \Omega}(\hat{r}_1, \hat{r}_2, \hat{r}) \, y_{q_1 q_2 \mu}(\hat{r}_1, \hat{r}_2) \, \tilde{\theta}^{J'M'}_{j_1' j_2' j_{12}' \Omega'}(\hat{r}_1, \hat{r}_2, \hat{r}),$$

$$(85)$$

and is given by[31]

$$\tilde{B}^{q_1 q_2 \mu}_{\beta \Omega \beta' \Omega'} = \delta_{JJ'} \delta_{MM'} \delta_{\Omega \Omega'} [2(1 + \delta_{\mu 0})]^{-\frac{1}{2}} [(2j_1 + 1)(2q_1 + 1)(2j_1' + 1)(2j_2 + 1)$$

$$\times (2q_2 + 1)(2j_2' + 1)(2j_{12} + 1)(2j_{12}' + 1)]^{\frac{1}{2}} (-1)^{j_{12}' + j_1' + j_2' + q_1 + q_2 + \Omega}$$

$$\times \begin{pmatrix} j_1 & q_1 & j_1' \\ 0 & 0 & 0 \end{pmatrix} \begin{pmatrix} j_2 & q_2 & j_2' \\ 0 & 0 & 0 \end{pmatrix} \sum_{q_{12}} (2q_{12} + 1) \begin{pmatrix} j_{12}' & q_{12} & j_{12} \\ -\Omega & 0 & \Omega \end{pmatrix}$$

$$\times \begin{pmatrix} q_{12} & q_1 & q_2 \\ 0 & \mu & -\mu \end{pmatrix} \begin{Bmatrix} j_{12}' & q_{12} & j_{12} \\ j_1' & q_1 & j_1 \\ j_2' & q_2 & j_2 \end{Bmatrix} [1 + (-1)^{q_1 + q_2 + q_{12}}], \quad (86)$$

where $\begin{Bmatrix} \cdots & \cdots & \cdots \\ \cdots & \cdots & \cdots \\ \cdots & \cdots & \cdots \end{Bmatrix}$ is a 9 - j symbol.[26] The laboratory-frame basis functions used for the wavefunction expansion are related to these body-frame functions by the transformation[31,32]

$$\theta^{JM}_{j_1 j_2 j_{12} \ell}(\hat{R}_1, \hat{R}_2, \hat{r}) = \sum_{\Omega} (2\ell + 1) \begin{pmatrix} \ell & j_{12} & J \\ 0 & \Omega & -\Omega \end{pmatrix} (-1)^{\Omega} \, \tilde{\theta}^{JM}_{j_1 j_2 j_{12} \Omega}(\hat{r}_1, \hat{r}_2, \hat{r}),$$

$$(87)$$

where $\begin{pmatrix} \cdots & \cdots & \cdots \\ \cdots & \cdots & \cdots \end{pmatrix}$ is a 3 - j symbol[26] (which is simply a re-phased and re-normalized Clebsch-Gordan coefficient), and the $B^{q_1 q_2 \mu}_{\beta \ell \beta' \ell'}$ which we need to evaluate are related to the body-frame integrals by

$$B^{q_1 q_2 \mu}_{\beta \ell \beta' \ell'} = (2\ell + 1)(2\ell' + 1) \sum_{\Omega \Omega'} \begin{pmatrix} \ell & j_{12} & J \\ 0 & \Omega & -\Omega \end{pmatrix} \begin{pmatrix} \ell' & j_{12}' & J' \\ 0 & \Omega' & -\Omega' \end{pmatrix} (-1)^{\Omega + \Omega'} \, B^{q_1 q_2 \mu}_{\Omega \ell \Omega' \ell'}.$$

$$(88)$$

Substituting (86) into (88) yields

$$B_{\beta\ell\beta'\ell'}^{q_1 q_2 \mu} = \delta_{JJ'} \delta_{MM'} (-1)^{j_{12}+j_1'+j_2'+q_1+q_2+J} [2(1+\delta_{\mu 0})]^{-\frac{1}{2}} [(2j_1+1)(2q_1+1)$$

$$\times (2j_1'+1)(2j_2+1)(2q_2+1)(2j_2'+1)(2j_{12}+1)(2j_{12}'+1)$$

$$\times (2\ell+1)(2\ell'+1)]^{\frac{1}{2}} \begin{pmatrix} j_1 & q_1 & j_1' \\ 0 & 0 & 0 \end{pmatrix} \begin{pmatrix} j_2 & q_2 & j_2' \\ 0 & 0 & 0 \end{pmatrix} \sum_{q_{12}} (2q_{12}+1)$$

$$\times \begin{pmatrix} \ell & \ell' & q_{12} \\ 0 & 0 & 0 \end{pmatrix} \begin{pmatrix} q_{12} & q_1 & q_2 \\ 0 & \mu & -\mu \end{pmatrix} \begin{Bmatrix} \ell & \ell' & q_{12} \\ j_{12}' & j_{12} & J \end{Bmatrix} \begin{Bmatrix} j_{12}' & q_{12} & j_{12} \\ j' & q_1 & j_1 \\ j_2' & q_2 & j_2 \end{Bmatrix}$$

$$\times [1 + (-1)^{q_1+q_2+q_{12}}] , \tag{89}$$

where $\begin{Bmatrix} \cdots & \cdots & \cdots \\ \cdots & \cdots & \cdots \end{Bmatrix}$ is a 6 - j symbol.[26] For the special case of $J = 0$, eq. (89) can be simplified to

$$B_{\beta\ell\beta'\ell'}^{q_1 q_2 \mu} = \delta_{j_{12}\ell} \delta_{j_{12}'\ell'} (1+\delta_{\mu 0})^{-\frac{1}{2}} [2(2j_1+1)(2q_1+1)(2j_1'+1)(2j_2+1)$$

$$\times (2q_2+1)(2j_2'+1)(2\ell+1)(2\ell'+1)]^{\frac{1}{2}} (-1)^{j_{12}+j_1'+j_2'+q_1+q_2}$$

$$\times \begin{pmatrix} j_1 & q_1 & j_1' \\ 0 & 0 & 0 \end{pmatrix} \begin{pmatrix} j_2 & q_2 & j_2' \\ 0 & 0 & 0 \end{pmatrix} \sum_m \begin{pmatrix} \ell & j_1 & j_2 \\ 0 & m & -m \end{pmatrix} \begin{pmatrix} \ell' & j_1' & j_2' \\ 0 & -\mu-m & \mu+m \end{pmatrix}$$

$$\times \begin{pmatrix} j_1' & q_1 & j_1 \\ -\mu-m & \mu & m \end{pmatrix} \begin{pmatrix} j_2' & q_2 & j_2 \\ \mu+m & -\mu & -m \end{pmatrix} . \tag{90}$$

In terms of Clebsch-Gordan coefficients eq. (90) is

$$B_{\beta\ell\beta'\ell'}^{q_1 q_2 \mu} = \delta_{j_{12}\ell} \delta_{j_{12}'\ell'} (-1)^{j_1'+j_2+\ell'+q_1+q_2+\mu} [(2j_2'+1)(2j_2+1)(1+\delta_{\mu 0})]^{-\frac{1}{2}}$$

$$\times [2(2\ell+1)(2\ell'+1)(2q_1+1)(2q_2+1)]^{\frac{1}{2}} (j_1 0 q_1 0 | j_1 q_1 j_1' 0)$$

$$\times (j_2 0 q_2 0 | j_2 q_2 j_2' 0) \sum_m (\ell 0 j_1 m | \ell j_1 j_2 m)(\ell' 0 j_1' -\mu-m | \ell' j_1' j_2' -\mu-m)$$

$$\times (j_1' -\mu-m q_1 \mu | j_1' q_1 j_1 -m)(j_2' \mu+m q_2 -\mu | j_2' q_2 j_2 m) . \tag{91}$$

The $C_{\gamma\gamma'}^{q_1 q_2 \mu}$ are calculated using a recently proposed optimized quadrature scheme using Gauss-ground-state[33] nodes. Seven points per vibrational coordinate were found sufficient to converge the final S matrix elements for the initial state $v_1 = v_2 = 1$, $j_1 = j_2 = 0$ to better than 1%. These convergence tests were performed with a basis with $v_1 + v_2 = 2$ and $j_1 + j_2 \leq 5$ at a

relative translational energy of 76 meV. N was 54 for these calculations.
Thus the integrand of eq. (82) is required at 49 points for each q_1, q_2, μ,
γ, γ', and r, and this was evaluated as discussed next.

The three-dimensional integral of eq. (83) must be evaluated numerically
in each sector for each of the 49 values of R_1 and R_2 required. The θ_1 and
θ_2 integrals were evaluated using N_q-point Gauss-Legendre quadrature and the
$\phi_1 - \phi_2$ integral was evaluated using N_q-point Gauss-Chebyshev quadrature, and
we took N_q to be $q_{max} + 1$ where q_{max} is the maximum value of q_1 and q_2 used
in eq. (78). Note that $0 \leq \mu \leq \min(q_1, q_2)$ where $\min(q_1, q_2)$ is the minimum
of q_1 and q_2. N_q ranged from 11 at small r to 4 for r greater than 24.5 a_0.
Thus the potential function must be evaluated $49N_q^3$ times per sector. This
is an important number because the evaluation of the potential at a given
geometry is very time consuming. As coded by us, the Poulsen-Billing-
Steinfeld[29] potential requires the evaluation of 3 exponentials, 23 additions
or subtractions, 36 multiplications, 1 division, 1 square root, and 6 raisings
to a power per potential evaluation. The Redmon-Binkley[28] potential is even
more complicated; our version requires 34 exponential evaluations, 303 additions or subtractions, 405 multiplications, 19 divisions, 5 square roots,
and 2 raisings to a power per potential evaluation. Note, however, that the
number of potential evaluations is independent of the number of channels in
the close-coupling expansion.

To complete the specification of our calculations it is necessary to
assign values to the number of channels N, i.e., eq. (9), and to the number
of terms in the potential expansion, eq. (78). We choose the number of terms
in eq. (78) by including all q_1, q_2, and μ with $q_1 + q_2 \leq q_{sum}$ and increasing
q_{sum} until eq. (78) accurately represents $V(\underset{\sim}{x},r)$ at a given value of r. In
our case accurately means that the expansion differ from $V(\underset{\sim}{x},r)$ on the order
of 1% for r > 4 a_0 and slightly more for r < 4 a_0. This yields $q_{sum} = 10$ at
small r and $q_{sum} = 3$ for r larger than 24.5 a_0. Equation (78) then contains
from 161 to 13 terms. It should be noted that a large number of the $B_{\beta\ell\beta'\ell'}^{q_1 q_2 \mu}$
are zero and to avoid storing up to $322N^2$ numbers, many of which are zero,
our program only stores the nonzero values of $B_{\beta\beta'\ell'}^{q_1 q_2 \mu}$, alternating with one
word which indicates the values of q_1, q_2, μ, β, ℓ, β', and ℓ'. This
requires approximately $30N^2$ words of storage for $q_{sum} = 10$, with the exact
coefficient of N^2 depending on N.

The fact that many $B_{\beta\ell\beta'\ell'}^{q_1 q_2 \mu}$ are zero does not necessarily translate into
zero values of $V_{nn'}$. In fact at small r, $V_{nn'}$ has no zero elements, but as
r increases and we decrease the number of terms in eq. (78), we do obtain
zeroes in the V matrix. Table 1 summarizes the percentage of zero elements
as a function of r.

Table 1. Potential and sector information

r (a_0)	N_q	Number of terms in eq. (78)	% zero elements in V matrix[a]	Number of sectors in this range[a]
3 to 6.06	11	161	0	52
6.06 to 6.54	10	125	0	8
6.54 to 7.02	9	95	1.1	8
7.02 to 7.56	8	70	3.8	9
7.56 to 11.0	7	50	9.8	51
11.0 to 12.6	6	34	20	12
12.6 to 24.7	5	22	34	51
24.7 to 150	4	13	54	100

[a]based on N = 530

There are many ways to choose N. In the present chapter, the only block of the Schroedinger equation that we attempt to solve is the one with $J = M = 0$, and the only initial state considered explicitly is the one with $v_1 = v_2 = 1$ and $j_1 = j_2 = 0$. As a consequence we need only consider the sub-block with $\eta = \zeta = +1$. Consider a total energy with the $v_1 = v_2 = 1$, $j_1 = j_2 = 0$ channel open, and the $v_1 = v_2 = 1$, $j_1 = 0$, $j_2 = 1$ channel closed. One possible choice for N is to include all open channels plus selected closed channels. However, this would lead to an extremely large value for N since the rotational energy spacing is so much smaller than the vibrational spacing. For the energy under consideration there are 1548 open channels with $J = M = 0$ and $\eta = \zeta = +1$, and this number increases rapidly as either J or the energy increases. We have found previously,[11b] for model diatom–diatom collisions with central potentials such that only vibrational energy and not rotational energy is transferred, that a good way to choose channels is to include all channels with $v_1 + v_2 \leq v_{sum}$, where v_{sum} is large enough to include at least one closed vibrational level of each molecule. We assume (based on experience[10] with rotational energy transfer in atom–molecule collisions) that highly closed rotational channels may also be excluded, and we also assume that channels with very high j_i, even when open, do not have significant dynamical coupling to an initial state with $j_1 = j_2 = 0$. Thus we finally arrive at a scheme in which we include all channels with $v_1 + v_2 \leq v_{sum}$, $j_1 + j_2 \leq j_{sum}$, where j_{sum} is allowed to depend on v_1 and v_2. In the final analysis we must converge the calculations with respect to v_{sum} and all the j_{sum}. If convergence is obtained the assumptions behind our channel-selection scheme do not affect the accuracy of the final results, but they do affect the rate of convergence with increasing N. Because the HF molecules are indistinguishable and we properly include the interchange symmetry, it is not a further restriction to generate the channels from a list in which $v_1 \leq v_2$ and $j_1 \leq j_2$ when $v_1 = v_2$. For the present chapter we consider four sizes of the basis set, with N = 55, 101, 400, and 530. The first two are small basis sets used only for debugging and timing analyses; the third and fourth are an attempt to test convergence. The 400-channel basis is obtained using $v_{sum} = 3$, $j_{sum} = 8$ for $v_1 + v_2 \leq 2$, and $j_{sum} = 6$ for $v_1 + v_2 = 3$, and the 530-channel basis is obtained using $v_{sum} = 3$, $j_{sum} = 9$ for $v_1 + v_2 \leq 2$, and $j_{sum} = 7$ for $v_1 + v_2 = 3$. (Another difference is that the 55, 400, and 530-channel runs are for the Redmon-Binkley potential with proper inclusion of interchange symmetry, whereas the 101-channel calculations are simplified as discussed in section VI.) None of these basis sets includes all open channels with $v_1 + v_2 \leq 1$; channels with high $j_1 + j_2$ are excluded. These open channels should not be too important since channels with intermediate $j_1 + j_2$ will have poor translational overlap with the initial state because of the large energy defect and coupling to channels with higher $j_1 + j_2$ require terms in eq. (78) with high q_1 and q_2, and these become relatively less important as q_1 and q_2 increase.

We integrate eq. (12) using a fixed stepsize of 0.06 a_0 for r less than 10 a_0 and then allow the stepsize to increase according to eq. (38) with EPS equal to 0.1 and h_{max} large enough to not limit $h^{(i+1)}$. By the time the calculation is stopped at r = 150 a_0, the stepsize has increased to about 3 a_0, requiring approximately 290 sectors.

V.   VECTORIZATION OF THE COMPUTER CODE

On the basis of the theory discussed in section II and the algorithm reviewed in section III, the solution of the close coupling equations may be broken into six segments, as described in Table 2.

A general name for the work of segment V is asymptotic analysis. In general it need only be performed in the last sector, but as a check that the integration has been carried out to large enough r it is usually performed more than once. As mentioned above, runs at second and subsequent energies require less work if appropriate matrices are saved from the calculation at the first energy. For a first energy run with N_s sectors for which the asymptotic analysis is performed N_a times, segments I, II, and IV are performed N_s times, segment III is performed $N_s - 1$ times, and segment V is performed N_a times. Segment VI contains some tasks performed only once, as well as some tasks performed N_s times or N_a times, but the tasks performed more than once all involve very little computational work compared to those singled out as segments I-V. For the various runs discussed here $N_s = 288$ or 291 and $N_a = 4$. Thus most of the computational work is in

Table 2. Segments of the computation

Segment	Principal subprogram	Task
I	POT	Calculation of $V_{nm}(r_C^{(i)})$
II	DCALC	Assembly and diagonalization of $\underset{\sim}{D}^{(i)}$ to obtain $\lambda_{nn}^{(i)}$ and $\underset{\sim}{T}^{(i)}$
III	TAUMTS	Calculation of $\underset{\sim}{T}(i-1,i)$ and/or its inverse
IV	RCALC	Calculation of $\underset{\sim}{R}_4^{(i)}$ and $P^{(i+1)}$
V	GENSCAT	Calculation of $\underset{\sim}{S}$ from $\underset{\sim}{R}_4^{(i)}$
VI	...	everything else

segments I–IV. For convenience in discussing the segments in the rest of this chapter, segments I–V will be named after their principal subprograms and segment VI will be called overhead.

Segment I involves the computation of a very large number, approximately $\frac{1}{2}N^2$, of multi-dimensional integrals, and segments II–V include a number of matrix operations for which the number of arithmetic steps becomes asymptotically proportional to N^3 or $[P^{(i)}]^3$ as the matrix orders are increased. Such matrix operations are called "N^3 steps". Most of the computer time goes into the potential calculation and these N^3 steps, and these are thus the operations for which vectorization offers the most potential benefits. We will next concentrate our attention on the N^3 steps. Table 3 lists the number and type of such steps for each segment. It is useful to comment briefly on the appearance of an inversion step in the table. In general whenever possible, it is more efficient to solve linear equations $\underset{\sim}{A}\underset{\sim}{x} = \underset{\sim}{B}$ directly than to perform the separate steps of inverting $\underset{\sim}{A}$ followed by the matrix multiplication of $\underset{\sim}{A}^{-1}$ times $\underset{\sim}{B}$ to give the solution $\underset{\sim}{x} = \underset{\sim}{A}^{-1}\underset{\sim}{B}$. However, eqs. (44) and (46) are of the form $\underset{\sim}{x} = \underset{\sim}{B}\underset{\sim}{A}^{-1}$. Although this is equivalent to $\underset{\sim}{x}^T = (\underset{\sim}{A}^T)^{-1}\underset{\sim}{B}^T$, we instead solve it by first inverting $\underset{\sim}{T}(i-1,i)$, which plays the role of $\underset{\sim}{A}$, and then form the matrix products $\underset{\sim}{B}_1\underset{\sim}{A}^{-1}$ for eq. (44) and $\underset{\sim}{B}_2\underset{\sim}{A}^{-1}$ for eq. (46). Note that $\underset{\sim}{B}_2$ is diagonal. Multiplying a matrix by a diagonal matrix is, like a matrix times a vector or a matrix plus a matrix, only an "N^2 step", and such operations are not included in Table 3. Figure 1 gives a flow chart of our program, showing the flow of the calculation between the various program segments discussed above. On the first energy, POT, DCALC, TAUMTS, and RCALC are called once per sector. For calculations at second energies, TAUMTS is called once in each sector

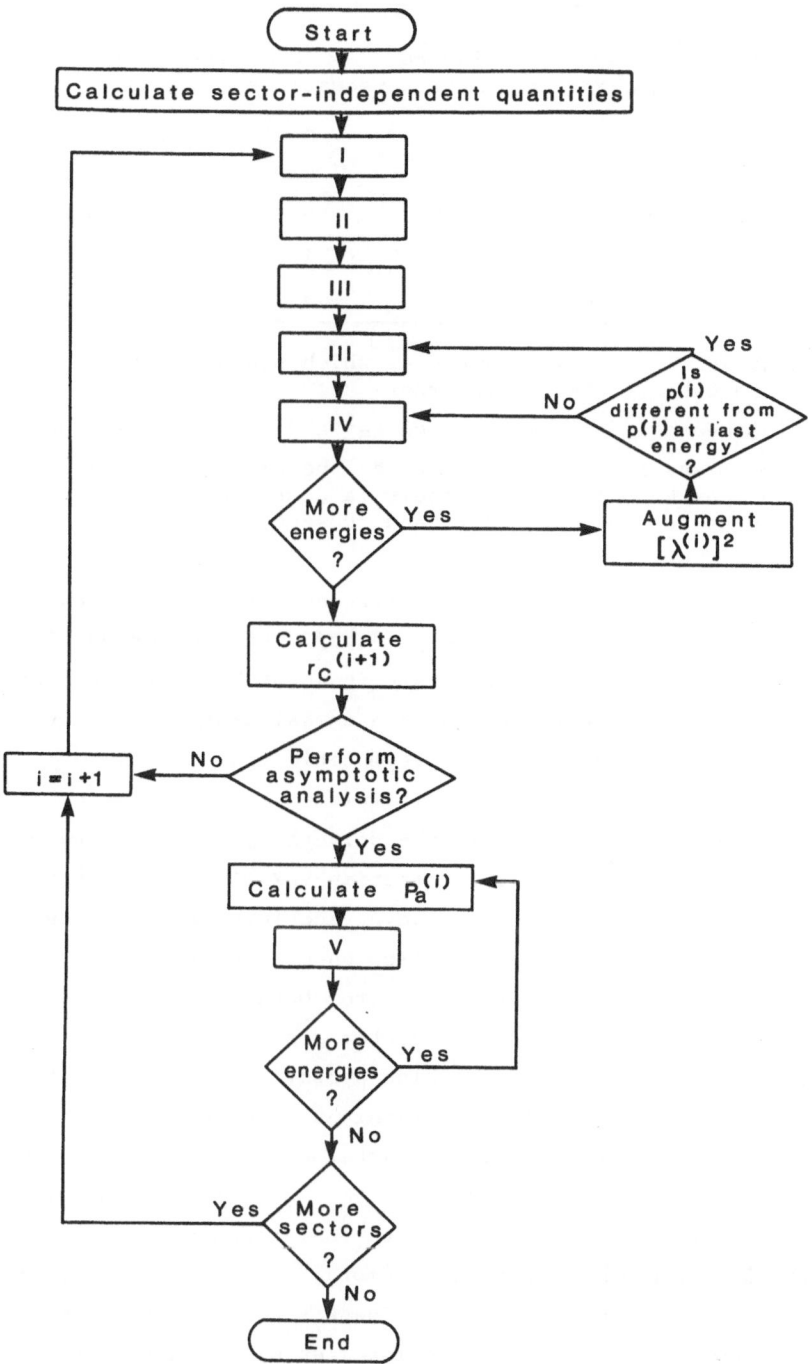

Fig. 1 Flow Chart

Table 3. Number and type of "N^3 steps" per program section[a]

	operation				
Segment	diagonalize	invert matrix	solve linear equations	multiply matrix by matrix	Equations
I. POT	0	0	0	0	...
II. DCALC	$(1,0)$[b]	0	0	0	(16)
III. TAUMTS	0	$[1,0]$[c]	0	$(1,0)$	(41),(46)
IV. RCALC	0	0	1	2	(44),(49)
V. GENSCAT	0	0	4	4	(21),(58),(60)

[a] significant operations are ones which scale for large matrix orders
as the cube of the matrix order

[b] $(1,0)$ means 1 for the first energy and 0 for subsequent energies

[c] $[1,0]$ means 1 for all segments at the first energy and for sectors
in which the number of channels is decreased at any energy, but 0
for other sectors at second and subsequent energies

for which $P^{(i)}$ at this energy is different from $P^{(i)}$ at the last energy,
and RCALC is called once in every sector. For the production runs $P^{(i)}$
differs from one sector to the next in about half of the sectors. At
selected distances near the end of each calculation, for every energy,
GENSCAT is called to perform an asymptotic analysis.

We now discuss the practical implementation of these calculations on
three computers. The first is the University of Minnesota Department-of-
Chemistry Digital Equipment Corporation VAX 11/780. This is a "minicomputer"
equipped with a scalar floating point accelerator and 4 Megabytes of physical
memory (equivalent to a half million 64-bit words). The second is the
Colorado State University (CSU) Control Data Corporation Cyber 205 equipped
with a scalar processor and two vector pipelines and two million 64-bit
words of physical memory. (We also performed some of our Cyber 205 calculations on similarly configured machines located at the Control Data Corporation offices in Arden Hills, Minnesota; timing analyses will be given only
for the CSU machine.) The third is the University of Minnesota University-
Computer-Center Cray-1, equpped with scalar and vector processors and one
million 64-bit words of physical memory, of which 0.91 Megawords are
available to an individual user. These second and third machines are "class
VI supercomputers". These three machines will henceforth be called simply
the VAX, the Cyber 205, and the Cray-1, respectively. The VAX and the
Cyber 205 are virtual-memory machines, while the Cray-1 is not. In our

discussion of implementing the codes we will pay special attention to utilization of the vector modes available on the two supercomputers.

There are two aspects of vectorizing a program. The first concerns memory usage. Since our calculations require the manipulation of matrices which are quite large, the finite amount of memory and mass storage available to us limits the size and number of matrices which can be used in a given calculation. Thus a considerable amount of effort was spent to modify our code to use the smallest amount of memory possible. This is especially important on the Cray-1 which is not a virtual memory machine. Our vectorized code on the Cray-1 holds only two $N \times N$ matrices in memory, with other $N \times N$ matrices and the B matrix elements stored on disk. This limits us to N less than about 640. It is possible to hold up to four $N \times N$ matrices in memory for N less than about 470, but to avoid presenting results with more than one Cray-1 vector version of our code, we present results here only for the version holding only 2 $N \times N$ matrices in memory. The second aspect of vectorization is to modify the program to use the supercomputers' vector capabilities. This can be done in several ways, as discussed next.

A portion of code that can be vectorized, that is, that which can be replaced with a vector instruction, is a DO-loop that performs exactly the same arithmetic operation on each element of a vector. Examples of DO-loops which cannot be vectorized include those involving recursion, subroutine calls, input or output statements, branch instructions, and most conditional statements. Different machines have different definitions of a vector. On the Cyber 205, a vector is a series of contiguous memory locations. The Cray-1 has 8 vector registers, each capable of containing 64 words (one word is 64 bits), and it can fill these registers with values from a series of memory locations, each incremented by a constant amount.

To understand how a vector machine achieves high-speed arithemtical throughput, it is necessary to first discuss how a scalar machine works. A typical operation, such as the multiplication of two numbers, can be broken down into a series of suboperations. The amount of time required by a suboperation is at least one minor cycle. On a scalar machine, each suboperation must wait until the preceding one has completed before it can start. Thus the multiplication of two numbers will take many minor cycles; the exact number of minor cycles will depend on whether or not the operands were initially in registers or in memory. In a vector machine, the units which perform the suboperations on vectors are more independent. The set of units that perform these suboperations on vectors are called pipelines and the units can be operating simultaneously on different operands, the different vector elements. It is then necessary to wait several minor cycles until the first element of a vector has received all of the sub-operations, and after that, the results of the operation on the following vector elements will come out at the rate of one per minor cycle. There are several ways to increase the speed even further. For example, one can add more pipelines, as on the Cyber 205 machine we used, which has two. Then, after the startup time, the time necessary for the first result to emerge, a result will be obtained from each pipeline every minor cycle. Another way to speed up the arithmetic is to perform more than one operation at a time; an example of this is multiplying a scalar times a vector and adding it to a different vector (a scalar is a vector of length 1). This is called linking (or chaining), and using it produces the results of two operations per pipe per minor cycle. On the Cyber 205 linking is restricted to two vector operands and each of the four units (add/subtract, multiply/divide, logical, and shift) in the CPU can be used only once. On the Cray-1, all 8 of the vector registers can be used, but again each unit in the CPU can be used only once. On the Cyber 205 it is also possible to use half precision (32 bits) rather than the normal 64-bit precision. In this case one obtains twice as many 32-bit results as 64-bit results. All of these techniques increase the startup time, so it is necessary to have vectors as long as possible to get the maximum speed. The maximum vector length on the Cyber 205 is 65,355. The Cray-1 uses only 64 elements of a vector at a given time, but it has no maximum vector length. In our calculations, we use linking whenever possible and we always use 64-bit precision. For the processes which take up most of the time in segments II-V, the vector length is always less than or equal to $P^{(i)}$. For some operations on the Cyber 205, such as setting a matrix equal to another matrix, we exceed the maximum vector length and have to split up the operation into parts, each of which does not exceed the maximum vector length. In our vectorized version of POT we use vector lengths of up to 1331, however, most of the time required is spent on operations on vectors of length 49.

As stated already above, most of the work in our calculations is spent on matrix manipulations. To take advantage of vectorization one can write utility routines which are vectorizable or use pre-existing library routines supplied by the system. On the Cyber 205 we have used the Math-Geophysical Vector library (MAGEV), which contains routines to diagonalize matrices (these are modifications of EISPACK routines) and solve linear equations and to perform other useful tasks. On the Cray-1, we have used the SCILIB library, which contains similar routines as the MAGEV library.

To help make the vectorization process more clear, we will now consider in detail how one would write a utility routine to multiply two

matrices. We first consider doing this on the Cyber 205, which has the
restriction that the components of the vectors be contiguous memory loca-
tions. A typical FORTRAN code to multiply an $N \times M$ matrix $\underset{\sim}{A}$ times an $M \times P$
matrix $\underset{\sim}{B}$ to produce an $N \times P$ matrix $\underset{\sim}{C}$ using the inner product method is

```
      DO 1 I=1,N
      DO 1 J=1,P
      C(I,J)=0.0
      DO 1 K=1,M
    1 C(I,J)=C(I,J)+A(I,K)*B(K,J).
```

In FORTRAN the elements of arrays are stored in memory in the order produced
by varying the first index most rapidly. Thus in the above example only the
elements of B are accessed sequentially from memory. Since the elements of
A are not accessed sequentially from memory, the inner-most loop cannot be
vectorized (this loop could be vectorized on the Cray-1). One possible solution is to store the transpose of the matrix A, but it is much more efficient to simply re-order the loops as follows (this is sometimes known as
the outer product method):

```
      DO 1 J=1,P
      DO 1 I=1,N
    1 C(I,J)=0.0
      DO 2 J=1,P
      DO 2 K=1,M
      DO 2 I=1,N
    2 C(I,J)=C(I,J)+A(I,K)*B(K,J).
```

Here the inner-most loop can be replaced with the linked vector instruction:
add element I of the vector starting at C(1,J) to the scalar B(K,J) times
element I of the vector starting at A(1,K) and store the result in element
I of the vector starting at C(1,J). Since this uses the linking capabilities of the Cyber 205, the matrix multiplication can be performed extremely
fast. (The method is not the only efficient way to vectorize a matrix
multiply on the Cyber 205, but is the one we use.)

In Cyber 205 FORTRAN there are two ways to indicate vector instructions. The first is to simply turn on the compiler option (after of course
modifying the program to allow vector instructions). There are two options,
V (for Vectorization) and U (for Unsafe vectorization). The U option causes
some classes of DO loops to be replaced with vector instructions which the
V option does not vectorize. We found that these options were not useful
to us because the V option will not vectorize DO loops with variable limits,
and the U option sometimes caused incorrect results. The second way to
indicate vector instructions uses an explicit notation. An example of

this notation is A(I,J,K;N) which indicates a vector containing N elements
starting with A(I,J,K). Using this notation the matrix multiply example
becomes

```
        DO 1 J=1,P
    1   C(1,J;N)=0.0
        DO 2 J=1,P
        DO 2 K=1,M
    2   C(1,J;N)=C(1,J;N)+A(1,K;N)*B(K,J).
```

On the Cray-1, vectorization is accomplished differently. Since the
Cray-1 can fill its vector registers from memory locations which are not
contiguous, the first as well as the second multiplication example can be
vectorized. We have found that, of the methods we tried, the fastest way
to multiply two matrices is to use the DO loops ordered as in the first
example but replacing the inner-most DO loop with a call to the Cray-1 dot
product function:

```
        DO 1 I=1,N
        DO 1 J=1,P
    1   C(I,J)=SDOT(M,A(I,1),N,B(1,J),1)
```

where it has been assumed that the array A has N as its first dimension.
Cray-1 FORTRAN does not contain a special explicit notation for vectors;
thus to make the compiler use vector instructions it is necessary to use
the compiler option V. However, unlike Cyber 205 FORTRAN, Cray-1 FORTRAN
allows for what are called compiler directives. These allow one to control
whether or not specific DO loops are vectorized.

In the final version of our scattering code on the Cyber 205 we use
in-line FORTRAN statements to perform matrix multiplications, and we call
routines from the MAGEV library to solve linear equations, invert matrices,
and diagonalize matrices. These routines use Gaussian elimination to solve
linear equations and modifications of EISPACK routines for matrix diagonalization. In comparison to our own vectorized FORTRAN utility routines to
perform these operations, the MAGEV linear equations solver (GEL) is
slightly faster and the MAGEV matrix diagonalization routines are significantly faster. The matrix diagonalization routines from the MAGEV library
are faster than our FORTRAN routines because matrix diagonalization (first
a Householder transformation to tridiagonal form followed by an implicit QL
diagonalization of the tridagonal matrix) involves many steps which cannot
be vectorized, and the MAGEV routines use many special calls (to what are
called STACKLIB routines) which cause the compiler to generate in-line
code which efficiently performs these non-vectorizable steps. We have
not included any STACKLIB routines in our own coded sections.

On the Cray-1, the final version of our scattering code uses in-line FORTRAN statements and the dot product function SDOT to perform matrix multiplications. To diagonalize the matrix $\underset{\sim}{D}$ we use the SCILIB version of the EISPACK routine RS, and for matrix inversion and linear-equation solving, we use our own vectorized version of a routine made up of LINPACK routines. Our routine consists of the LINPACK routines SGEFA, SGEDI, and SGESL combined together and modified to avoid all unnecessary subroutine calls. The original LINPACK routines use BLAS (Basic Linear Algebra Subroutines) routines to perform the innermost DO loop operations. In our routine, we found it faster to substitute FORTRAN statements to perform the operations done by the BLAS routines rather than to call the BLAS routines contained in the SCILIB library. In contrast to the Cyber 205, the SCILIB EISPACK routines are only slightly faster than using our own vectorized FORTRAN versions. The SCILIB inversion and linear-equation-solver subprogram MINV is faster than our own vectorized FORTRAN routine (based on modified LINPACK routines as discussed above); however we do not use it because the SCILIB version requires matrix input in an augmented matrix form that is inconvenient and also since the ratio of the time required by MINV to the time required by our vectorized FORTRAN routine was approaching one as the matrix size is increased.

VI. EXECUTION TIME ANALYSIS

In this section we present actual execution times for various runs and tasks within runs and for various computers and versions of our code. First we consider a special set of early runs with $N = 101$ for which we made the most detailed timing comparisons. These calculations used the simpler Poulsen-Billing-Steinfeld[29] potential with a smaller number of terms in eq. (78); in particular, we restricted q_1 and μ to 0 and q_2 to be less than or equal to 2. These calculations also differ from the production runs in several other respects; the most important of which are: we used 27-point quadrature (3 points per dimension) to evaluate eq. (83), the vibrational dependence was expanded in a Taylor series, keeping up to quadratic terms in each oscillator coordinate for a total of nine terms; we did not enforce interchange symmetry, we restricted the basis to functions with $j_1 = 0$ (so that the potential is effectively spherically averaged over the first rotor's orientation), and we used 4 sectors and multiplied the times obtained by the appropriate factors to mimic a calculation using $N_s = 291$. In the second part of this section we consider calculations with the Redmon-Binkley[28] surface with $N = 55$, 400, and 530 and with the production values of the numerical parameters, e.g., $N_q = 4$ to 11, depending on r.

Table 4. CPU times (in sec) for various versions of the six program segments on various machines. The times are based on N (number of channels) = 101, N_s (number of sectors) = 291, and N_a (number of asymptotic analyses) = 4.

Machine[a]	Version	Tasks done only at 1st E		Tasks done every E				Totals	
		POT	DCALC	TAUMTS (1st E)[b]	TAUMTS (2nd E)[c]	RCALC[d]	GENSCAT[d]	1st E[b]	2nd E[c]
VA	scalar	3170	18800	14600	4740	29200	894	66800	34800
CY	scalar	69.0(46)[e]	654(29)	427(34)	145(33)	889(33)	27.8(32)	2070(32)	1060(33)
CY	scalar with compiler vectorization (options V and U)	69.0(46)	634(29)	425(34)	141(34)	889(33)	27.4(33)	2060(32)	1060(33)
CY	explicitly vectorized	69.0(46)	81(232)	29(490)	8.8(538)	43(684)	4.45(201)	226(294)	56(628)
CR	scalar	95(33)	551(34)	536(27)	86(55)	1250(23)	35.2(25)	2470(27)	1370(25)
CR	scalar with compiler vectorization	98(32)	269(70)	386(38)	12(393)	1110(26)	28.2(32)	1890(35)	1150(30)
CR	explicitly vectorized	130(24)	117(160)	63(233)	13(375)	116(252)	5.58(160)	432(154)	134(261)

[a] VA: VAX 11/780; CY: Cyber 205; CR: Cray-1

[b] time for first energy

[c] time for second energy or any subsequent energy

[d] time for these tasks is same for 1st or any other energy

[e] numbers in parentheses are the ratio of the VAX time to the time of that particular entry

Table 4 presents the execution times for each of the segments I–V for the 101-channel runs. The times for segment VI (overhead) are not given in this table because they comprise less than 1% of the total time. As stated above, it is assumed that the integration range may be divided into 291 sectors, but an asymptotic analysis to calculate the scattering matrix need be performed 4 times (the fourth to actually obtain the scattering matrix and the first three as convergence checks of the maximum r value r_{max}). For the 101-channel test runs, the segment POT was not vectorized; it was only modified as required to interface with the other program modifications. Table 4 shows that running in scalar mode on the Cyber 205 results in a speedup factor of about 32 as compared to our VAX for both first and second energies. On the Cray-1 this speedup factor is a factor of 26. Vectorizing our program by just turning on the compiler vectorization options (options V and U on the Cyber 205 and option V on the Cray-1) produces very little difference on the Cyber 205, while on the Cray-1 some program sections are significantly speeded up, resulting in an overall speedup factor compared to our VAX of 35 times for first eneriges and 30 times for second energies. Our utility routines to perform diagonalization, determine inverses, and solve linear equations are speeded up significantly on the Cray-1, while our routines to multiply matrices were not speeded up at all, since they were written in a manner which did not allow the compiler to use vector instructions. Explicitly vectorizing our code as described above results in substantial speed increases for both the Cyber 205 and Cray-1. The speedup factor for the Cyber 205 over the VAX increases to 294 times for first energies and 628 times for second energies. The large speedup for second energies is mainly due to the fact that the processes which vectorize relatively less well on the Cyber 205 (the potential evaluation and matrix diagonalization) are not performed for second energies. The situation is similar on the Cray-1 where the speedup factor over the VAX increases to 154 times for first energies and 261 times for second energies. The reason why TAUMTS at second energies is slightly faster in the compiler vectorized version of our code than the explicitly vectorized version is that a different routine to invert matrices is used. We use the slightly slower inverter because it is the routine we have chosen to use to solve linear equations. Some program segments are speeded up even more, e.g., we achieve a factor of 684 for the RCALC segment on the Cyber 205; this indicates that other problems or other algorithms for this problem might be speeded up even more than the cases reported here. Thus it is possible to achieve very large speedup factors if the effort is made to make efficient use of the vector capabilities of a machine.

As discussed in section IV, the Poulsen-Billing-Steinfeld potential is simpler than the Redmon-Binkley one, and as mentioned in an earlier part of this section, we restricted q_{1max}, μ, and q_{2max} to 0, 0, and 2 for test runs with this potential, and the angular integration of eq. (83) was carried out by 27-point rather than N_q^3-point quadrature for the 101-channel timing runs. Nevertheless, the POT segment still consumes about 30% of the total time when the rest of the code is vectorized. In our production runs we used the Redmon-Binkley potential, larger values of q_{max}, and N_q^3-point quadratures with $N_q = 4$–11 (see Table 1); these improvements, which are necessary for good accuracy, resulted in POT requiring much more time in production runs. Thus we spent extra effort to vectorize our final potential routine for these runs, as described next.

The vectorization of the POT segment proceeded as follows. The first part of the potential matrix evaluation, which is independent of sector, is the calculation of the $B_{\beta\ell\beta'\ell'}^{q_1 q_2 \mu}$, via eq. (91). This is the most time consuming part of the calculations with the versions of the code used for the timing analyses presented here, and it cannot easily be vectorized because most of the $B_{\beta\ell\beta'\ell'}^{q_1 q_2 \mu}$, require different numbers of terms in the sum of eq. (91) and our routines to calculate 3j symbols (or Clebsch-Gordan coefficients) are not easily vectorizable. The sector-dependent parts of POT are also time consuming, but are more amenable to vectorization, as discussed next.

We calculate the $v_{q_1 q_2 \mu}$ of eq. (78) by performing the three-dimensional quadrature of eq. (83). In carrying out this integration we evaluate the potential function at all N_q^3 different orientations with fixed R_1, R_2, and r using vector arithmetic prior to doing any of the quadratures. This requires considerable storage, but it enables us to use fairly large vector lengths, equal to N_q^3, which varies from 1331 at small r, and decreasing to 64 at large r. Note that these vector lengths are independent of the number of channels. The sum in the three-dimensional integral to get the $v_{q_1 q_2 \mu}$ is done using the vector dot product functions. To do this we write

$$v_{q_1 q_2 \mu}(R_1, R_2, r) = \sum_{n_1=1}^{N_q} \sum_{n_2=1}^{N_q} \sum_{n_3=1}^{N_q} \omega_{n_1} \omega_{n_2} \omega'_{n_3} y^*_{q_1 q_2 \mu}(\theta_{n_1}, \theta_{n_2}, \phi_{n_3})$$

$$\times V(\theta_{n_1}, \theta_{n_2}, \phi_{n_3}, R_1, R_2, r) \tag{92}$$

$$= \sum_{i=1}^{N_q^3} W_i^{q_1 q_2 \mu} V(\underset{\sim}{x}_i, r) \tag{93}$$

where ω_i, ω'_i, θ_i and ϕ_i are the appropriate quadrature weights and nodes, and $W_i^{q_1 q_2 \mu}$ is a product of three weights times the expansion function, and we use the vector dot product function (SDOT on the Cray-1, Q8SDOT on the Cyber 205) to evaluate eq. (93). The $W_i^{q_1 q_2 \mu}$ depend on the value of N_q and thus on the value of r for which eq. (93) is evaluated; they are calculated once and stored for each r range given in Table 1. As for the evaluation of the potential function, the three-dimensional integral in eq. (83) is independent of the total number of channels. The vector length used in this part of the calculation is again N_q^3, which varies from 1331 to 64.

Our routine then loops through the channel pairs and calculates the $C_{\gamma\gamma'}^{q_1 q_2 \mu}$, the $V_{\alpha\alpha'}$ via eq. (80), and finally the $V_{nn'}$ using eq. (77). The calculation of the $C_{\gamma\gamma'}^{q_1 q_2 \mu}$ requires a 2-dimensional integral and this too is done using the vector dot product functions; that is, we write

$$C_{\gamma\gamma'}^{q_1 q_2 \mu}(r) = \sum_{i=1}^{7} \sum_{j=1}^{7} \tilde{\omega}_i^{v_1 j_1 v_1' j_1'} \tilde{\omega}_j^{v_2 j_2 v_2' j_2'} v_{q_1 q_2 \mu}(x_i, x_j, r) \tag{94}$$

$$= \sum_{k=1}^{49} \tilde{w}_k^{\gamma\gamma'} v_{q_1 q_2 \mu}(x_{1k}, x_{2k}, r) \tag{95}$$

and evaluate eq. (95) using the Cray-1 or Cyber 205 dot product function. The $\tilde{\omega}_i^{v j v' j'}$ are independent of r and are computed once before the start of the R matrix propagation. The $\tilde{w}_k^{\gamma\gamma'}$ are calculated from the $\tilde{\omega}_i^{v j v' j'}$ at each sector for each $\gamma\gamma'$ pair. We compute the $\tilde{w}_k^{\gamma\gamma'}$ on the Cyber 205 by first using the vector Q8VGATHR function to form two vectors of length 49 containing the appropriate $\tilde{\omega}_i^{v j v' j'}$, then by multiplying these two vectors. The $\tilde{w}_k^{\gamma\gamma'}$ are recomputed rather than stored because storing them would involve, for the 530-channel case, an array of length 11.6 million words, whereas the amount of storage required for the $\tilde{\omega}_i^{v j v' j'}$ is only 3.4 thousand words The integrals in eq. (95) must be performed for each $q_1 q_2 \mu$ with nonzero $B_{\beta\ell\beta'\ell'}^{q_1 q_2 \mu}$ for each of the $\beta\ell\beta'\ell'$ pairs. The number of these integrations is thus proportional to N^2, and depends on the value of r, because the number of terms in eq. (78) depends on the value of r (see Table 1). On the Cyber 205, in the large channel limit where the evaluation of eq. (92) is negligible, these procedures result in a total speedup factor of about 3 for the vector version of segment POT as compared to the scalar version. On the Cray-1, the speedup factor is about 2.

We now consider some computer times for cases with $N = 400$ and 530. Table 5 gives the times for calculations of 7 energies. (For $N = 530$ on the Cray-1 these times are estimated from a two-energy run.) The Cyber 205 uses 70% of the time required by the Cray-1 for $N = 400$ and 50% of the time for $N = 530$. Also included in Table 5 is the cost of the calculations in

Table 5. Execution times for $N = 400$ and $N = 530$ for 7 energies

Machine	CPU hours	billing units (equivalent hours)
$N = 400$[a]		
Cray-1	11.7	18.1
Cyber 205	7.6	9.2
$N = 530$[b]		
Cray-1	21.0	31.8
Cyber 205	9.4	15.5

[a] $N_s = 288$

[b] $N_s = 291$

Table 6. Breakdown of execution times (in hours) into program sections for $N = 400$ and $N = 530$ on the Cray-1

Program section	$N = 400$[a]	$N = 530$[b]
first energy		
Angular integrals[c] ($B_{\beta \ell \beta' \ell'}^{q_1 q_2 \mu}$)	1.2	2.4
Potential function evaluation[d]	0.2	0.2
Radial integrals[e]	1.4	2.2
Subtotal for POT[f]	2.8	4.8
DCALC	1.1	1.8
TAUMTS	0.6	1.2
RCALC	1.0	1.6
Total[g]	5.5	9.4
second or subsequent energy		
TAUMTS	0.1	0.3
RCALC	1.0	1.6
Total[g]	1.1	1.9

[a] $N_s = 288$

[b] $N_s = 291$

[c] eq. (91)

[d] Time spent in subprogram FUN, which evaluates the interaction potential

[e] eq. (95)

[f] The evaluation of eq. (92) for all $q_1 q_2 \mu$ and all sectors is included in this total but requires only 28 seconds.

[g] The time for GENSCAT is included but is less than 2 minutes. Overhead is also included but is less than 1%.

equivalent hours. The difference between the CPU time and the equivalent
time is due to charges for memory usage, input/output requests, permanent
file usage, and, on the Cyber 205, paging to virtual memory. On the Cyber
205, to minimize page fault charges and turnaround time, we mapped all of
our large arrays onto large (65,356 word) pages. We let the Cyber 205
operating system schedule all the page requests.

Table 6 shows a breakdown of the total times into those for individual
program segments and subsegments for first and second energies on the Cray-1.
For first energies for both N = 400 and 530, POT requires about half the
total time, while in POT, about half of its total time is spent evaluating
the sector-independent coefficients $B_{\beta\ell\beta'\ell'}^{q_1 q_2 \mu}$. It should be noted that after
these calculations were performed, we were able to implement a new algorithm
for which the calculation of the $B_{\beta\ell\beta'\ell'}^{q_1 q_2 \mu}$ requires about a factor of 4 less
time. Our original routine for calculating the $B_{\beta\ell\beta'\ell'}^{q_1 q_2 \mu}$ used eq. (91) and
evaluated the Clebsch-Gordan coefficients using recursion over one of the
j indices.[34] The faster version evaluates the Clebsch-Gordan coefficients
using recursion over one of the m indices.[34] This saves time since most of
the Clebsch-Gordan coefficients required are in a sum over one of the m
indices, thus the recursion needs to be performed only a maximum of 6 times
as compared to $2 + 4(m_{max} - m_{min})$ times for the original version, where m_{max}
and m_{min} are the range of m in eq. (91). For consistency the CPU times in
Tables 5 and 6 are based on the slower version.

The bulk of the remaining time in POT is taken up with the 2-dimensional
integrals of eq. (95). This is largely as a consequence of the facts that
(i) we have not assumed the separability of rotation and vibration and so
our vibrational eigenfunctions depend on the rotational state, and (ii) we
have retained a very large number of terms in eq. (78) to accurately represent a realistic diatom-diatom potential. The seven points per radial coordinate is essentially optimal[33] and use of any less optimal numerical quadrature in this step would greatly slow the calculations. The program sections DCALC, TAUMTS, and RCALC all require about the same amount of time,
about one hour for N = 400 and 1.5 hours for N = 530. GENSCAT and overhead
require a very small fraction of the total time. For second energies, the
majority of computer time is spent on RCALC.

Table 7 shows times per sector on the Cray-1 for N = 55, 101, 400, and
530. We see that, on this computer, the time required scales approximately
as $N^{\frac{1}{2}}$ for N going from 55 to 101, N^3 for N going from 101 to 400, and N^2 for
N going from 400 to 530. On the Cyber 205 the approximate scaling for N
going from 400 to 530 is N^1.

Table 7. Execution times (in seconds) per sector for N =
          55, 101, 400, and 530 on the Cray-1

| | Program section | | | |
N	DCALC	TAUMTS (1st E)	RCALC	GENSCAT
55[a]	0.074	0.043	0.060	0.41
101[b]	0.33	0.15	0.27	1.1
400[a]	14.	7.6	12.	8.8
530[a]	22.	14.	19.	23.

[a]Redmon–Binkley potential

[b]Poulsen–Billing–Steinfeld potential

VII. RESULTS

We have computed probabilities for the transitions

$$2HF(v=1,j=0) \rightarrow HF(v_1'=0,j_1') + HF(v_2'=2,j_2') \tag{96}$$

for several energies for N = 500, 530, and 694. However, the results are
not converged with respect to increasing N. More complete discussion of the
physics will be possible when convergence is achieved.

VIII. CONCLUSIONS

We have successfully used supercomputers to set up and solve quantum
mechanical scattering problems representing diatom–diatom collisions with
up to 694 coupled equations. These problems are already considerably larger
than the biggest calculations performed so far without supercomputers, and
thus we may say that the use of supercomputers has definitely extended our
capability. Converged solutions of the vibration–vibration energy transfer
problem studied here will require even more coupled equations than considered
here. We anticipate though that convergence, at least for the prototype case
of total angular momentum zero, may be achievable by a combination of (i)
further code optimizations, (ii) possible use of algorithms that make more
efficient use of vector processors, (iii) use of even more powerful supercomputers, and/or (iv) longer runs.

The speed enhancements we have achieved with respect to the most popular
minicomputer, the VAX 11/780 with floating point acceloerator, are very significant. For the total problem, based on 101-coupled-equations test runs
and for a case in which seven energies are calculated simultaneously, the
speed enhancement corresponds to a factor of 490 on the Cyber 205 and a

factor of 223 on the Cray-1. Some individual program segments are speeded up even more, for example a factor of 684 on the Cyber 205 for a segment involving linear-equation solving and matrix multiplication. We did not run more than 101 coupled equations on the minicomputer but the supercomputers become relatively even more efficient as the equation order and hence vector length increases.

These early calculations were often frustrating and slow because of the difficulties of using new systems, remote usage, and access. At present though many of the problems have been or are being ironed out, and the future looks rosy for wide progress, by other groups as well as our own, for solving very-large-scale quantum mechanical scattering problems on these machines.

IX. ACKNOWLEDGMENTS

We are happy to acknowledge several individuals and grants for contributing to and supporting this research. Nancy Mullaney Harvey and Devarajan Thirumalai collaborated on earlier applications of R matrix propagation to easier problems. Robert W. Numrich of Control Data Corporation provided the crucial encouragement (translate: early funding) that started us on the road to the present vector computations. Computer time was provided by Control Data Corporation, first by grant no. 82-CSUD3 for use of the Cyber 205 at Colorado State University and later by making additional computer time available at Arden Hills, and by the Minnesota Supercomputer Institute, by a grant of time on the University of Minnesota Cray-1. The work at Control Data Corporation was facilitated by both Robert W. Numrich and David Antongiovanni and the work at the University of Minnesota was facilitated by Michael Skow, acting director of the University Computer Center, and his staff. Our departmental minicomputer was purchased with funds obtained in part from the National Science Foundation, and our work on molecular scattering calculations has been supported for many years by the National Science Foundation, most recently by grant no. CHE83-17944. The continued and augmented support for the present project by the chemical physics program officers of the National Science Foundation is greatly appreciated.

REFERENCES

1. H. F. Schaefer III, Interaction potentials I: Atom-molecule potentials, in: "Atom-Molecule Collision Theory", R. B. Bernstein, ed., Plenum Press, New York (1979), chap. 2.
2. G. F. Adams, G. D. Bent, R. J. Barlett, and G. D. Purvis, Calculation of potential energy surfaces for HCO and HNO using many-body methods, in: "Potential Energy Surfaces and Dynamics Calculations", D. G. Truhlar, ed., Plenum Press, New York (1981), chap. 5; M. S. Gordon, Potential energy surfaces in excited states of saturated molecules, ibid., chap. 7; K. Morokuma and S. Kato, Potential energy characteristics for chemical reactions, ibid., chap. 10; T. H. Dunning, Jr.,

S. P. Walsh, and A. F. Wagner, Theoretical studies of selected reactions in the hydrogen-oxygen system, _ibid._, chap. 14.

3. K. Morokuma, S. Kato, K. Kitaura, S. Obara, K. Ohta, and M. Hanamura, Potential energy surfaces of chemical reactions, _in_: "New Horizons of Quantum Chemistry", P.-O. Löwdin and B. Pullman, eds., D. Reidel, Dordrecht, Holland (1983), chap. 16; W. Kolos, Ab initio methods in calculations of intermolecular interaction energies, _ibid._, chap. 17.

4. W. A. Lester, Jr., Calculation of cross sections for rotational excitation of diatomic molecules by heavy particle impact: Solution of the close-coupling equations, _Meth. Comp. Phys._ 10:211 (1973); W. A. Lester, Jr., Coupled-channel studies of rotational and vibrational energy transfer by collision, _Adv. Quantum Chem._ 9:199 (1975); W. A. Lester, Jr., The N coupled-channel problem, _in_: "Dynamics of Molecular Collisions, Part A", W. H. Miller, ed., Plenum Press, New York (1976), chap. 1.

5. J. C. Light, Inelastic scattering cross sections I: Theory, _in_: "Atom-Molecule Collision Theory", R. B. Bernstein, ed., Plenum Press, New York (1979), chap. 6.

6. D. Secrest, Rotational excitation I: The quantal treatment, _in_: "Atom-Molecule Collision Theory", R. B. Bernstein, ed., Plenum Press, New York (1979), chap. 8.

7. R. Bellman, R. Kalaba, and G. M. Wing, Invariant imbedding and the reduction of two-point boundary value problems to initial value problems, _Proc. Natl. Acad. Sci._ 46:1646 (1960); R. Bellman, R. Kalaba, and M. Prestrud, "Invariant Imbedding and Radiative Transfer in Slabs of Finite Thickness", American Elsevier, New York (1963).

8. A. Degasperis, Generalization of the phase method to multi-channel potential scattering, _Nuovo Cimento_ 34:1667 (1964); M. E. Riley and A. Kuppermann, Vibrational energy transfer in collisions between diatomic molecules, _Chem. Phys. Lett._ 1:537 (1968); B. R. Johnson and D. Secrest, Quantum-mechanical calculations of the inelastic cross sections for rotational excitation of para and ortho H_2 upon collision with He, _J. Chem. Phys._ 48:4682 (1968).

9. J. C. Light and R. B. Walker, An R matrix approach to the solution of coupled equations for atom-molecule reactive scattering, _J. Chem. Phys._ 65:4272 (1976); E. B. Stechel, R. B. Walker, and J. C. Light, R-matrix solution of coupled equations for inelastic scattering, _J. Chem. Phys._ 69:3518 (1978).

10. N. A. Mullaney and D. G. Truhlar, The use of rotationally and orbitally adiabatic basis functions to calculate rotational excitation cross sections for atom-molecule collisions, _Chem. Phys._ 39:91 (1979).

11. (a) N. Mullaney Harvey and D. G. Truhlar, The use of vibrationally adiabatic basis functions for inelastic atom-molecule scattering, _Chem. Phys. Lett._ 74:252 (1980); (b) D. Thirumalai and D. G. Truhlar, Rapid convergence of V-V energy transfer calculated using adiabatic basis functions and an accurate two-state model for low-energy resonant V-V energy transfer, _J. Chem. Phys._ 76:5287 (1982).

12. N. A. Mullaney and D. G. Truhlar, Rotationally and orbitally adiabatic basis sets for electron-molecule scattering, _Chem. Phys. Lett._ 58:512 (1979); D. G. Truhlar, N. M. Harvey, K. Onda, and M. A. Brandt, Applications of close coupling algorithms to electron-atom, electron-molecule, and atom-molecule scattering, _in_: "Algorithms and Computer Codes for Atomic and Molecular Quantum Scattering Theory", Vol. I, L. Thomas, ed., National Resource for Computation in chemistry, Lawrence Berkeley Laboratory, Berkeley, CA (1979), chap. 14.

13. E. P. Wigner, Resonance reactions, _Phys. Rev._ 70:606 (1946); E. P. Wigner and L. Eisenbud, Higher angular momenta and long range interaction in resonance reactions, _Phys. Rev._ 72:29 (1947); P. G. Burke and W. D. Robb, The R-matrix theory of atomic processes, _Adv. At._

Mol. Phys. 11:143 (1975); R. W. Numrich and R. G. Kay, Dissociation
dynamics of collinear triatomic systems by the R-matrix method, J.
Chem. Phys. 70:4343 (1979); J. Gerratt and I. D. L. Wilson, L^2 R-
matrix studies of molecular collision processes; Energy dependence
of $\sigma_{vj \to v'j'}$ for ^{4}He + H$_2$, Proc. Roy. Soc. Lond. Ser. A 372:219 (1980).

14. D. J. Diestler and V. McKoy, Quantum mechanical treatment of inelastic
collisions. II. Exchange reactions, J. Chem. Phys. 48:2951 (1968);
D. G. Truhlar and A. Kuppermann, Exact and approximate quantum
mechanical reaction probabilities and rate constants for the collinear
H + H$_2$ reaction, J. Chem. Phys. 56:2232 (1972).

15. A. Askar, A. Cakmak, and H. Rabitz, Finite element methods for reactive
scattering, Chem. Phys. 33:267 (1978).

16. I. H. Zimmerman, M. Baer, and T. F. George, F + H$_2$ collisions on two
electronic potential energy surfaces: Quantum mechanical study of
the collinear reaction, J. Chem. Phys. 71:4132 (1979); M. Baer,
Quantum mechanical treatment of electronic transitions in atom-
diatom exchange collisions, Ber. Bunsenges. Physik. Chem. 86:448
(1982); M. Baer, Quantum mechanical treatment of electronic transitions in atom-molecule collisions, Top. Curr. Phys. 33:117 (1983).

17. T. G. Schmalz, E. B. Stechel, and J. C. Light, Time independent quantum
theory of electron transfer collisions using a nonorthogonal basis
and R-matrix propagation, J. Chem. Phys. 70:5640 (1979); B. C.
Garrett, M. J. Redmon, D. G. Truhlar, and C. F. Melius, Ab initio
treatment of electronically inelastic K + H collisions using a direct
integration method for the solution of the coupled-channel scattering
equations in electronically adiabatic representations, J. Chem. Phys.
74:412 (1981); B. C. Garrett and D. G. Truhlar, The coupling of electronically adiabatic states in atomic and molecular collisions,
Theor. Chem. Adv. Perspectives 6A:215 (1981); J. Gerratt, R-matrix
theory of charge transfer, Phys. Rev. A 30:1643 (1984).

18. G. C. Schatz and A. Kuppermann, Quantum mechanical reactive scattering
for three-dimensional atom plus diatom systems. I. Theory, J. Chem.
Phys. 65:4642 (1977); R. E. Wyatt, Direct-mode chemical reactions
I: Methodology for accurate quantal calculations, in: "Atom-Molecule
Collision Theory", R. B. Bernstein, ed., Plenum Press, New York
(1979), chap. 17.

19. D. L. Moores, Applications of the close-coupling method to electron
molecule scattering, in: "Electron-Molecule and Photon-Molecule
Collisions", T. Rescigno, V. McKoy, and B. Schneider, eds., Plenum
Press, New York (1979), chap. 1; M. A. Morrison, The coupled-
channels integral-equations method in the theory of low-energy
electron-molecule scattering, ibid., chap. 2.

20. D. Secrest, Theory of rotational and vibrational energy transfer in
molecules, Annu. Rev. Phys. Chem. 24:379 (1973); M. Faubel and J. P.
Toennies, Scattering studies of rotational and vibrational excitation of molecules, Adv. At. Mol. Phys. 13:229 (1977).

21. M. H. Alexander, Close-coupling studies of rotationally inelastic HF-HF
collisions at hyperthermal energies, J. Chem. Phys. 73:5735 (1980).

22. H. Rabitz, The dimensionality and choice of effective hamiltonians for
molecular collisions, J. Chem. Phys. 63:5208 (1975); D. G. Truhlar,
Recent progress in atomic and molecular collisions and the interface
with electronic structure theory, Int. J. Quantum chem. Symp. 17:77
(1983).

23. M. H. Alexander and A. E. DePristo, Symmetry considerations in the
quantum treatment of collisions between two diatomic molecules, J.
Chem. Phys. 66:2166 (1977).

24. P. Pechukas and J. C. Light, On the exponential form of time displacement operators in quantum mechanics, J. Chem. Phys. 44:3897 (1966);

P. Chang and J. C. Light, Exponential solution of the Schrödinger equation: Potential scattering, J. Chem. Phys. 50:2517 (1969); J. C. Light, Quantum calculations in chemically reactive systems, Meth. Comp. Phys. 10:111 (1971).

25. J. N. Murrell and K. S. Sorbie, New analytic form for the potential energy curves of stable diatomic states, J. Chem. Soc. Faraday Trans. II 70:1552 (1974).

26. A. R. Edmonds, "Angular Momentum in Quantum Mechanics", Princeton University Press, Princeton, NJ (1960).

27. D. U. Webb and K. N. Rao, Vibration rotation bands of heated hydrogen halides, J. Mol. Spectry. 28:121 (1968).

28. M. J. Redmon and S. B. Binkley, to be published.

29. L. L. Poulsen, G. D. Billing, and J. I. Steinfeld, Temperature dependence of HF rotational relaxation, J. Chem. Phys. 68:5121 (1978).

30. G. Gioumousis and C. F. Curtiss, Molecular collisions. II. Diatomic molecules, J. Math. Phys. 2:96 (1961).

31. J. M. Launay, Molecular collision processes I. Body-fixed theory of collisions between two systems with arbitrary angular momenta, J. Phys. B 10:3665 (1977).

32. A. E. DePristo and M. H. Alexander, Relationships among the coupled-states, P-helicity decoupling and effective potential methods, Chem. Phys. 19:181 (1977).

33. D. W. Schwenke and D. G. Truhlar, An optimized quadrature scheme for matrix elements over the eigenfunctions of general anharmonic potentials, Comp. Phys. Commun., in press.

34. K. Schulten and R. G. Gordon, Exact recursive evaluation of 3j- and 6j-coefficients for quantum-mechanical coupling of angular momenta, J. Math. Phys. 16:1961 (1975).

PHONON CALCULATIONS IN COVALENT SEMICONDUCTORS USING A VECTOR COMPUTER

P.E. Van Camp and J.T. Devreese

University of Antwerp (RUCA and UIA)
Antwerp, Belgium

I. INTRODUCTION

In the past decade two ab-initio methods have been developed to calculate phonon dispersion curves of covalent semiconductors: the dielectric screening formalism and the total energy difference method. In the first method linear response theory is applied in order to evaluate the effect of the ionic displacements on the electronic system. The electronic structure is calculated in the framework of the local density approximation. Starting from a local ionic pseudopotential a self-consistent pseudopotential band calculation is performed to find the electron wave functions and energies. The lattice constant is not taken from experiment but obtained from minimalization of the total crystalline energy with respect to ionic displacements. The convergence of the phonon frequencies as a function of the number of reciprocal lattice vectors used in the Hamiltonian and in the linear response matrices is investigated in detail. Because the programs to make the above described calculations take a large amount of computer time the codes have been vectorized during the past year. They were executed on a CDC Cyber 205 (1-pipe, 2 M words) giving a gain in time (with respect to the scalar mode) by a factor 5. This number refers to the whole program. In some parts the factor was as high as 45.

II. THEORY

In the dielectric screening formalism the total electronic energy of the crystal in an arbitrary configuration is expanded in a Taylor series with respect to the ionic displacements from equilibrium. The ionic displacement is equivalent to an externally applied potential δV^{EXT}. The charge density induced by this potential, $\delta\rho$, is given in linear response theory by:

$$\delta\rho = \chi \; \delta V^{EXT} \qquad (1)$$

Eq. (1) is written in matrix notation, i.e. $\delta\rho$ and δV^{EXT} are column vectors and χ (the density response) is a matrix. The density response matrix determines the inverse dielectric matrix ε^{-1} through the relation:

$$\varepsilon^{-1} = 1 + v_c \chi \tag{2}$$

where v_c is the Coulomb interaction potential. The dielectric matrix ε itself is defined as the factor of proportionality between the total potential and the externally applied potential as seen by a test charge:

$$\varepsilon = 1 - v_c \tilde{\chi} (1 - v_{xc} \tilde{\chi})^{-1} \tag{3}$$

Here v_{xc} is the Fourier transform of the exchange-correlation potential and $\tilde{\chi}$ is the polarizability matrix given by [1]:

$$\tilde{\chi}(\vec{q},\vec{G},\vec{G}') = \frac{1}{\Omega} \sum_{\ell m} \frac{\eta_\ell - \eta_m}{E_\ell - E_m} <\ell | e^{-i(\vec{q}+\vec{G})\vec{r}} | m> \, .$$

$$<m | e^{i(\vec{q}+\vec{G}')\vec{r}} | \ell> \tag{4}$$

with Ω the crystal volume.

The dynamical matrix of the system determines the frequencies of the ionic vibrations. This matrix contains, apart from the ionic potentials, the density response matrix.

The computational scheme is therefore as follows: first compute the polarizability matrix (Eq. (4)). Then construct the dielectric matrix (Eq. (3)), the inverse of which is used to calculate the density response matrix. The only input for the computation of the polarizability matrix are the electron energies and wave functions. These are obtained from a local self-consistent band calculation.

In short the calculation can be described as follows:
1. Start with an ionic potential. In the present work the Topp-Hopfield [2], the Appelbaum-Hamann [3] and the Schlüter-Chelikowsky-Louie-Cohen [4] expressions are used. Only the first one is truly ab-initio since the second and the third take already crystalline properties into account.
2. Using a starting pseudopotential one solves the Schrödinger equation and forms the electronic charge density.
3. Subsequently one evaluates the Hartree and the exchange-correlation potentials. Together with the ionic potential this forms the new pseudopotential.
4. Go back to 2., but first replace the old pseudopotential by the new one. If the difference between two consecutive calculated charge densities is less than a prechosen value the iteration is terminated.
5. Calculate the total energy of the crystal for the given lattice parameter.
The whole calculation (step 1-5) is repeated for several lattice parameters and the total energy is minimalized to give the equilibrium lattice constant [5].

III. COMPUTATIONAL PROCEDURE

The most time comsuming part of the program is the calculation of the polarizability matrix (Eq. (4)). Therefore this part was first vectorized and the method is described below. In the pseudopotential scheme the electron wave functions are expanded in plane waves:

$$\psi_{\vec{k}n}(\vec{r}) = \frac{1}{\sqrt{\Omega}} \sum_{G} C_{\vec{k}}^{n}(\vec{G}) \, e^{i(\vec{k}+\vec{G})\cdot\vec{r}} \tag{5}$$

where $\vec{k}n$ are the electron wave vector and band index and $\vec{G}$ is a reciprocal lattice vector. Typically the sum in Eq. (5) is expanded using up to 193 plane waves. The expansion coefficients $C_{\vec{k}}^{n}(\vec{G})$ are the eigenvectors of the Hamiltonian matrix H:

$$H \, C = E \, C \tag{6}$$

and the electron energies are the eigenvalues E. Maximally the dimension of the eigenvalue problem is then 193.

Inserting the expansion (Eq. (5)) into Eq. (4) leads to:

$$\tilde{\chi}(\vec{q},\vec{G},\vec{G}') = \frac{1}{\Omega} \sum_{\vec{k}nn'} \frac{\eta_{\vec{k}n} - \eta_{\vec{k}+qn'}}{E_{\vec{k}n} - E_{\vec{k}+qn'}} \, T_{\vec{k}\ \vec{k}+q}^{nn'}(\vec{G}) \, T_{\vec{k}\ \vec{k}+q}^{nn'}(\vec{G}') \tag{7}$$

with:

$$T_{\vec{k}\vec{k}'}^{nn'}(\vec{G}) = \sum_{G} C_{\vec{k}}^{n}(\vec{G}') \, C_{\vec{k}'}^{n'}(\vec{G}'-\vec{G}) \tag{8}$$

Now the sum in Eq. (8) looks like a scalar product but the products are <u>not</u> taken component by component. This is due to the fact that the elements of the second vector ($C_{\vec{k}'}^{n'}(\vec{G}'-\vec{G})$) are not to be used in the same sequence as those of the first vector ($C_{\vec{k}}^{n}(\vec{G}')$), except when $\vec{G}$ = zero. This means that in order to obtain vector speeds in executing the summation in Eq. (8) one has to rearrange the sequence of the elements of the second vector. This was realized in the following way:

1. Construct the difference vector $\vec{G}'-\vec{G}$. Note that each element of this vector is a (three-dimensional) vector itself.
2. Because we are working with a limited (up to 193) set of reciprocal lattice vectors one must check if all the differences $\vec{G}'-\vec{G}$ belong to this set. This is most easily done by using a vector relational and a bit vector.
3. Next calculate the address of each element $\vec{G}'-\vec{G}$. Put address equal zero if $\vec{G}'-\vec{G}$ does not belong to the set considered.
4. Subsequently gather the elements using the index list.
5. Then the dot product, controlled by the bit vector, is calculated using a special call (Q8SDØT) [6].

Table I. Time ratio (scalar time/vector time) for matrix
diagonalization and inversion versus the matrix
dimension.

dimension	diagonalization	inversion
25	2.7	8.6
50	3.9	15.7
75	5.2	20.7
100	6.5	25.2
125	7.9	28.5
150	9.4	37.9
175	11.7	45.4

Although in programming the above cited steps the code is
expanded from 10 to 45 lines the execution speed is reduced by
a factor of 10. In the next stage various other parts of the
program (e.g. matrix diagonalization, matrix inversion, etc.)
have been vectorized. In table I time ratios (scalar
time/vector time) for matrix inversion and diagonalization are
shown for different dimensions of the matrices considered.

The run time of the complete phonon program is mainly
determined by the dimension of the Hamiltonian matrix H (which
gives the electronic structure) and of the dielectric matrix
$\varepsilon(\vec{q},\vec{G},\vec{G}')$ (for the linear response theory). For the highest
dimensions used (193 for H and 181 for ε) the program needed
4400 CPU-secs (for one phonon wave vector) on a one-pipe
Cyber-205. Needless to say that executing such a job on a
serial computer would be highly impractical.

IV. RESULTS

A. Silicon

One of the main objectives of our research in the present
stage is the investigation of the convergence of the calculated
phonon frequencies in terms of the number of plane waves used
in the basis (see Eq. 5)) and of the number of reciprocal
lattice vectors taken into account in the linear response
theory (i.e. the dimension of the dielectric matrix). All the
convergence tests discussed below were performed for Silicon,
using the Schluter-Chelikowsky-Louie-Cohen ionic potential.
Minimalization of the total energy with respect to the lattice
spacing yields a lattice constant of 5.373 Å to be compared with
the experimental value of 5.429 Å [7].

1. Convergence of the phonon frequencies in terms of the
dimension of ε. In table II results for the phonon frequencies
as a function of the dimension of $\varepsilon(\vec{q},\vec{G},\vec{G}')$ are given. It is
noted that the convergence is relatively slow.

To include effects of reciprocal lattice vectors beyond
181 we added a diagonal block to the dielectric matrix, the
results of which will be given below.

Table II. Phonon frequencies of Si in terms of the dimension of the dielectric matrix $\varepsilon(\vec{q},\vec{G},\vec{G}')$, together with the experimental values (in THz).

dimension $\varepsilon(\vec{q},\vec{G},\vec{G}')$	Γ	TO(X)	LOA(X)	TA(X)
89	22.40	19.31	14.41	6.11
113	18.18	16.75	13.97	3.68
137	18.74	17.28	14.36	4.38
169	18.77	17.22	14.39	4.75
181	18.67	17.12	14.33	4.55
Experiment [8]	15.5	13.9	12.3	4.5

2. Convergence of the phonon frequencies in terms of the dimension of H. In table III results for the phonon frequencies of Si as a function of the dimension of the Hamiltonian matrix H are given.

As can be seen from the table the convergence is relatively slow. Therefore, as a first step, an extrapolation was made using:

$$\omega = A \frac{1}{N} + B \qquad (9)$$

where the phonon frequency ω is taken to depend linearly on $\frac{1}{N}$ with N the number of plane waves used in the Hamiltonian. The extrapolated values using Eq. (9) with the results of 137 and 150 plane waves are also given in table III.

The complete phonon dispersion curves of Silicon are shown in figure 1. An overall agreement with experiment is obtained with an average discrepancy of 14%.

Table III. Phonon frequencies of Si in terms of the dimension of the Hamiltonian matrix H, together with the extrapolated and experimental values (in THz).

dimension H	Γ	TO(X)	LOA(X)	TA(X)
137	18.89	17.50	14.51	4.39
150	18.74	17.28	14.36	4.38
169	18.54	17.07	14.20	4.39
193	18.46	17.05	14.29	4.51
Extrapolation	17.10	14.93	12.81	4.16
Experiment [8]	15.5	13.9	12.3	4.5

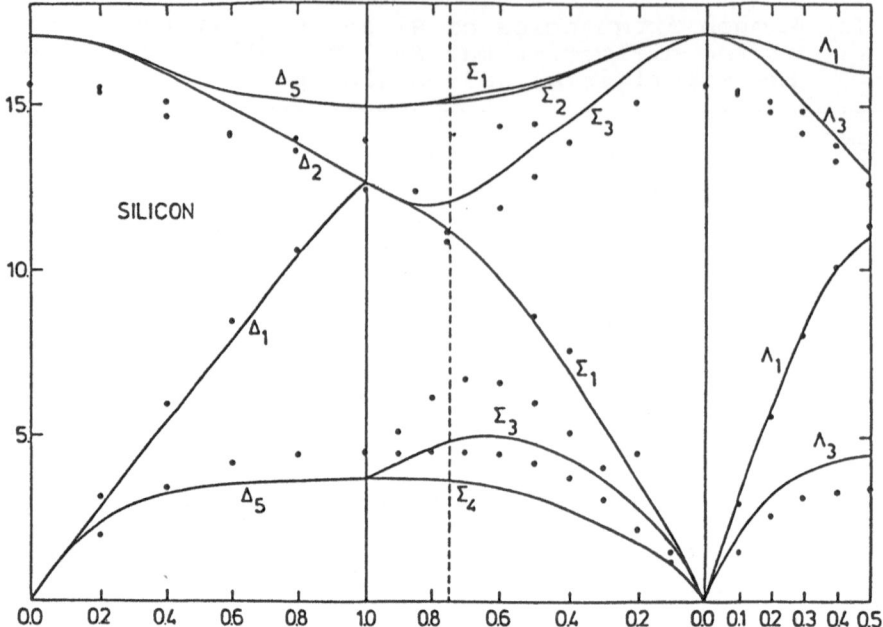

Figure 1. Calculated phonon frequencies for Silicon in Δ-, Λ-
and Σ-directions (in THz). The experimental data
are also shown [8].

B. Germanium

Recently the methods described above have also been
applied to Germanium. As ionic potential the form proposed by
Topp and Hopfield [2] is used. The calculated lattice constant
is 5.517 Å to be compared with the experimental value of 5.652
Å. The calculated phonon frequencies, together with the
experimental values, are given in table IV in the Γ- and
X-point.

The calculated values quoted in table IV are obtained
using a dimension of 193 for the Hamiltonian matrix and of
181 for the dielectric matrix $\varepsilon(\vec{q},\vec{G},\vec{G}')$.

In table V runtimes for the complete phonon program are
given for different dimensions of the Hamiltonian matrix H
and of the dielectric matrix ε. The value between parentheses
is an estimation.

Table IV. Phonon frequencies of Ge together with the experimental values (in THz).

	Γ	TO(X)	LOA(X)	TA(X)
calculation	9.90	9.40	7.68	2.70
experiment [8]	9.13	9.26	7.21	2.40

Table V. Runtime (in sec) of the phonon programs as a function of the dimension of the Hamiltonian H and of the dielectric matrix $\varepsilon(\vec{q},\vec{G},\vec{G}')$.

dimension H	dimension ε	runtime (sec)
169	137	2242.0
169	169	3176.1
169	181	(3575.0)
193	137	2827.3
193	169	3909.9
193	181	4397.4

V. CONCLUSIONS

It is shown that for the calculation of the phonon dispersion relations of covalent semiconductors in the dielectric screening formalism the use of a pipelined computer is indispensable.

As far as the physics is concerned the authors were led to draw the following conclusions.
1. Consistency on three levels is necessary to get meaningful results:
   A. Minimalization of the total crystal energy with respect to the lattice spacing.
   B. The same ionic potential should be used in the Hamiltonian and in the dynamical matrix.
   C. Exchange-correlation effects should be treated in the same way in the Hamiltonian and in the linear response theory.
2. It is very important to use a basis consisting of a sufficiently large number of plane waves. For the polarizability exactly the same basis should be used (i.e. all calculated conduction bands should be taken into account).

Acknowledgments

The numerical work was performed on the CDC Cyber-205 in Karlsruhe (F.R.G.) with a grant from the "Supercomputer Project" 1983-84 of the N.F.W.O. (National Fund for Scientific Research, Belgium). The authors also would like to thank Dr. V.E. Van Doren who was involved in previous stages of the research.

REFERENCES

1. S. Adler, Phys. Rev. 126, 413 (1962).
   N. Wiser, Phys. Rev. 129, 62 (1963).
2. W.C. Topp, J.J. Hopfield, Phys. Rev. B7, 1295 (1973).
3. J.A. Appelbaum, D.R. Hamann, Phys. Rev B8, 1777 (1973).
4. M. Schlüter, J. Chelikowsky, S. Louie, M.L. Cohen, Phys. Rev. B12, 4200 (1975).

5. P.E. Van Camp, V.E. Van Doren, J.T. Devreese, Phys. Rev. Lett. 42, 1224 (1979); Phys. Rev. B24, 1096 (1981); Phys. Rev. B25, 4270 (1982); in "Ab-Initio Calculation of Phonon Spectra", eds. J.T. Devreese, V.E. Van Doren, P.E. Van Camp, Plenum, New York (1983); in "Proceedings of the 6th General Conference of the European Physical Society", Prague, to be published (1984).

6. CDC-Cyber 200 Fortran Version 2 Reference Manual, publication number 60485000, Sunnyvale, Ca. (1981).

7. J. Donahue, "The Structure of the Elements", Wiley, New York (1972).

8. G. Nilsson, G. Nelin, Phys. Rev. B6, 3777 (1972).

PIECEWISE POLYNOMIAL HYLLERAAS CONFIGURATION INTERACTION WAVEFUNCTIONS

FOR HELIUM

D. W. Zaharevitz,* H. J. Silverstone* and D. M. Silver+

Department of Chemistry,* The Johns Hopkins University
Baltimore, Maryland 21218, and Applied Physics Laboratory+
The Johns Hopkins University, Laurel, Maryland 20707

ABSTRACT

The scientific problem treated in this work is that of describing the electronic structure of atoms and molecules with sufficient accuracy to be useful for predictions of chemical properties. The present activity consists of deriving computational solutions of the many-body Schrodinger equation for small chemical species. Two specific objectives are to explore the consequences of an explicit correlation term in the electronic wavefunction and to examine the effectiveness of certain mathematical forms as an expansion basis. The magnitude of the computational problem is massive and the existence of supercomputers like the CYBER 205 helps to bring these problems into a feasible range.

I. INTRODUCTION

The most widely used technique for representing electronic wavefunctions is "configuration interaction" (CI). CI trades accuracy for flexibility: the accuracy of CI energies has a practical limit[1] of 0.0005 eV. To go beyond this limit requires explicitly correlated wavefunctions (i.e., which depend on the interelectronic coordinate r_{12}, such as in the Hylleraas and the CI-Hylleraas methods.[2]

One "benchmark" for helium calculations is the value obtained by Pekeris[3] using a Hylleraas wavefunction[4] whose basis consisted of 1078 functions of the form, $(r_1+r_2)^i(r_1 - r_2)^j r_{12}^k \exp[-c(r_1 + r_2)]$, with $i + j + k < 21$. Pekeris' value was -2.903724375 a.u., whose error is 2×10^{-9}

a.u. A second benchmark is the Frankowski-Pekeris[5] calculation, whose main difference from the Pekeris calculation was the use of powers of $\log(r_1+r_2)$: the result, -2.9037243770326 a.u.; the error, 2×10^{-12} a.u. A third and final benchmark is a slight modification of the Frankowski-Pekeris calculation by Freund, Huxtable, and Morgan,[6] who obtained -2.90372437703407 a.u.

A methodological limitation of these three calculations is that they are difficult to apply to other atoms because of the problem of evaluating the multi-electron integrals that arise. For molecules the problem is exacerbated by the existence of multi-center multi-electron integrals involving exponentials.

The present work addresses the integrals problem by choosing a piecewise polynomial Hylleraas configuration interaction (PP-HY-CI) basis rather than the exponentials in the Pekeris expansion. More precisely, the PP-HY-CI wavefunction is constructed to have the form:

CI-wavefunction + r_{12} x CI-wavefunction′

in which the radial parts of the orbitals are piecewise polynomials.[1,7,8] For atoms, the radial integrals are reduced to localized, disjoint contributions, which are considerably fewer in number than would arise from bases of Slater or Gaussian functions of the same size.

In scope the PP-HY-CI wavefunction is similar to the Pekeris wavefunction, but with a distribution of exponentials. One aim of this work is to examine the rate of convergence of such expansions and to evaluate the need for additional logarithmic factors.

The next Section presents details of the PP-HY-CI method. Section III discusses some computational aspects of the problem. In Section IV, numerical results are given for the ground state of the helium atom. The final section contains some observations and conclusions.

II. THE PP-HY-CI METHOD

The HY-CI method has been developed principally by Hagstrom and coworkers.[2] The wavefunction, specialized for the ground state of helium, has the form

$$\psi = \psi_{CI} + r_{12} \times \psi'_{CI} \tag{1}$$

where

$$\psi_{CI} = \sum_{n_1 n_2 \ell m} C_{n_1 n_2 \ell m} R_{n_1 \ell}(r_1) R_{n_2 \ell}(r_2) Y_{\ell m}(\theta_1 \phi_1) Y^*_{\ell m}(\theta_2 \phi_2) \qquad (2)$$

and where the $Y_{\ell m}$ are the usual spherical harmonics. ψ'_{CI} is similar to ψ_{CI}, but with different expansion coefficients, $C'_{n_1 n_2 \ell m}$. The radial orbitals $R_{n \ell}$ are here expanded on a piecewise polynomial basis set.[1]

In particular, the piecewise polynomials are fixed by the "mesh" defining the subdivision of the r-axis into finite elements, the degree of the polynomials, and the degree of continuity at the mesh points. The most convenient organization of the raw polynomials is into "mesh-oriented unit-nth-derivative" (MOUND) functions which have the property that each basis function and its first s derivatives are zero at every mesh point, except for one derivative, say the nth, whose value is unity at only one mesh point, say r_p:

$$(d/dr)^k \chi_{np}(r)\Big|_{r=r_m} = \delta_{kn} \delta_{pm}, \quad (k \leqslant s). \qquad (3)$$

These MOUND functions are given explicitly by a formula based on Hermite's interpolation formula:[9]

$$\chi_{np}(r) \equiv 0, \quad (r \leqslant r_{p-1}, \; r \geqslant r_{p+1})$$

$$= \sum_{k=n}^{s} \binom{2k-n}{k} \frac{(-1)^n}{(2k-n)! (r_p - r_{p-1})^{2k+1-n}} \qquad (4)$$

$$\times [k(r-r_{p-1})^{k+1}(r_p - r)^k - (k-n)(r - r_{p-1})^k (r_p - r)^{k+1}],$$

$$(r_{p-1} \leqslant r \leqslant r_p), \qquad (5)$$

$$= \sum_{k=n}^{s} \binom{2k-n}{k} \frac{1}{(2k-n)! (r_{p+1} - r_p)^{2k+1-n}}$$

$$\times [k(r - r_p)^k (r_{p+1} - r)^{k+1} - (k-n)(r - r_p)^{k+1}(r_{p+1} - r)^k],$$

$$(r_p \leqslant r \leqslant r_{p+1}). \qquad (6)$$

If $R(r)$ is any PP function of degree $2s+1$, then the expansion coefficients

of R(r) on the MOUND basis are just the values of R and its first s
derivatives at the N mesh points,

$$R(r) = \sum_{n=o}^{s} \sum_{p=o}^{N} R^{(n)}(r_p) \chi_{np}(r).$$

(7)

III. COMPUTATIONAL ASPECTS

With the form of the wavefunction specified by Eqs. (1)-(7), the
problem is to calculate the expansion coefficients, $C_{n_1n_2\ell m}$ and
$C'_{n_1n_2\ell m}$, and the expectation value for the energy. First one must calculate the matrix elements of the Hamiltonian H with respect to the
$R_{n_1\ell}R_{n_2\ell}$ configurations and $r_{12}R_{n_1\ell} R_{n_2\ell}$ "r_{12}-configurations." The
formulas are similar to those given in Ref. 1, Appendix A, but require here
the expansion of r_{12} in spherical harmonics. Next the H matrix is transformed to an orthogonalized basis, and the lowest root is determined by
standard methods.[10,11] The transformation step is the most consumptive of
computational effort and was accomplished by adaptations of a method described by Bender[12] involving successive quarter transformations.

The calculations were executed on the CYBER 205 at Colorado State
University. For most of the work, double precision (128 bit word length)
was required to eliminate numerical differencing arising from the inherent
over-completeness (near linear dependence) of the HY-CI wavefunction. In
selected subroutines, FORTRAN code was modified in order to permit vectorization of some DO loops. The overall increase in performance (i.e., decrease in execution time) from the original code to the final "optimized"
version was a factor of two. Performance increases in selected subroutines
were on the order of three to five.

The computer programs used in this work were originally written for
use on a DEC 10 computer. For some small-scale comparison runs, the
increase in execution speed was on the order of a factor of 50 faster on
the CYBER 205 over the DEC 10. Larger-scale comparisons would be expected
to yield even higher ratios of performance. Two additional aspects of the
CYBER 205 made its use advantageous over the DEC 10: namely, the existence
of 128 bit double precision (mentioned above) and virtual memory which
facilitated the scaling-up of the calculations for higher-order cases.

Memory management became important during a matrix transformation
procedure for the larger calculations performed. Since the matrix is
hermitian, only a lower triangle of the matrix elements had been stored and

an indexing array had been used to identify terms. When the matrices became large (1800 x 1800), the efficiency of the virtual memory manager was thwarted because entire columns of the matrix were required in sequence. These were obtained by accessing an element per column until the column containing the diagonal element, then proceeding down that column. The solution was to produce a copy of the matrix that included the redundancies so that all elements of any given column would be available together. This yielded an efficient execution.

IV. NUMERICAL RESULTS

The calculated ground state energies of helium for PP-HY-CI wavefunctions with various polynomial degrees, mesh points, and maximum angular momentum quantum numbers are given in Table I. The maximum angular momentum quantum numbers are designated S, P, D and F. In each column, the energies (going downward) approach an "S-limit", "P-limit", etc. The convergence of these results is illustrated further in Table II where differences between these calculated energies and the "extrapolated energy" of Frankowski-Pekeris[5] are tabulated.

Table I. Calculated ground state energy of helium
for PP-HY-CI wavefunctions[a]

Mesh [b]	S	P	D	F
3 x 5	−2.902576	−2.903627	−2.903664	−2.9036669
5 x 3	−2.90343133	−2.90371793	−2.90372043	−2.90372058
5 x 4		−2.90372134		
5 x 6		−2.90372414		
7 x 3	−2.90349584	−2.9037242176	−2.903724350	−2.9037243515
7 x 4	−2.90349783	−2.9037242611	−2.9037243741	
7 x 5	−2.90349807	−2.9037242665	−2.9037243766	
7 x 6	−2.90349823	−2.9037242679	−2.9037243769	
7 x 7	−2.90349827			

[a] Energies in a.u.

[b] Mesh is defined as the polynomial degree $(2s + 1)$ x the number of mesh points (N).

Table II. Difference in energy between the calculated PP–HY–CI results and the extrapolated Frankowski-Pekeris[a] energy for helium.[b]

Mesh[c]	S x 10^4	P x 10^7	D x 10^{10}	F x 10^{10}
3 x 5	11.5	973.77	599,770	570,000
5 x 3	2.93	64.47	39,470	38,000
5 x 4		30.33		
5 x 6		23.34		
7 x 3	2.29	1.60	270	254
7 x 4	2.27	1.16	29.5	
7 x 5	2.26	1.11	4.05	
7 x 6	2.26	1.09	1.37	
7 x 7	2.26			

[a] Ref. 5

[b] Energies in a.u.

[c] Mesh is defined as the polynomial degree (2s + 1) x the number of mesh points (N).

Table III. Comparison of selected configuration interaction energies for the helium atom[a]

Energy	(year)	Author	[Ref.]	Basis[b]
−2.8992	(1928)	Hylleraas	[13]	SEL
−2.90170	(1963)	Davidson	[14]	NRO–SEL
−2.90322	(1966)	Ahlrichs et al.	[15]	NRO–STO
−2.90338	(1965)	Green et al.	[16]	STO
−2.90344	(1958)	Tycko et al.	[17]	SEL
−2.90370	(1979)	Carroll et al.	[1]	NRO–PP

[a] Energies in a.u.

[b] Definitions: SEL is single exponential Laguerre, NRO is natural radial orbital, STO is Slater-type orbital, PP is piecewise polynomial.

Table IV. Comparison of selected non-configuration interaction energies for the helium atom[a]

Energy	(year)	Author	[Ref.]	Basis[b]	
-2.90324	(1929)	Hylleraas	[4]	HY	
-2.903498	(1984)	This work		PP-HY-CI	S-limit
-2.903624	(1976)	Morrell, et al.	[18]	HYPER-CO	
-2.9037225	(1957)	Kinoshita	[19]	K	
-2.9037239	(1963)	Davidson	[14]	K	
-2.90372424	(1976)	Sims, et al.	[2]	HY-CI	
-2.90372427	(1984)	This work		PP-HY-CI	P-limit
-2.90372435	(1963)	Scherr-Knight	[20]	PT	
-2.903724375	(1959)	Pekeris	[3]	HY-P	
-2.90372437616	(1962)	Schwartz	[21]	HY-P	
-2.9037243769	(1984)	This work		PP-HY-CI	D-limit
-2.9037243770326	(1966)	Frankowski-Pekeris	[5]	HY-P	
-2.90372437703407	(1983)	Freund, et al.	[6]	HY-P	

[a]Energies in a.u.

[b]Definitions: HY is "Hylleraas," PP-HY-CI is piecewise polynomial Hylleraas configuration interaction, HYPER-CO is hyperspherical coordinates, K is "Kinoshita,", PT is perturbation theory, HY-P is "Hylleraas-Pekeris."

Representative lists of total energies for helium are given in Table III for configuration interaction[1,13-17] and in Table IV for non-configuration interaction[2-6,14,18-21] approaches. The earliest CI calculation is that of Hylleraas,[13] who quickly thereafter initiated non-configuration interaction techniques.[4] Comparison of the two Tables indicates the slow convergence of the CI expansion for helium: the most accurate CI results[1] do not even reach the P-limit of the non-CI method.

V. OBSERVATIONS AND CONCLUSIONS

The convergence of the PP-HY-CI calculations for the energy of helium as a function of maximum angular momentum quantum number, L, appears to be on the following order: the S-limit residue is 10^{-4} a.u., the P-limit residue is 10^{-7} a.u., the D-limit residue is 10^{-10} a.u. An extrapolation to the F-limit gives an estimate of 10^{-12} a.u. or better for this residue. Thus, the F-limit results might exceed the quality of the Frankowski-Pekeris[5] limit. One implication of this situation is that it may not be necessary to include logarithmic factors in the HY-CI expansion to achieve the Frankowski-Pekeris result. The intrinsic flexibility arising from the disjoint nature of the piecewise polynomial basis functions is responsible.

A further observation from Tables I and II is that the wavefunction needs to be saturated with functions of low angular momentum quantum number, S and P, before the functions with higher L values can make a contribution in their own realm. This is especially evident for the D component of the wavefunction for each mesh represented, and the 5 x 3 mesh indicates that the effect is also present for the F components.

Work in progress is extending this work to an application to the hydrogen negative ion (H⁻). The wavefunction for H⁻ needs to be more diffuse and extend to larger radial distances than needed for He. The initial indication is that the piecewise polynomial basis functions can yield this flexibility. The final accuracy of the H⁻ calculation is expected to be uniform with that found for He.

In principle, the configuration interaction technique offers an expansion of the wavefunction in a complete space of many-electron functions. Hence the addition of the r_{12} - CI factor, the second term in Eq. (1), produces an expansion that is over-complete. One advantage is faster convergence than for CI alone but linear dependence is a consequence. In the present work, this aspect led to the need for high precision numerical techniques.

The characteristics of piecewise polynomial basis functions combined with the Hylleraas configuration interaction expansion offer a useful approach to the determination of electronic wavefunctions. The ability of this approach to achieve high accuracy for the energy of helium and other two-electron systems makes it necessary to re-examine arguments related to the functional forms needed for a complete description of a wavefunction.

ACKNOWLEDGMENT

This work was supported in part by Control Data Corporation through a grant of computer time on the CYBER 205 at Colorado State University, Project #83CSU26, and in part by the Department of the Navy, Naval Sea Systems Command under Contract N00024-83-C-5301.

REFERENCES

1. D. P. Carroll, H. J. Silverstone, and R. M. Metzger, "Piecewise Polynomial Configuration Interaction Natural Orbital Study of 1^2s Helium," J. Chem. Phys. 71, 4142 (1979).

2. J. S. Sims, S. A. Hagstrom, and J. R. Rumble, Jr., "Combined Configuration-Interaction-Hylleraas-Type Wave-Function Study of the Ground State of the Beryllium Atom," Int. J. Quantum Chem. 10, 853 (1976) and references therein.

3. C. L. Pekeris, "1^1s and 2^3s State of Helium," Phys. Rev. 115, 1216 (1959).

4. E. Hylleraas, "Neue berechnung der energie des heliums im grundzustande, sowie des tiefstens terms von ortho-helium," Z. Physik 54, 347 (1929).

5. K. Frankowski and C. L. Pekeris, "Logarithmic Terms in the Wavefunctions of the Ground State of 2-Electron Atoms," Phys. Rev. 146, 46 (1966).

6. D. E. Freund, B. D. Huxtable, and J. D. Morgan III, "Variational Calculations on the Helium Isoelectronic Sequence," (preprint, 1983).

7. J. L. Gázquez and H. J. Silverstone, "Piecewise polynomial electronic wavefunctions," J. Chem. Phys. 67, 1887 (1977).

8. H. J. Silverstone, D. P. Carroll and D. M. Silver, "Piecewise Polynomial Basis Functions for Configuration Interaction and Many-Body Perturbation Theory Calculations. The Radial Limit of Helium," J. Chem. Phys. 68, 616 (1978).

9. Ch. Hermite, "Sur La Formule d'Interpolation de Lagrange," J. reine angew. Math. 84, 70 (1878).

10. I. Shavitt, EIGEN, Program 172.1, Quantum Chemistry Program Exchange (Indiana University, Bloomington, Indiana, 1970).

11. B. T. Smith, J. M. Boyle, B. S. Garbow, Y. Ikebe, V. C. Klema, and C. B. Moler, "Matrix Eigensystem Routines - EISPACK Guide," (Springer, Berlin 1974).

12. C. F. Bender, "Integral Transformation. A Bottleneck in Molecular Quantum Mechanical Calculations," J. Comput. Phys. 9, 547 (1972).

13. E. Hylleraas, "Uber den Grundzustand des Helium Atoms," Z. Physik 48, 469 (1928).

14. E. R. Davidson, "Natural Expansion of Exact Wavefunctions. III. The Helium-Atom Ground State," J. Chem. Phys. 39, 875 (1963).

15. R. Ahlrichs, W. Kutzelnigg and W. A. Bingel, "On the Solution of the Quantum Mechanical 2-Electron Problem by Direct Calculation of the Natural Orbitals. III. Refined Treatment of the Helium Atom and the Helium-Like Ions," Theor. Chim. Acta 5, 289 (1966).

16. L. C. Green, E. K. Kolchin and N. C. Johnson, "Wave Functions for the Excited States of Neutral Helium," Phys. Rev. A 139, 373 (1965).

17. D. H. Tycko, L. H. Thomas and K. M. King, "Numerical Calculation of the Wave Functions and Energies of the 1^1s and 2^3s States of Helium," Phys. Rev. 109, 369 (1958).

18. G. O. Morrell, Jr. and R. G. Parr, "Estimates of Helium ℓ-Limit Energies," J. Chem. Phys. 71, 4139 (1979).

19. T. Kinoshita, "Ground State of the Helium Atom," Phys. Rev. 105, 1490 (1957).

20. C. W. Scherr and R. E. Knight, "2-Electron Atoms. III. A Sixth Order Perturbation Study of the 1^1s Ground State," Rev. Mod. Phys. 35, 436 (1963).

21. C. Schwartz, "Ground State of the Helium Atom," Phys Rev. 128, 1146 (1962).

DYNAMICS OF VIBRATIONAL PREDISSOCIATION

OF VAN DER WAALS MOLECULES

Lawrence L. Halcomb [a] and Dennis J. Diestler

Department of Chemistry
Purdue University
West Lafayette, Indiana 47907

INTRODUCTION

A van der Waals molecule (VDWM) consists of atoms or molecules held together by intermolecular attractions [1]. These relatively weak interactions, on the order of 10-100 wavenumbers, are responsible for such phenomena as deviation of real gases from ideality and condensation of molecules into liquids and solids. One class of VDWM's that has attracted attention is the triatomic complex X···BC, where X represents a rare gas atom. Of particular interest is the mechanism by which vibrational energy in the chemical bond BC is transferred to the van der Waals bond X···B, resulting in dissociation into X and BC. Both experimental [1,2] and theoretical [3-5] investigations of this process of <u>vibrational predissociation</u> (VP) have been carried out.

We recently proposed a dynamical theory [6,7] that allows a system being treated by classical mechanics to interact with a system being treated by quantum mechanics. Here, we apply this "hemiquantal" approach to the VP of the triatomic VDWM, using as an example $He \cdots I_2 (B^3 \pi)$. In this case, the motion of the helium is quantal in nature, while the vibration of the diatomic iodine is classical.

The hemiquantal equations (HQE) for this system consist of a partial differential equation (PDE) for the "quantal" subsystem coupled to two ordinary differential equations (ODE's) that describe the "classical" subsystem. The integration of the PDE is accomplished using second-order time differencing in conjunction with a fast Fourier transform (FFT) scheme [8] while the ODE's are integrated using a predictor-corrector technique. Since the initial conditions for the classical subsystem are not unique, many such systems must be solved and the results averaged. The use of a vector processor such as the CYBER 205 tremendously simplifies the computations, especially if these systems are integrated simultaneously.

The next Section contains the necessary background and a brief derivation of the HQE. The numerical algorithm along with the steps necessary to implement it on the CYBER 205 is presented next. In the following Section we apply the algorithm to the VDWM $He \cdots I_2 (B^3 \pi)$. Finally, the modifications necessary to integrate many systems at the same time are discussed.

THEORETICAL TREATMENT

The general theoretical problem is to describe the VP of a VDWM on a single electronic potential-energy surface. We specifically restrict ourselves to VDWM's of the form X···BC, where X is a rare gas atom.

Background

The physical quantity of interest is the rate of the collision-free unimolecular decay,

$$X \cdots BC \; (\nu) \longrightarrow X + BC \; (\nu' < \nu) \; ,$$

where ν and ν' are the initial and final vibrational quantum numbers of BC. The (first-order) rate constant k_ν is defined through the quantity

$$D_\nu(t) = \exp\left[-k_\nu t \right] \; , \tag{1}$$

which is the probability of dissociation at time t. Experimentally, VP is detected by means of supersonic jet and picosecond laser techniques [1,2]. The quantity connecting experiment and theory is the decay half-width of the fluorescence excitation spectrum, which is given by

$$\Gamma_\nu = \hbar k_\nu / 2 \; . \tag{2}$$

Even though the complex has six degrees of freedom (exclusive of the center of mass), most calculations have been performed using fewer because of the computational complexity of the problem. Indeed, an exact quantum-mechanical treatment of the full 3-D problem appears prohibitive. Models involving two degrees of freedom include the collinear and "T-shaped" models. The collinear molecule is depicted in Figure 1; q is the distance between the atom X and the center of mass of BC, while Q is the BC separation. In the T-shaped variety, the rare gas atom is constrained to move on a line perpendicular to the interatomic axis of BC. Experimental evidence [1,2] indicates that triatomic VDWM's are nonlinear and could exist as a hybrid of the two forms. Past theoretical approaches have been based on both models but the only exact quantum-mechanical calculation [4] of the VP dynamics has been performed on the collinear model. For this reason, we also adopt this model.

For the collinear model we can write the Hamiltonian as

$$\mathcal{H} = \frac{p_Q^{\;2}}{2\mu_{BC}} + U_{BC}(Q) + \frac{p_q^{\;2}}{2\mu_{X,BC}} + V_{XB}(Q,q) \; , \tag{3}$$

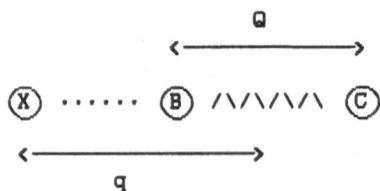

Figure 1. Collinear triatomic van der Waals molecule

where the reduced masses μ_{BC} and $\mu_{X,BC}$ are given by

$$\mu_{BC} = \frac{m_B m_C}{m_B + m_C} \quad \text{and} \quad \mu_{X,BC} = \frac{m_X(m_B + m_C)}{m_X + m_B + m_C} . \tag{4}$$

The operators $\hat{q}$ and $\hat{Q}$ correspond to q and Q respectively. The potential-energy functions are chosen to be of the Morse type:

$$U_{BC}(Q) = D_{BC}\left\{ \exp\left[-2\alpha_{BC}(Q-Q_o)\right] - 2\exp\left[-\alpha_{BC}(Q-Q_o)\right] \right\} \tag{5}$$

$$V_{XB}(q,Q) = D_{XB}\left\{ \exp\left[-2\alpha_{XB}[q-\gamma Q-(q-\gamma Q)_o]\right] - 2\exp\left[-\alpha_{XB}[q-\gamma Q-(q-\gamma Q)_o]\right] \right\}. \tag{6}$$

In Eq. (6) the constant γ is defined by $\gamma = m_C/(m_B + m_C)$.

Hemiquantal Equations

In this treatment, we assume that the internal vibration Q of the diatomic molecule BC behaves classically and that the relative coordinate q is quantal in nature. We begin by writing Heisenberg's equations of motion for the operators $\bar{Q}$ and $\bar{\mathcal{P}}_Q$. For $\bar{Q}$ we have

$$\dot{\bar{Q}} = -\frac{i}{\hbar}\left[\bar{Q}, \frac{\bar{\mathcal{P}}_Q^2}{2\mu_{BC}}\right] = \frac{\bar{\mathcal{P}}_Q}{\mu_{BC}} , \tag{7}$$

where the overbarred operators are in the Heisenberg picture. Similarly, for $\bar{\mathcal{P}}_Q$ we have

$$\dot{\bar{\mathcal{P}}}_Q = -\frac{\partial U_{BC}(\bar{Q})}{\partial \bar{Q}} - \frac{\partial V_{XB}(\bar{Q},\bar{q})}{\partial \bar{Q}} . \tag{8}$$

Assuming that the initial state of the system is separable, i.e.

$$|\Psi(0)\rangle = |\phi(0)\rangle \, |\Psi(0)\rangle , \tag{9}$$

where $|\phi(0)\rangle$ and $|\Psi(0)\rangle$ are the initial states of the classical and quantal subsystems respectively, we can take the partial trace of Eqs. (7) and (8) to obtain, for example,

$$\langle \Psi(0)| \dot{\bar{\mathcal{P}}}_Q |\Psi(0)\rangle = \langle \Psi(0)| -\frac{\partial U_{BC}(\bar{Q})}{\partial \bar{Q}} - \frac{\partial V_{XB}(\bar{Q},\bar{q})}{\partial \bar{Q}} |\Psi(0)\rangle . \tag{10}$$

We now make the crucial approximation, which requires the classical degree of freedom to follow __precisely__ its classical trajectory. By setting

$$U_{BC}(\bar{Q}) = U_{BC}(\langle\bar{Q}\rangle) \, \mathbb{1} \tag{11}$$

and

$$V_{XB}(\bar{Q},\bar{q}) = V_{XB}(\langle\bar{Q}\rangle,\bar{q}) \, \mathbb{1} , \tag{12}$$

where $\mathcal{I}$ is the identity operator, we can write the expectation values of $\dot{\bar{Q}}$ and $\dot{\bar{\mathcal{P}}}_Q$ as

$$\dot{Q}(t) \equiv \langle\langle\Psi(0)|\dot{\bar{Q}}|\Psi(0)\rangle\rangle = - \frac{\langle\langle\Psi(0)|\bar{\mathcal{P}}_Q|\Psi(0)\rangle\rangle}{\mu_{BC}} \equiv \frac{P_Q(t)}{\mu_{BC}} \tag{13}$$

$$\dot{P}_Q(t) = - \frac{\partial U_{BC}(Q)}{\partial Q} + \langle\Psi(t)| - \frac{\partial V_{XB}(Q,q)}{\partial Q} |\Psi(t)\rangle . \tag{14}$$

Note that the last term has been rewritten in the Schrödinger picture.

For the description of the wavepacket corresponding to the relative motion of X, we first write the "hemiquantal" evolution equation

$$\dot{\mathcal{U}} = - \frac{i}{\hbar} \left[\frac{p_q^2}{2\mu_{X,BC}} + V_{XB}(Q,q) \right] \mathcal{U} . \tag{15}$$

Applying Eq. (15) to the ket $|\Psi(0)\rangle$, we obtain

$$\frac{\partial}{\partial t} |\Psi(t)\rangle = - \frac{i}{\hbar} \left[\frac{p_q^2}{2\mu_{X,BC}} + V_{XB}(Q,q) \right] |\Psi(t)\rangle . \tag{16}$$

Eqs. (13)-(16) comprise a representationless form of the HQE for this system.

Recasting Eq. (16) in the q representation, we obtain (dropping the subscript on V_{XB})

$$\frac{\partial}{\partial t} \Psi(q,t) = - \frac{i}{\hbar} \left[\frac{-\hbar^2}{2\mu_{X,BC}} \frac{\partial^2}{\partial q^2} + V(Q,q) \right] \Psi(q,t) . \tag{17}$$

Similarly, the other HQE are (dropping the subscripts on P_Q and U_{BC}):

$$\dot{Q} = P/\mu_{BC} \tag{18}$$

$$\dot{P} = - \frac{\partial U(Q)}{\partial Q} + \int_{R(q)} \Psi^*(q,t) \left[- \frac{\partial V(q,Q)}{\partial Q} \right] \Psi(q,t) \, dq ; \tag{19}$$

R(q) is the range of q. Eq. (17) is strikingly similar to the time-dependent Schrödinger equation, the only difference being the appearance of the classical variable Q. The other HQE [Eqs. (18) and (19)] have the appearance of Hamilton's equations, with the addition of a force term that arises from the classical-quantal coupling.

The HQE describe the time evolution of the wavefunction $\Psi(q,t)$ and the coordinates Q and P. If the complex is considered to be dissociated when $q \geq q_D$, the dissociation probability at time t is given by

$$D(t) = \int_{q \geq q_D} |\Psi(q,t)|^2 \, dq , \tag{20}$$

where $\Psi(q,t)$ is assumed to be normalized. The rate coefficient [see Eq. (1)] is calculated from the relation

$$k_\nu = - \frac{\partial \ln[1-D(t)]}{\partial t} , \tag{21}$$

i.e. the negative slope of a plot of $\ln[1-D(t)]$ versus t.

A crucial aspect of the problem is the choice of initial conditions. Since hemiquantal mechanics requires separable initial conditions, a "mixed" state cannot be used. We assume an initial (ground-state) Morse wavefunction [9] that corresponds to the van der Waals interaction given by Eq. (6), namely

$$\Psi(q,0) = N \, e^{- \frac{b+1}{2} \, e^{-\alpha_{XB}(q-q_o)}} \left[(b+1) \, e^{-\alpha_{XB}(q-q_o)} \right]^{b/2} , \tag{22}$$

where

$$b = \frac{\sqrt{8\mu_{X,BC} D_{XB}}}{\hbar \, \alpha_{XB}} - 1 . \tag{23}$$

Since the diatomic is assumed to be in vibrational state ν initially, the initial energy contained in BC can be calculated from the following expression for the energy levels of the Morse potential:

$$E_\nu = - \frac{D\hbar}{\kappa^2} \left(\kappa - \nu - \frac{1}{2} \right)^2 , \tag{24}$$

where $\kappa \equiv (2\mu D)^{1/2}/\hbar\alpha$. The corresponding initial values of Q and P (i.e. the initial phase of the classical subsystem) must satisfy the relation

$$E_\nu = \frac{1}{2\mu} P^2 + U(Q) . \tag{25}$$

Since an infinite number of phases fulfill this condition, it is necessary to solve the HQE for a representative number of these phases and then average the results.

NUMERICAL INTEGRATION OF THE HEMIQUANTAL EQUATIONS

The HQE are unusual in that a PDE [Eq. (17)] is coupled to two ODE's [Eqs. (18) and (19)]. We have devised an atypical algorithm that melds a novel approach to the numerical integration of the PDE with rather more standard methodology for the ODE's.

Algorithm

Recently, a new technique for solving the time-dependent Schrödinger equation was proposed [8]. This method uses a numerical Fourier transform to approximate spatial derivatives and a second-order differencing scheme for time derivatives. We have chosen the Fourier method for two principal reasons:

(1) Although employed in other fields, the technique has not, to our knowledge, been used for a meaningful quantum-mechanical computation. Therefore, its viability in this context should be examined.

(2) The fast Fourier transform (FFT) algorithm [10,11] is highly vectorizable and a CYBER 205 FFT subroutine is available in the Control Data Math/Geophysical Vector (MAGEV) library [12].

The Fourier method is based on the relation between the Fourier transform of the derivative of a function and the Fourier transform of the function itself. If we define the Fourier transform of a function f(q) as

$$F(k) \equiv \mathcal{F}\{f(q)\} \equiv \int_{-\infty}^{\infty} f(q) \, e^{-i2\pi kq} \, dq \; , \tag{26}$$

then, for example, the second derivative can be expressed as

$$\frac{\partial^2 f(q)}{\partial q^2} = \mathcal{F}^{-1}\{ \, -(2\pi k)^2 \, F(k) \, \} \; . \tag{27}$$

The discrete Fourier transform of a function g(q) known on a grid of evenly spaced points $n\Delta q$ is given by [10,11]

$$G\left(\frac{m}{N\Delta q} \right) = \sum_{n=0}^{N-1} g(n\Delta q) \, e^{-i2\pi mn/N}, \quad m = 0, \ldots, N-1 \tag{28}$$

with inverse

$$g(n\Delta q) = \frac{1}{N} \sum_{m=0}^{N-1} G\left(\frac{m}{N\Delta q} \right) e^{i2\pi mn/N} \; , \quad n = 0, \ldots, N-1 \; . \tag{29}$$

We can numerically approximate the second spatial derivative of $\Psi(q,t)$ by taking the FFT of the (discrete) function $\Psi(q,t)$, multiplying the transformed function by a vector of the proper wavenumbers, and finally, taking the inverse FFT of this result.

The time derivatives are approximated with the second-order differencing scheme

$$\frac{\partial \psi^n}{\partial t} = \frac{\psi^{n+1} - \psi^{n-1}}{2\Delta t} \; , \tag{30}$$

where Δt is the time step and ψ^n is the (discretized) wavefunction at time $n\Delta t$. Higher-order differencing schemes could be used if desired [8].

We now proceed to Eqs. (18) and (19) for the classical subsystem. Unfortunately, these ODE's cannot be economically integrated with the same time step and difference scheme employed for the PDE. To circumvent this problem, we used a predictor-corrector algorithm of the following form [assuming a generic system of equations $x(t) = f(x,t)$] [13]:

predictor:
$$p^{n+1} = x^{n-1} + 2\Delta t f^n \tag{31}$$

modifier:
$$m^{n+1} = p^{n+1} + \frac{4}{5}(\, c^n - p^n \,) \tag{32}$$

$$\dot{m}^{n+1} = f[\, m^{n+1}, (n+1)\Delta t \,] \tag{33}$$

corrector:
$$c^{n+1} = x^n + \frac{\Delta t}{2} (\dot{m}^{n+1} + f^n) \qquad (34)$$

modifier:
$$x^{n+1} = c^{n+1} - \frac{1}{5} (c^{n+1} - p^{n+1}) . \qquad (35)$$

There are two such sets of equations, one for the coordinate Q and one for the momentum P.

Before the algorithm can be used, it is necessary to obtain estimates of the wavefunction and phase at the first time point from the initial conditions. An Euler approximation [13] was found to be adequate for the wavefunction, i.e.

$$\psi^1 = \psi^0 + (\Delta t)(-iH\psi^0) , \qquad (36)$$

where we have adopted units such that $\hbar$ appears as one. The function $-iH\psi^0$ is the right-hand side of Eq. (17) at time t= 0. A more accurate approximation must be used for Q(Δt) and P(Δt). We employed the second-order Runge-Kutta algorithm [13]

$$K_1 = \Delta t \ f^n \qquad (37)$$

$$K_2 = \Delta t \ f[\ x^n + K_1/2, \ (n+1)\Delta t \ ] \qquad (38)$$

$$x^{n+1} = x^n + (K_1 + K_2)/2. \qquad (39)$$

Implementation on the CYBER 205

We assume that a discrete grid of N points has been chosen in q-space (with spacing Δq) and that the system of equations is to be integrated using a time step of Δt. We introduce the following vector notation:

(1) $\vec{q}$ is the vector of grid points; $(\vec{q})_j = j\Delta q$, j= 0, ..., N-1.

(2) $\vec{k}$ is the vector of wavenumbers corresponding to $\vec{q}$. The discrete Fourier Transform exhibits "conjugate symmetry" [10,11] about N/2, i.e., $(\vec{k})^T = (1/N\Delta q)[\ 0,1,2,...,N/2,-(N/2-1),-(N/2-2),...,-1 \ ]$.

(3) $\vec{\psi}^n$ is composed of values of $\Psi(q,t)$ at time nΔt.

(4) $\vec{V}^n$ consists of the values of the coupling potential V(Q,q) at time nΔt. (This vector is not constant since the coordinate Q changes at every time point.)

(5) $\vec{F}^n$ consists of the values of the force $-\partial V(Q,q)/\partial Q$ at time nΔt.

The algorithm for the integration of the PDE consists of evaluating $\ddot{\vec{\psi}}^n$ and $\vec{V}^n$, constructing the vector $-iH\vec{\psi}^n$, and then calculating $\vec{\psi}^{n+1}$. Each of the "complex" vectors has the structure of $\vec{\psi}^T = [Re\{\vec{\psi}\}, Im\{\vec{\psi}\}]$. The integration of the PDE can be diagramed as follows:

$$\ddot{\vec{\psi}}^n = FFT^{-1}\left[\begin{bmatrix} -(2\pi\vec{k})^2 \\ -(2\pi\vec{k})^2 \end{bmatrix} * FFT\left[\vec{\psi}^n\right]\right]$$

Compute $\vec{V}^n$
(discussed below)

$$Re\{-iH\vec{\psi}^n\} = -(2\mu_{X,BC})^{-1} Im\{\ddot{\vec{\psi}}^n\} + \vec{V}^n * Re\{\vec{\psi}^n\}$$

$$Im\{-iH\vec{\psi}^n\} = (2\mu_{X,BC})^{-1} Re\{\ddot{\vec{\psi}}^n\} - \vec{V}^n * Im\{\vec{\psi}^n\}$$

$$\vec{\psi}^{n+1} = \vec{\psi}^{n-1} + (2\Delta t)(-iH\vec{\psi}^n)$$

permute pointers of $\vec{\psi}^{n+1}$, $\vec{\psi}^n$ and $\vec{\psi}^{n-1}$

The FFT routine in the library MAGEV [12] is entitled "FFT1D". The routine is extremely fast, taking 457 μs to do an FFT of a complex vector of length 512 (real and imaginary parts separated). The routine uses trigonometric tables that are built using a separate call and requires a scratch area the same size as the input vector. Complete use and timing information can be found in the library documentation.

The above diagram exhibits several noteworthy features: (1) the vector additions and multiplications are made efficient by using the entire vector $\vec{\psi}$ as often as possible; (2) the conjugation of "H$\vec{\psi}$" is performed essentially for free since linked triads are used in the process, and (3) the result vector is not moved; only pointers are permuted using simple register-to-register operations.

Actually, we can get by with only two wavefunction vectors since the wavefunction at time $(n+1)\Delta t$ can replace the wavefunction at time $(n-1)\Delta t$. Thus, only an interchange of pointers is necessary. In fact, storage for many of the vectors can overlap. For example, the scratch area required by FFT1D can also be used for the potential and force vectors. Although storage is not crucial now, it could become so when modifying this scheme to integrate sets of HQE at the same time, a topic to be covered below.

We now turn to the evaluation of the derivatives $\dot{Q}$ and $\dot{P}$. In order to evaluate the integral in Eq. (19), the magnitude of the wavefunction at each grid point must be calculated. This is accomplished in two vector operations, i.e., $|\vec{\psi}|^2 = Re\{\vec{\psi}^2\} + Im\{\vec{\psi}^2\}$. Once $\vec{F}$ is computed, the value of the integral (using the trapezoidal rule) can be written as $I = \Delta q(|\vec{\psi}|^2 \cdot \vec{F})$. Since $\Psi(q,t)$ is negligible at the edge of the grid, the contribution of the endpoints is ignored. The dot product is taken using the hardware instruction "Q8SDOT".

The remainder of this derivative evaluation involves only scalar manipulations. It is worthwhile to note that many of these operations can

280

overlap with the vector instructions since they can be performed in registers. This section of code is executed twice, since the predictor-corrector scheme in Eqs. (31)-(35) requires two derivative evaluations.

The computation of the potential vector at each time point is carried out as follows. We wish to calculate a vector of the form

$$V(\vec{r}) = D\left[e^{-2\alpha(\vec{r}-r_o)} - 2e^{-\alpha(\vec{r}-r_o)} \right] , \tag{40}$$

where $\vec{r} = \vec{q} - \gamma Q$, and $\vec{r}_o = (\vec{q} - \gamma Q)_o$. If we define the vectors

$$\vec{V}_1 = De^{-2\alpha(\vec{q}-r_o)} \tag{41}$$

$$\vec{V}_2 = 2De^{-\alpha(\vec{q}-r_o)} , \tag{42}$$

we can write $V(Q,\vec{q})$ as

$$\vec{V} = e^{\alpha\gamma Q} (e^{\alpha\gamma Q} \vec{V}_1 - \vec{V}_2) . \tag{43}$$

This calculation takes two vector operations per time step (a linked triad and a multiplication). The force vector is calculated similarly.

Mechanics

The program first performs all necessary setup (pointer assignment, calculation of trigonometric tables, etc.). The main loop consists of reading initial Q and P values, calculating the initial wavefunction and "starting" values, and then integrating the differential equations. This process is continued for each set of initial Q and P values. At selected time points , the probability of dissociation is stored. [The integral D(t) in Eq. (20) is evaluated in the same manner as I above.] When the integrations are complete, the curves for D(t) corresponding to initial phases are averaged to obtain ⟨D(t)⟩. The rate is obtained by linear regression from the slope of a plot of ln[1-⟨D(t)⟩] versus t. The vibrational energy in the I_2 bond [Eq. (25)] is also monitored in the simulation. This allows us to watch the flow of vibrational energy from the classical subsystem. Additional features of the program include simultaneous I/O for following the evolution of the wavepacket and a mechanism for checking the conservation of total energy and total probability at user-specified intervals.

Table 1.   Units

Frequency	$3.33 \times 10^{12} \text{ s}^{-1}$
Energy	$3.52 \times 10^{-15} \text{ erg}$
Time	$0.30 \times 10^{-12} \text{ s}$
Length	$1.38 \times 10^{-10} \text{ cm}$
Mass	1.00 amu

APPLICATION TO HeI$_2$(B^3π)

The VDWM HeI$_2$(B^3π), discovered in the mid 1970's, was the first example of a polyatomic molecule containing helium bound in the ground electronic state. The molecule has since become the subject of numerous experimental investigations [1,2] and, as a result, most theoretical studies [3-5]. Previous dynamical studies based on the collinear model include the quasiclassical calculations of Woodruff and Thompson [3] and the quantum-mechanical calculations of Viswanathan et al. [4]. These studies are the classical and quantal analogs of the present work.

For the computations, we adopted the units and Morse parameters given in Tables 1 and 2. Numerical experiments were performed to optimize the grid and the time step size. It was found that the second spatial derivative is poorly approximated on the edge of the grid and a portion of the grid was allocated to compensate for this effect. A grid 30 units in length covered by 512 Fourier points was used in the simulations; the "physical" portion of the grid is given by R(q)= [0,26]. The time step was chosen by fixing Q and P, integrating the PDE, and checking the conservation of total probability. For the present problem, a time step of 0.0005 was sufficient to obtain constancy to ten decimal places.

Once the time step was selected, the classical equations were integrated using the predictor-corrector technique [Eqs. (31)-(35)]. Conservation of total energy was checked to insure that the time step was appropriate. In order to obtain initial conditions for the iodine molecule, the energy was fixed at the vibrational state of interest and Q and P values were chosen on a uniform time grid over the period of the I$_2$ oscillator.

For each set of initial Q and P values, the HQE were integrated from 24,000-30,000 time steps (for lower vibrational states, it is necessary to integrate the HQE further in time). It was found that the results of 40-50 such integrations must be averaged to obtain reliable dissociation curves. So that comparisons may be made with previous work, simulations were performed for the iodine vibrational states 22 $\leq$ v <32.

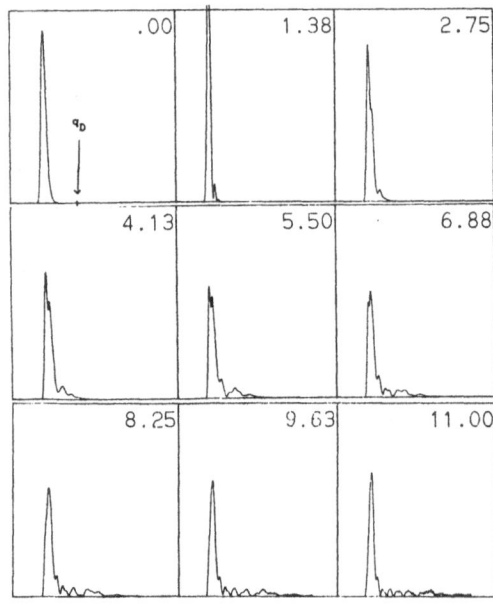

Figure 2. Sample wavepacket evolution for an initially compressed iodine molecule in vibrational state v= 32.

Table 2. Physical Parameters

Morse Parameter	He-I	I-I
D	0.7910	277.42
α	1.623	2.436
r_o	2.909	2.193

$$\mu_{HeI_2} = 3.941$$
$$\mu_{I_2} = 63.450$$
$$q_D = 9.50$$

Since it is not appropriate to enumerate fully the results here, we are including complete results in a companion paper [14]. As a representative example, we consider the case $\nu = 32$. The evolution of $|\Psi(q,t)|^2$ for a sample trajectory is shown in Figure 2. This wave packet is atypical in that the iodine molecule is initially compressed with no momentum, giving rise to a "violent" wave packet. Special note should be taken of the "rippling" in the last two frames of Figure 2. This effect is due to reflection from the edge of the grid. It is necessary to choose a grid large enough to prevent this unphysical "backwash."

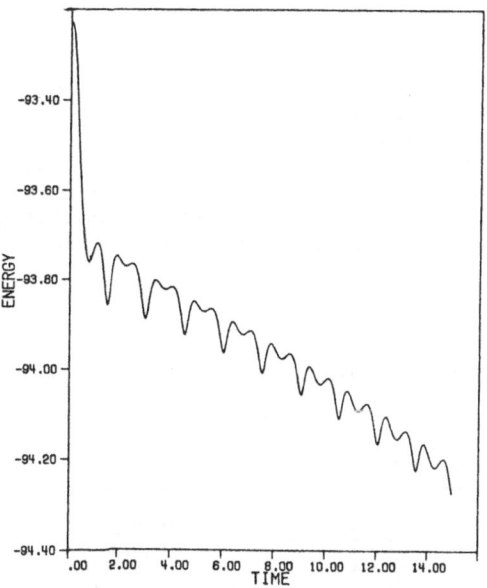

Figure 3. E(t) for a sample trajectory ($\nu = 32$).

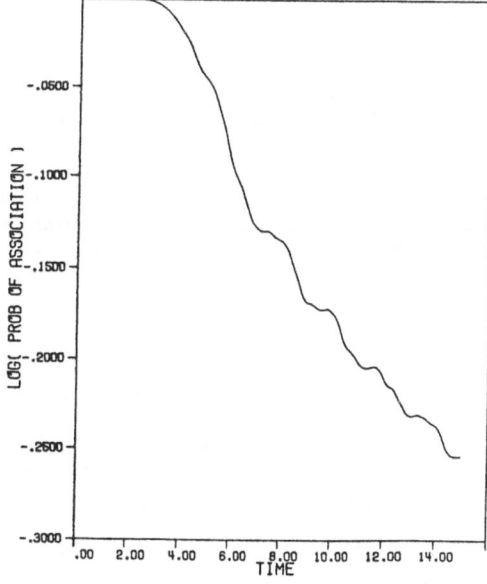

Figure 4. ln[1-D(t)] for a sample trajectory ($\nu = 32$).

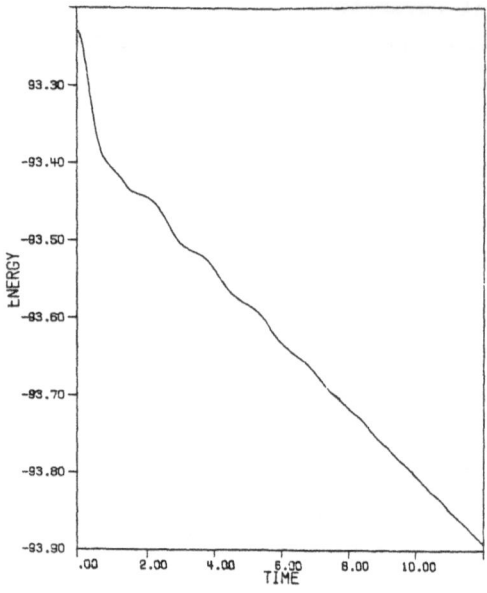

Figure 5. ⟨E(t)⟩ for 50
trajectories (ν = 32).

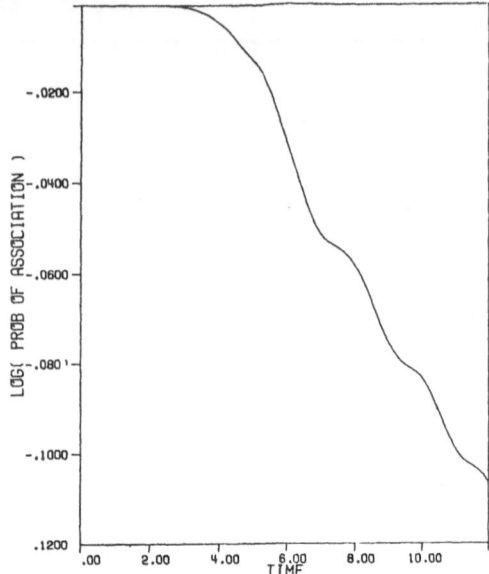

Figure 6. ln[1-⟨D(t)⟩] for
50 trajectories (ν = 32).

In Figures 3 and 4, plots of vibrational energy in the I_2 bond and ln[1-D(t)] versus time are shown for the same trajectory. After a sharp dip, E(t) oscillates as it decays. The large amplitude oscillations correspond to the frequency of the iodine oscillator. The results of averaging 50 such trajectories can be seen in Figures 5 and 6. The curves have smoothed considerably and are clearly linear.

Linear regression was performed on the ln[1-⟨D(t)⟩] curves to obtain the slope; an arbitrary "induction" time at the beginning of the simulation was ignored. For the case ν = 32, various induction times were used to explore the effect on the slope. For induction times from 5 to 10 time units, the decay half-width Γ decreases, but the total variation on this interval is less than 10% of the experimental value. The correlation coefficients also decrease but are always well above 0.98. In fact, for induction times from 6-8 this coefficient is over 0.99. The final value chosen was 6.0 (adding or subtracting one unit changes Γ by less than 2%). A similar process of elimination was used for the other vibrational states. The hemiquantal and some experimental results are summarized in Table 3. As can be seen, they are in good agreement.

Table 3. Rates and Decay Half-Widths for He$\cdots$I$_2$(B^{3}Π)

ν	k_ν expt[1] ($\times 10^{10} s^{-1}$)	k_ν ($\times 10^{10} s^{-1}$)	Γ_ν (cm^{-1})
22	1.7	1.92	0.051
24	2.1	2.26	0.060
26	2.6	2.52	0.067
28	–	3.09	0.082
30	–	3.54	0.094
32	–	4.07	0.108

In order to realize the maximal performance of the CYBER 205, vectors that are frequently used should be as long as possible. Since each trajectory is independent, it would be desirable to redesign the algorithm so that all trajectories are integrated at once. This leads to vector lengths as long as $2 \times 512 \times 50 = 51,200$ for the present case. In this new "parallel" algorithm, the wavefunction vectors have the structure

$$\vec{\Psi}^T = \left[\ \mathrm{Re}\{\vec{\Psi}_1\}, \ldots, \mathrm{Re}\{\vec{\Psi}_{50}\}, \mathrm{Im}\{\vec{\Psi}_1\}, \ldots, \mathrm{Im}\{\vec{\Psi}_{50}\} \ \right] . \tag{44}$$

The former scalars such as Q and P simply become vectors of length 100, i.e., all 50 values of Q followed by all 50 values of P.

The subroutine FFT1D can perform many FFT's at once but the input vector must be stored rowwise rather than columnwise. Hence, the input and output vectors must be transposed. However, the time saved by doing the large number of FFT's at once more than compensates for the extra time spent doing the transposes. The routine TPMOV in MAGEV was used to perform the transposes. The potential vectors cannot be evaluated in "parallel" economically. Therefore, at each time step, these vectors are computed separately and used to build a larger vector so that advantage can be taken of its longer length in subsequent multiplications. Apart from the transpositions and the separate evaluations of the potential vectors, the integration of the PDE is accomplished as before.

Almost all of the computations involved in the calculation of the derivatives of Q and P as well as the predictor-corrector manipulations can now be accomplished using vectors of length 100. The magnitudes of all the wavefunctions can be formed at once but the force vectors and the integrals must still be calculated separately.

In Table 4, the speeds of the vector operations in the individual and "parallel" schemes are compared. (In both cases, vector operations comprise over 90% of the CPU time.) The "parallel" algorithm is almost twice as fast, with most of the gain coming from the evaluation of the spatial derivative.

Table 4.   Comparison Timings (Vector Operations)

Calculation	Individual (μs)	"Parallel" (μs)
$\ddot{\vec{\Psi}}$	937	480
$\vec{V}$	14	14
$\vec{\Psi}^{n+1}$	39	31
PDE(total)	990	525
$|\vec{\Psi}|^2$	17	15
$\vec{F}$	14	14
$\vec{F} \cdot |\vec{\Psi}|^2$	13	13
other	0	23
ODE(total)	88	107
Total	~1.08 ms	~0.63 ms

CONCLUSION

In this paper we have presented one application of a "hemiquantal" mechanics. While this methodology considerably simplifies the treatment of degrees of freedom that behave classically, a price must be paid in the need to integrate and average over many sets of initial conditions. Even so, problems of higher dimensionality become feasible since it is easier to compute classical trajectories than to propagate multidimensional wavefunctions.

We have used in this work a Fourier technique for approximating spatial derivatives that had not been previously applied to large-scale quantal calculations. The method looks quite promising even in view of the increased number of grid points required, especially since an FFT can be so rapidly performed on a vector processor. We have also demonstrated that the CYBER 205 can be used even more economically by carrying out computations in "parallel," even when the individual computation is highly vectorized. The individual calculation itself may be performed more rapidly either by decreasing the number of grid points or by using a higher-order time-differencing scheme and then increasing the time step.

ACKNOWLEDGEMENTS

We are grateful to Control Data Corporation and to Purdue University for the support of this work. We would also like to thank Carol Phife for proofreading and improving the manuscript.

REFERENCES

a) Control Data Corporation PACER Postdoctoral Fellow.

1. See, for example, D. H. Levy, Adv. Chem. Phys. $\underline{47}$, 323 (1981) and references therein.

2. D. H. Levy, Science $\underline{214}$, 263 (1981).

3. S. B. Woodruff and D. L. Thompson, J. Chem. Phys. $\underline{71}$, 376 (1979).

4. R. Viswanathan, L. M. Raff, and D. L. Thompson, J. Chem. Phys. $\underline{77}$ 3939 (1982).

5. (a) See, for example, J. A. Beswick and J. Jortner, Adv. Chem. Phys $\underline{47}$, 363 (1981), and references therein; (b) M. Aguado, J. Salgado, G Serrono, P. Mareca, P. Villarreal, and G. Delgado-Barrio, J. Mol Struct. $\underline{93}$, 325 (1983); (c) M. Aguado, P. Villareal, G. Delgado Barrio, P. Mareca, and J. A. Beswick, Chem Phys. Letters. $\underline{102}$, 22 (1983); (d) G. Delgado-Barrio, P. Villareal, P. Mareca,, and G Albelda, J. Chem. Phys. $\underline{78}$, 280 (1983).

6. L. L. Halcomb, Doctoral Dissertation, Purdue University, 1984.

7. D. J. Diestler, J. Chem. Phys. $\underline{78}$, 2240 (1983).

8. D. Kosloff and R. Kosloff, J. Comp. Phys. $\underline{52}$, 35 (1983).

9. E. A. McCullough Jr. and R. E. Wyatt, J. Chem. Phys. $\underline{54}$, 3578 (1974)

10. N. J. Nussbaumer, <u>Fast Fourier Transform and Convolution Algorithms</u> (Springer-Verlag, New York, 1981).

11. E. O. Brigham, <u>The Fast Fourier Transform</u> (Prentice-Hall, Englewood, NJ, 1974).

12. Math/Geophysical Vector Library, Version 3.1, Control Data Corporation, 1983.

13. J. S. Rosko, <u>Digital Simulation of Physical Systems</u> (Addison-Wesley, Reading, MA, 1972), Chap. 5.

14. L. L. Halcomb and D. J. Diestler, to be published.

IMPLEMENTING CONSTRAINED LEAST-SQUARES REFINEMENT
OF HELICAL POLYMERS ON A VECTOR PIPELINE MACHINE

R. P. Millane, M. A. Byler and Struther Arnott

Department of Biological Sciences
Purdue University
West Lafayette, Indiana, USA

1. INTRODUCTION

X-ray crystallography is the most powerful and widely used method of determining molecular structures at atomic resolution. A crystal of the material is exposed to a monochromatic X-ray beam and the *diffraction pattern* formed by the scattered X-rays recorded on a film or by an electronic detector. The crystal diffracts the X-rays only in certain directions so that the diffraction pattern consists of spots which are called *reflections* whose amplitudes (the square-root of the diffracted energy) are called *structure amplitudes*. Measurement of the structure amplitudes allows the structure of the molecule (the positions of the atoms) forming the crystal to be determined by standard (although not necessarily straightforward) X-ray crystallographic techniques[1].

Many polymers such as nucleic acids are long thin molecules which do not form extensive regular crystals and hence are not suitable for conventional crystallographic analysis. However, diffraction studies have played an important role in the determination of the structures of DNA and a wide range of other linear biopolymers because specimens of these materials can be prepared in which the molecules have their long axes approximately parallel and are laterally organized into very small regions of three-dimensional crystallinity. The azimuthal orientations of these crystallites about the direction parallel to the long axes of the molecules are however random. Diffraction patterns from these *fibrous* specimens often contain sufficient information to make structural analysis possible. However, the number of measured diffraction data is usually less than the number of independent atoms in the structural unit of the molecule, and they rarely extend beyond 3Å resolution, making determination of the structure from the diffraction data alone a very under-determined problem. The diffraction data must be supplemented by other independent information if structure determination is to be possible. The additional information available is of a stereochemical nature.

Linear polymers invariably display some helical symmetry and considerable simplification results when the molecule is approximately regular and can be represented by a single helical repeating unit rather than a complete turn of the helix. Further simplification results from the assumption that bond lengths and angles in the polymer are fixed at the values observed in monomers (determined using conventional crystallographic analysis). This reduces the solution of the polymer structure to determining the conformation angles about all (or some) of the bonds. It is possible to prepare what we call a *linked-atom* description of the molecule in which interatomic relationships are described in terms of bond lengths, bond angles and conformation angles. The linked atom description, together with the assumption of helical symmetry, results in a significant decrease in the number of parameters and improvement in the data-to-parameter ratio. A further very useful source of stereochemical information is the requirement that the structure exhibit no over-short non-bonded interatomic distances. Such short distances that do exist are called *contacts*, and the requirement that they be no less than minimum allowed distances provides additional data.

Examination of the diffraction pattern usually allows the helical symmetry of the molecule and the possible crystal packing arrangements to be determined. Using this and chemical information on the molecule, and knowledge of the molecular arrangements stereochemically possible, one can build-linked atom molecular models and incorporate them into plausible crystal structures. Each of these models is refined by adjusting chosen variable parameters in such a way that the stereochemical and diffraction properties calculated from the model, accord with those observed as well as possible, in a least-squares fashion. Figures of merit, which describe the stereochemical acceptibility of the model and the agreement between the calculated and measured diffraction intensities, can be calculated for each of the refined models to determine if one of them is clearly superior to the others.

These concepts were first applied to helical polymers in the 1960s[2] and the refinement programs extended and generalized since, resulting in a *linked-atom least-squares* (LALS) refinement system[3] which is generally applicable to a wide range of complex molecular structures and packing arrangements. This system is in use at Purdue and a number of laboratories throughout the world, and has been used successfully to investigate the structures of a large number of fibrous materials[3,4]. Until 1982 LALS was run on the CDC 6500/6600 system at Purdue. Although it has been possible to examine many interesting structures, we could not model low symmetry structures with large repeated motifs which exist in nucleic acids in some crystal packing arrangements. However, implementation of LALS on a CDC Cyber 205 recently installed at Purdue has allowed us to examine nucleic acid structures much more complex than was previously possible[5]. We describe here vectorization of LALS to take advantage of the vector processing and other special features of the Cyber 205 to obtain optimum performance.

In section 2 we outline the strategy of the LALS system and describe its major components. Vectorization of the three major components of LALS is described in section 3 together with results showing the improvements in performance obtained in each component. When refining large structures, memory swapping can be a more severe limitation than speed and an aspect of memory management is described in section 4. In section 5 we present examples illustrating the overall increase in performance of LALS when refining structures of different sizes, and discuss the implications of this work and further prospects.

2. THE LINKED-ATOM LEAST-SQUARES REFINEMENT SYSTEM

The strategy of linked-atom least-squares refinement is to refine a molecular and crystal structure model (i.e. adjust the atom positions) such that the structure amplitudes predicted by the model match those observed as well as possible while maintaining optimum stereochemistry. The model is adjusted by making changes to a set of chosen linked-atom parameters (bond angles and conformation angles) and parameters defining the positions and orientations of the molecules in the unit cell of the crystal. Refinement is achieved by minimizing a cost function Ω given by

$$\Omega = \sum_m \omega_m ({}_oF_m - F_m)^2 + \sum_m k_m (c_m - d_m)^2 + \sum_m \lambda_m G_m .$$
$$= \quad X \quad + \quad C \quad + \quad L$$

(1)

The term X involves the differences between the observed ${}_oF_m$ and calculated F_m structure amplitudes. The term C involves *restraints* in which the quantities c_m are driven towards desired values d_m. LALS accommodates a variety of quantities which can be incorporated as restraints[3], however almost all the computational load involves contacts between non-bonded atoms and so only these are described here. In this case the c_m are the close non-bonded interatomic distances and the d_m are the minimum acceptable distances. The ω_m and k_m are weights which are chosen to be inversely proportional to the estimated variances of the measured structure amplitudes or desired interatomic distances. In a polymer structure there usually exist relationships, which we call *constraints*, between parameters which must be satisfied exactly. These are inflexible requirements such as, for example, that one helical turn of the molecule must be continuous with the next turn, and that sets of atoms forming chemical ring systems are closed. LALS provides for a variety of types of constraint[3] which are incorporated in the term L in (1). The constraints G_m are satisfied when $G_m = 0$ and the λ_m are Langrange multipliers. The computational load involved in calculating constraints is small and so they are not discussed further here.

Since Ω is not a linear function of the varied parameters, denoted by p_m, minimization of (1) cannot be achieved immediately. If the starting model is close to that which minimizes Ω, then (1) can be linearized by expanding it as a Taylor series to first order in the differences Δp_m between the old and new (better) parameter values[2]. The *shifts* Δp_m to be applied to the old parameters can then be found by requiring that

$$\frac{\partial \Omega}{\partial \Delta p_m} = 0 \qquad and \qquad \frac{\partial \Omega}{\partial \lambda_m} = 0 .$$

(2)

Equation (2) can be written as the system of linear equations[2]

$$\mathbf{M}^T \mathbf{M} \mathbf{S} = \mathbf{A} \mathbf{S} = \mathbf{M} \mathbf{H}$$

(3)

where the elements of the matrix S are the parameter shifts Δp_m and the λ_m, the elements of the vector H are the $({}_oF_m - F_m)$, $(c_m - d_m)$ and G_m, and the elements of the matrix M are the partial derivatives of the F_m, c_m and G_m with respect to the parameter shifts multiplied by the square root of the corresponding weights. A is the normal matrix equal to $\mathbf{M}^T \mathbf{M}$ and the superscript T denotes transposition. Solution of the normal equations (3) provides the required parameter shifts. Since the Taylor series is truncated to first order, applying the shifts to the parameters will not minimize Ω immediately and the process must be iterated until the Δp_m are so small that Ω does not change significantly. Once convergence is obtained for each model considered, they can be compared using figures of merit as described in section 1.

A flow chart of the operation of LALS is shown in Fig. 1. Each iteration of least squares refinement (a *cycle*) forms the main loop in the program. Typically, up to 15 cycles are required for convergence of a well behaved model. The solution of a structure may involve examining several molecular models each in one of several crystal packing arrangements. In addition, each molecular model may need to be refined under several different restraint conditions to determine the correct domain of some of the conformation angles. Hence the solution of one structure can involve over a thousand cycles of least-squares refinement. One cycle of refinement of a complex low symmetry nucleic acid structure may require hundreds of seconds of CPU time on the CDC 6600 system so that examination of many of these structures is prohibitively expensive even on large scalar computers. The tremendous speed increase obtainable by vectorization of the most time consuming sections of the program has made examination of these structures possible and relatively inexpensive[5].

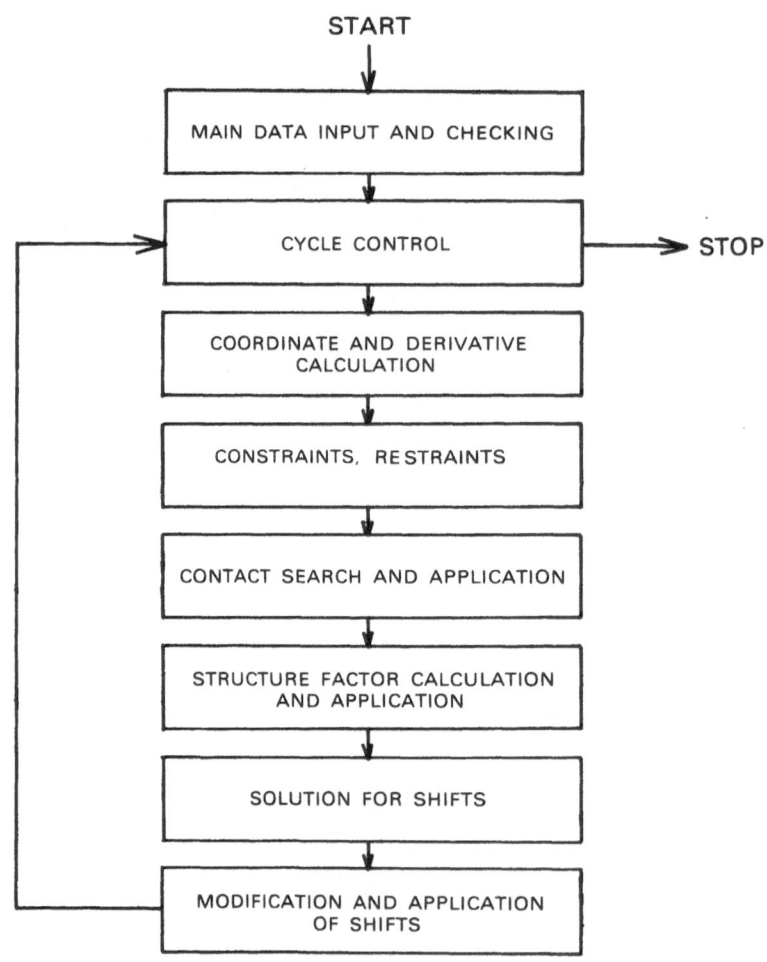

Fig. 1. Flow chart of the operation of LALS.

3. VECTORIZATION

3.1 Approach

There are two possible aproaches to vectorizing a large complex program written originally for a scalar machine. The algorithms can be rearranged in such a way that they are suited to vector processing and the problem recoded. Alternatively, the overall structure of the original program can be retained, and time-consuming areas of the code located and modified in a sequential fashion to make use of the vector processing facilities. We chose the later approach as the former would have required a large programming effort (approximately one man-year) and would not have provided a short term solution for current jobs which require excessive CPU time. Using the later approach we achieved a dramatic improvement in performance with an effort of approximately three man-months, and provided useable partially vectorized code during the modification period. LALS is a 7000 line program and although the primary algorithms are fairly straightforward, the program is quite complex because of the wide variety of structures and refinement options it accommodates. It is this complexity that makes a complete recoding of the program so time consuming.

We found that using the Fortran compiler option for implicit vectorization and the "VAST"[6] automatic vectorizing system provided minimal improvement in performance and they were not implemented in the end. These two utilities can only vectorize relatively simple structures although our experience with VAST is not extensive. Use of the CDC "PROFILE" utility to locate time consuming sections of the program was invaluable throughout the vectorization process. Attempting to locate these sections by examining the source code can be unreliable. Frequent use of PROFILE is essential, as the proportion of time used by different sections changes after vectorization. PROFILE was run to find the most time-consuming subroutines which were then profiled individually to locate the time consuming groups of statements. Innermost loops were vectorized if possible followed by outer loops. When innermost loops cannot be vectorized, it is sometimes possible to rearrange the order of loop nesting to make a vectorizable loop the inner loop. Use of temporary vectors and rearrangements of array storage (column-wise or row-wise) was often necessary in order to incorporate vectorization.

Calculation of the atomic coordinates is described in the next subsection. Calculation of the structure amplitudes and contacts and their derivatives, and solving the normal equations usually consumes at least 95% of the CPU time for a cycle and vectorization of these three sections is described in the following three subsections.

3.2 Atomic Coordinate Calculation

The description of the atomic coordinates in terms of the linked-atom parameters is summarized as follows. The Cartesian coordinates P_n of the nth atom of a *base residue* are related to those of the $(n-1)^{th}$ atom by

$$P_n = P_{n-1} + A_n B_n \tag{4}$$

where the vector B_n is a function of the bond length from atom $(n-1)$ to atom n, the matrix A_n is a function of the bond angle and conformation angle to atom n from atoms $(n-1)$ and $(n-2)$ respectively and the direction of the bond between atoms $(n-1)$ and $(n-2)$. Defining a coordinate system relative to the first three atoms allows the coordinates of the remaining atoms to be determined by repeated application of (4). The atomic coordinates of each base residue are calculated at the beginning of each cycle (Fig. 1) from the linked-atom parameters from the previous cycle or, for the first cycle, from the input parameters, and are stored for use during the cycle. The derivatives of the Cartesian coordinates of each atom with respect to each parameter are readily calculated using (4) and are also stored to use when calculating the elements of the normal matrix as described in the following subsections. Since the number of atoms in the base residues is usually quite small, this section has not yet been vectorized.

The unit cell contains a number of *molecules* in different positions. Each molecule consists of residues replicated by screw rotations along the helix axis. The coordinates Q of an atom in the mth residue of a *generated molecule* are calculated by

$$Q = W(R_m EP + T_m) + U \qquad (5)$$

where P are the coordinates in the base residue, E is a rotation matrix which aligns the helix axis parallel to the z-axis, R_m and T_m are the rotation and axial shift respectively for the mth helix symmetry operation, W is a z-axis rotation matrix and U is a vector which positions the molecule in the unit cell. A number of base residues may be defined, and a number of molecules can be generated in different positions using the same or different base residues. The atomic coordinates in residues of the generated molecules are calculated from the base residues as they are needed in the routines described in the subsections below. Since there are usually many residues in the unit cell (up to 60) the number of atoms is large so that the calculation of (5) is vectorized. This is easily done by replicating (using the GATHER instruction) the base residue coordinates and the elements of the matrices and vectors in (5) into vectors corresponding to the atoms of the generated molecules.

3.3 Structure Amplitude Calculation

Calculation of the structure factors and their derivatives (with respect to the varied parameters) usually consumes 30-60% of the time required for one cycle with the scalar program. The mth structure amplitude F_m (of which there may be as many as 1500) is given by

$$F_m = \left| \sum_j F_{mj} \right| = \left| \sum_j o_j f_{mt_j} \exp[i2\pi(hx_j + ky_j + \ell z_j)] \right| \qquad (6)$$

where o_j is the occupancy of atom j, t_j is the type (carbon, oxygen, etc.) of atom j, f_{mt} is the scattering factor of an atom of type t, the *Miller indicies* (h,k,l) of the reflection are uniquely related to m, and (x_j, y_j, z_j) are the Cartesian coordinates of atom j. The sum is over all atoms j in the unit cell (which may number up to 3000). Atoms defined by the user to be excluded from the structure are recorded using a control vector and are excluded from the vectors of atom types, occupancies and coordinates using the COMPRESS instruction. The coordinates of all the atoms in the unit cell are calculated using a vectorized form of (5) as described in section 3.2. The vectors of atom types and occupancies are used to generate (using the GATHER instruction) a vector containing the product of the occupancy and the scattering factor of each atom in the unit cell. The calculation (6) can then be fully vectorized using linked triads (chaining[7, 8]), and the VCEXP, SSUM and VCABS instructions. The speeds of the scalar and vector codes for calculating the structure amplitudes are compared in Fig. 2 for different numbers of atoms. For more than 1000 atoms the vector code is more than 12 times faster than the scalar code.

The derivatives of the structure amplitudes with respect to the varied parameters are given by

$$\frac{\partial F_m}{\partial p_n} = \sum_j \left\{ \frac{\partial F_m}{\partial x_j} \frac{\partial x_j}{p_n} + \frac{\partial F_m}{\partial y_j} \frac{\partial y_j}{\partial p_n} + \frac{\partial F_m}{\partial z_j} \frac{\partial z_j}{\partial p_n} \right\} \tag{7}$$

where the sum is over all the atoms in the unit cell. The derivative of F_m with respect to a Cartesian coordinate of the jth atom involves only F_m and F_{mj} which is defined in (6) and is obtained as an intermediate result when computing F_m. The derivatives of the molecule atomic coordinates with respect to the base residue atomic coordinates, i.e. the derivatives of the transformation (5), are also used. Most of these operations are easily vectorized using the derivatives of the coordinates with respect to the parameters which are stored appropriately in a vector as described in section 3.2. The increase in speed obtained by vectorizing the derivative calculation is similar to that obtained for the amplitude calculation.

The structure amplitude derivatives are added into the elements A_{mn} of the normal matrix as

$$A_{mn} \leftarrow A_{mn} + \sum_k \omega_k \frac{\partial F_k}{\partial p_m} \frac{\partial F_k}{\partial p_n} \tag{8}$$

where the sum is over all the reflections. This avoids storing the large matrix **M**. Equation (8) is a matrix-matrix product which is vectorized over m to provide maximum performance[8]. The product **M H** in (3) is evaluated at the same time and vectorized in a similar fashion. The whole

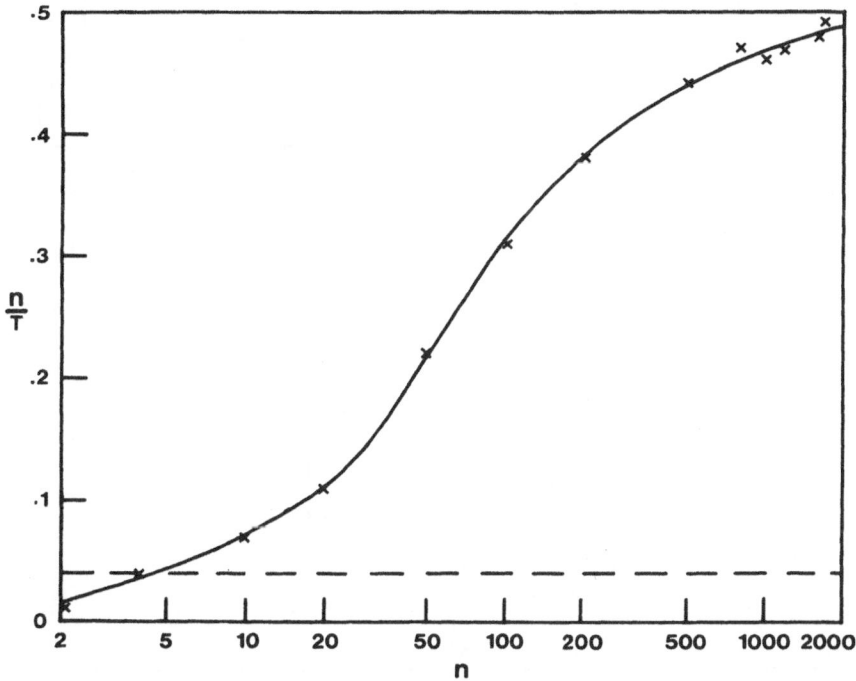

Fig. 2. Comparison of the speeds (n/T) in μs^{-1}, for calculation of the structure amplitudes using the scalar (---) and vector (——) codes, where T is the CPU time for n atoms.

structure amplitude routine is 17 times faster with the vector code
than with the scalar code for a large structure (structure III
described in section 5).

3.4 Contacts

The contact routines search for pairs of atoms which are closer than
their minimum allowable distances and calculate the derivatives with respect
to the parameters which are added into the normal matrix. The potential
number of interatomic distances is very large because intra-molecular, intermolecular and inter-unit-cell contacts need to be considered. A brute force
search of all possible interatomic distances would be a large computational
task in many cases, and a more efficient approach is possible. Many
contacts are identical because of symmetry properties of the molecules and
the crystal lattice, and these need be included only once and weighted by
their multiplicities. The user specifies the residue and molecule pairs to
be searched for contacts in a particular problem using a knowledge of the
symmetry and crystal packing. This minimizes the search to the necessary
residue pairs. A flow chart of the contact calculation is shown in Fig. 3.

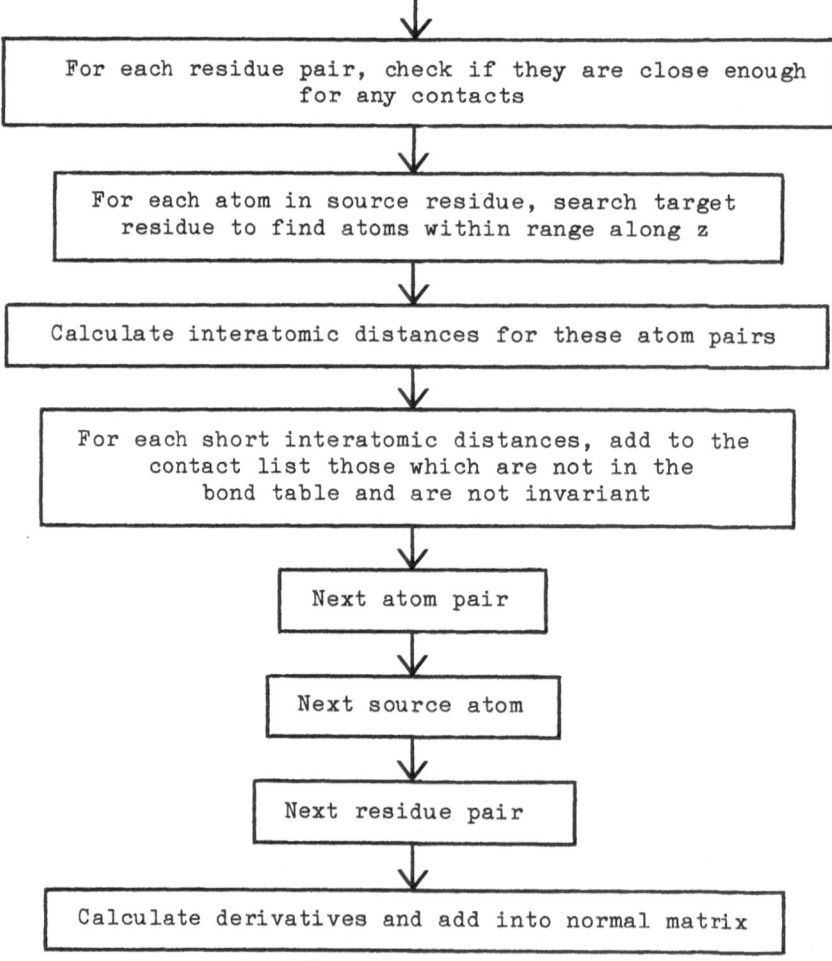

Fig. 3. Flow chart of the contact calculation.

Table 1. Comparisons of CPU Times Using Scalar and
Vector Codes for Contact Calculations of
Various Sizes

Atoms Searched	Contacts	CPU Time (s)	
		Scalar	Vector
60,000	44	1.1	0.5
120,000	138	2.5	0.9
249,000	436	7.4	2.1
638,000	769	12.9	3.5
860,000	1363	21.8	5.9

A table is first set up outside the cycle loop listing atoms which are bonded or are bonded through a third atom since short distances between such atoms are not considered contacts. In each cycle, the atoms of each residue considered are sorted by their z-coordinates. The sorted coordinate list is used to determine if contacts between residue pairs are possible, based on the residue separation. If contacts are possible, then for each atom in one residue (the "source" residue), the differences in the z-coordinates between it and the atoms of the other residue (the "target" residue), are calculated using vector instructions. The resulting vector is searched using the SLT and SGE instructions to locate a set of target atoms which may involve contacts with the source atom. The distances between the source atom and these target atoms are calculated using vector instructions. A list of contacts is accumulated by examining each atom pair to see if it is in the bonding table described above or if the contact is invariant with respect to the varied parameters. The derivatives with respect to the parameters are calculated in a similar manner to the structure factor derivatives using an expression analgous to (7), and added into the normal matrix in the same way as described in section 3.3.

The contact calculation is quite complex, however the most time-consuming sections could be vectorized and the vector comparison instructions were useful. A comparison of the performance of the contact routines before and after vectorization is shown in Table 1. Since these routines use vectors of a variety of lengths there is no obvious measure of the "size" of the contact calculation but we have chosen to use the number of atom pairs searched and number of contacts to indicate the sizes of the problems in Table 1. Inspection of Table 1 shows an increase in speed of vector code versus the scalar code by a factor between 2 and 3.6, depending on the size of the problem.

3.5 Solution of the Normal Equations

Solution of the normal equations (3) for the parameter shifts is given formally by

$$S = A^{-1} M H \tag{9}$$

and A is generated in the routines described above. Solution of the normal equations typically involved about 30% of the time for one cycle with the scalar program. It is usually convenient to choose a set of parameters and/or constraints which involve redundancies. The normal matrix will then be poorly conditioned leading to instability in the solution. However, in a physically meaningful system, the instability arises from consistent redundancies and can be eliminated by isolating and removing the singularities of **A** using singular value decomposition[9]. This involves decomposing the (symmetric) normal matrix into the eigenvalue-eigenvector form

$$A = V D V^T \tag{10}$$

where V is an orthogonal matrix (whose rows are the eigenvectors of A) and D is a diagonal matrix whose elements are the eigenvalues of A. The number of redundancies can usually be easily determined from the geometry of the problem and the singularities will be manifest as this same number of very small eigenvalues (relative to the others). Calculation of the inverse of the normal matrix is stabilized by setting these small eigenvalues to zero and, using the modified set of eigenvalues and eigenvectors, the inverse is given by

$$A^{-1} = V D^{-1} V^T \ . \tag{11}$$

The eigenvalues and eigenvectors are calculated by first reducing the normal matrix to tridiagonal form by a sequence of orthogonal similarity transformations using Householder's alogrithm[10], and then computing the eigenvalues and eigenvectors of the tridiagonal matrix using the implicit QL algorithm[11]. These two steps are carried out using the "EISPACK"[12] routines "TRED2" and "IMTQL2" respectively. TRED2 was vectorized by B. F. Putnam at the Purdue University Computer Center and IMTQL2 by the authors. The shifts are then computed using (9).

Vectorization of all the routines associated with the solution of (3) is relatively straightforward. These routines require many matrix multiplications involving both row and column access. The number of column accesses is maximized to allow vectorization, and where row accesses are necessary, the GATHERP instruction is used to store frequently used rows in temporary vectors and the SDOT instruction for dot products.

The performance of the scalar and vector EISPACK routines are compared in Fig. 4. The vectorized TRED2 runs 19 times faster than the scalar version for 500 x 500 matrices, and is slower only for matrices smaller than 8 x 8. IMTQL2 runs nearly seven times faster than the scalar code for 500 x 500 matrices when vectorized and is slower only for matrices smaller than 12 x 12. Further vectorization of IMTQL2 may be possible. Fig. 4 suggests that greater improvements in performance are likely for larger matrices.

A vector-scalar crossover length[13] is incorporated at a number of places where vectors may be either long or very short depending on the particular problem. Both scalar and vector codes are incorporated, the scalar code being executed if the vector is shorter than the crossover, and the vector code otherwise. Experiments showed that the optimum crossover length is almost always between 10 and 15 which is similar to that found by Reynolds and Lester[13].

Table 2. Comparison of CPU Times Using Scalar and
Vector Codes for Solution of the Normal
Equations

Matrix Size	CPU Time (s)	
	Scalar	Vector
89	1.3	0.3
212	17.8	1.9
411	175.0	10.5

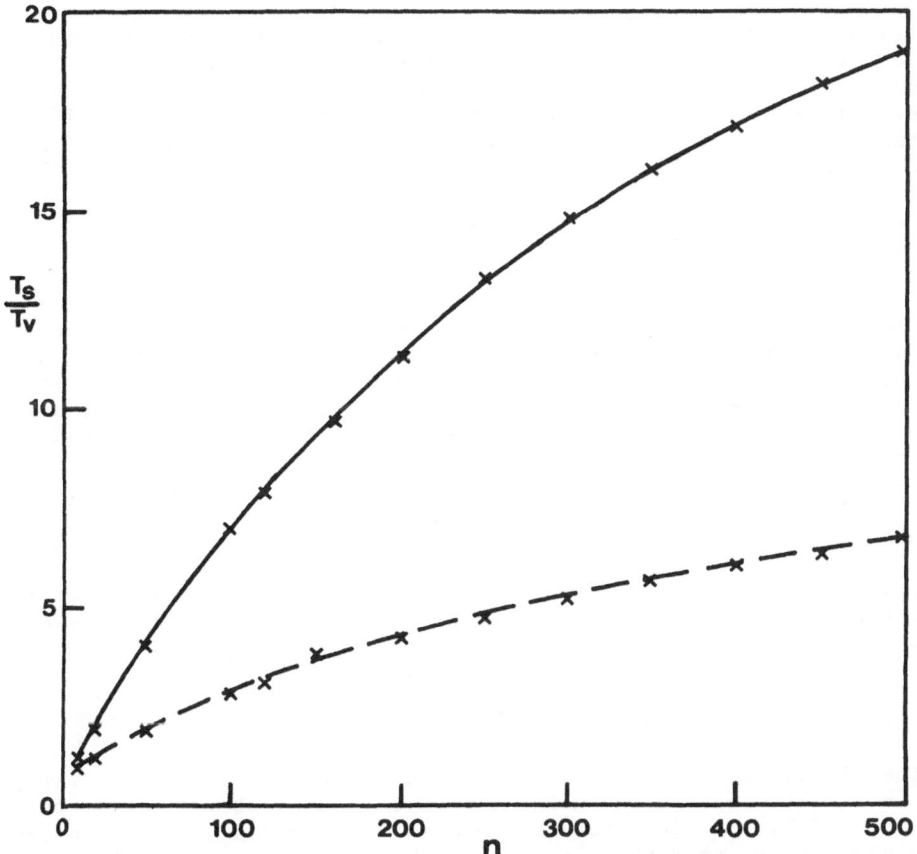

Fig. 4. Ratio of the scalar (T_s) to vector (T_v) CPU
times for execution of the EISPACK matrix
routines TRED2 (——) and IMTQL2 (---) versus
matrix size (n).

The total CPU times involved in all the routines for computing the solution to the normal equations for three different matrix sizes are compared, for the scalar and vector codes, in Table 2. The vector code is faster than the scalar code by factors of 4.3 and 17 for the smallest and largest normal matrices respectively.

4. MEMORY MANAGEMENT

The original version of LALS was an overlaid program run on a "real" memory system (CDC 6600), and as such no attention was paid to the order in which array elements were accessed. Reorganization of the algorithms or array storage was often possible to ensure access by columns to minimize page faults (and often to allow vectorization also). However, sometimes different sections of the program required access to the same array by both rows and columns. For a large array which will not fit into actual memory while it is being used, row accessing can cause a disasterous number of page faults. We found in a few cases that large arrays which are accessed often by both rows and columns could be processed more efficiently by storing a copy of the transposed array, allowing vectorization and efficient access of the rows of the original matrix.

Unless care is taken, even transposition of the array can cause an unacceptable number of page faults. We reduced paging during transposition of large arrays by using a type of block column algorithm[14] in which a block of k columns of one matrix is held in actual memory by accessing the block by rows, while the columns of the other (transposed) array are accessed one by one. It is necessary to allow two pages of the second matrix to reside in actual memory at one time to prevent part of the block from being paged out. The size of the block is then given by

$$[km/p] + [2n/p] = q - 2 \tag{12}$$

where the array has dimensions m x n, p is the number of array elements on a page, q is the number of pages available in actual memory, and $[x]$ denotes the greatest integer $\leq$ x. The number of pages available is determined dynamically using a system call. We use large pages (p = 65,000) so that p >> m and n, and the number of page faults for the transposition is given approximately by

$$mn(n+k)/pk \approx m^2n^2/p^2q \quad . \tag{13}$$

This gives a tremendous reduction in the number of page faults over that for straightforward sequential transposition, and a significant reduction over that using a submatrix approach[14].

5. DISCUSSION

The CPU times for one cycle of refinement for structures of various sizes on the 6600 system and using the scalar and vector codes on the 205 are compared in Table 3. The increase in speed of the 205 vector code over the scalar code varies from three times for the small structure (I) to 14 times for the large structure (III). The increase in speed and convenience and the decrease in cost is impressive. Our experience indicates that the difference in speeds on the 205 (scalar code) and the 6600, depends on the size of the job, and this difference for structure I shown in Table 3 is more marked than usual. Note also that the 6500 is typically about half the

Table 3. Comparisons of CPU Times for one Refinement Cycle
of Structures of Various Sizes Using Vector and
Scalar Codes

Structure	Atoms	Parameters	Data	Reflections	CPU Time (s)		
					6600	205 scalar	205 vector
I	60	89	241	100	48.0	6.0	2.2
II	223	212	581	266	-	51.0	6.9
III	386	411	754	414	-	275.0	21.5

Table 4. Comparisons of Proportions of CPU Times for one
Cycle Spent on the Major Routines Before and
After Vectorization

Code	Fraction of CPU Time (%)			
	Structure Amplitudes	Contacts	Solution of Normal Equations	Other
Scalar	48	16	35	1
Vector	31	33	28	8

speed of the 6600. The medium size structure (II) could have been run on
the 6600 (it was not because of the large amount of memory required) but
would have been quite expensive. Structure III would have been
prohibitively expensive and impracticable to run on the 6600 and is very
expensive using the 205 scalar code, but is quite practical to run using
the vector code.

Table 4 shows the proportions of CPU time spent in the most time
consuming sections of the program before and after vectorization. This
shows that the fraction of time for the most time consuming sections of the
program is reduced while the fraction for the less time consuming sections
is increased. This is what is required for vectorization to produce a
maximum increase in speed.

Although a more efficient program could be obtained by recoding the
whole program for a vector computer, we feel that most of the increase in
performance obtainable has been achieved by our approach at a much lower
cost. There are more sections of the program which can be vectorized and we
plan to implement some of these where we consider that the increased
performance justifies the effort involved.

The many special instructions available on the 205 proved extremely useful in vectorization. The need to gather together array elements not continguous in memory limited the utilization of the pipeline architecture. Although we could have used half-precision variables for much of the calculations, this was usually not worthwhile because many of the special instructions we used could not be used with half-precision arguments.

We thank R. Chandrasekaran for discussion, Beth Holle for word processing, Kathy Shuster for photography, and the National Science Foundation for support (PCM-81-02810).

REFERENCES

1. G. H. Stout and L. H. Jensen, "X-ray Structure Determination," MacMillan, London (1968).
2. S. Arnott and A. J. Wonacott, The refinement of the crystal and molecular structures of polymers using X-ray and stereochemical constraints, *Polymer*, 7:157 (1966).
3. P. J. C. Smith and S. Arnott, LALS: A linked-atom least-squares reciprocal-space refinement system incorporating stereochemical restraints to supplement sparse diffraction data, *Acta Cryst.*, A34:3 (1978).
4. S. Arnott, Twenty years hard labour as a fiber diffractionist, *in*: "Fiber Diffraction Methods," A. D. French and K. H. Gardner, eds., ACS Symposium Series Vol. 141, American Chemical Society, Washington, D. C. (1980).
5. R. Chandrasekaran, S. Arnott, R. G. He, R. P. Millane, H. S. Park, L. C. Puigjaner and J. K. Walker, More complex DNA structures, *J. Macromolecular Science - Physics*, B24 (1985), in press.
6. "VAST Automatic Vectorizor User Guide," Control Data Corporation, Publication No. 84002690 (1983).
7. R. W. Hockney and C. R. Jesshope, "Parallel Computers," Arrowsmith, Bristol (1981).
8. J. J. Dongarra, F. G. Gustavson and A. Karp, Implementing linear algebra algorithms for dense matrices on a vector pipeline machine, *SIAM Rev.*, 26:91 (1984).
9. C. L. Lawson and R. J. Hanson, "Solving Least Square Problems," Prentice-Hall, New Jersey (1974).
10. J. H. Wilkinson, Householder's method for symmetric matrices, *Num. Math.*, 4:354 (1962).
11. R. S. Martin and J. H. Wilkinson, The implicit QL algorithm, *Num. Math.*, 12:377 (1968).
12. B. T. Smith, J. M. Boyle, B. S. Garbow, Y. Ikebe, V. C. Klema and C. B. Moler, "Matrix Eigensystem Routines - EISPACK Guide," Lecture Notes in Computer Science, Vol. 6, Springer-Verlag, N. Y. (1974).
13. P. J. Reynolds and W. A. Lester, Chemical application of diffusion quantum Monte Carlo, CYBER 200 Applications Seminar, NASA Conference Publication 2295, p. 103 (1984).
14. J. J. Du Croz, S. M. Nugent, J. K. Reid and D. B. Taylor, Solving large full sets of linear equations in a paged virtual store, *ACM Trans. Math. Softw.*, 7:527 (1981).

Determinant, 197
Diagonalization, 219, 225, 242, 243, 258
Dielectric
  matrix, 258, 260
  screening, 255
Diffraction pattern, 289, 290
Diffusion, 29, 41-50
Dillner wing, 164
Dirac equation, 64-65
Dirichlet problem, 189
Dissociation probability, 274, 276
Dot product, 2, 3, 243, 247, 257, 280, 289
Double precision, 266
Drag, 160

Eddy
  coefficient, 147
  flux, 148
Eigenfunctions
  quantum mechanical operators, 230
Eigenvalue problem, 1, 2, 242, 257, 298
  generalized, 1, 2, 33 200
  Lanczos method, 4
  quantum mechanics, 220
Eigenvalues, 6, 225
  Chebyshev method, 44
  double, 6
  quantum mechanics, 230
  shifting the spectrum, 3-4
Eigenvectors, 257
Electronic
  structure, 215, 263
  wavefunction, 257, 263
Energy transfer, 215
Euler
  approximation, 279
  equations, 159, 164, 179
Exchange-correlation, 256

Fast Fourier transform, 104, 115, 273, 285, *see also* Fourier transform
Fill strategy
  conjugate gradient, 20-21
  incomplete Choleski, 19
Finite difference, 19
Finite element, 86, 87, 91, 265
  conjugate gradient, 19
  Galerkin method, 197, 206-207
  integral equations, 124-125
  Lanczos method, 5
Finite volume, 164
Flight characteristics, 159
Flow simulation, 88, 159
Fluid structure, 117-120
Fourier transform, 256, 277-288, *see also* Fast Fourier transform

Frontal method, 197, 198, 201

Galerkin method, 197, 201, 206-207
Gather, 41, 47, 53-56, 65, 77, 142, 247, 294
Gauge transformation, 65
Graphics, 86, 87, 116
Gravity wave, 148

Half precision, 64, 66, 102, 135, 240
Hamiltonian, 218, 255, 257, 266
Harmonic oscillator, 225, 228
Heat conduction, 164
Heisenberg picture, 275
Helium, 263, 264, 267
Hemiquantal equations, 273, 276
Hopf bifurcation, 200
Householder transformation, 242, 298
Hydrogen flouride, 217, 227
Hydrogen negative ion, 270
Hydrostatic equation, 147
Hylleraas wavefunction, 263
Hyper-line method, 29-30, 47
Hypernetted chain, 117

Imaging, 85, 86
Implicit methods, 180
In-core solution, 15, 180, 197, 201
Input/output, 15-16, 62-64, 136, 159
Integral equations, 122
Integrals, multidimensional, 229-233 236, 264
Interpolation
  Hermite, 265
  weather data, 137-138
  transfinite, 164
Invariant imbedding, 216
Inviscid flow, 190
Iterative methods
  acceleration by conjugate gradient, 92
  conjugate gradient, 58
  cyclic Chebyshev, 29-30
  incomplete Choleski conjugate gradient, 22
  Jacobi, 92
  Lanczos, 1, 3, 14-15
  overlapping block, 91
  subspace, 1, 10-15

Jacobi matrix, 45, 122, 187, 189, 198, 199, 206

Kuroshio current, 145

Lagrange multipliers, 291
Laminar flow, 160
Lanczos method, 1, 3, 14-15
Laplacian, 218
Large page, 1, 3, 102, 135, 177, 218, 249